Protocol Management in
Computer Networking

For a listing of recent titles in the *Artech House Telecommunications Library,* turn to the back of this book.

Protocol Management in Computer Networking

Philippe Byrnes

Artech House
Boston • London

Library of Congress Cataloging-in-Publication Data
Byrnes, Philippe.
    Protocol management in computer networking / Philippe Byrnes.
        p.   cm. — (Artech House telecommunications library)
    Includes bibliographical references and index.
    ISBN 1-58053-069-9 (alk. paper)
    1. Computer network protocols.   I. Title.   II. Series.
TK5105.55.B97   2000
004.6'2—dc21   99-052313
                                                                           CIP

British Library Cataloguing in Publication Data
Byrnes, Philippe
    Protocol management in computer networking — (Artech House
    telecommunications library)
    1. Computer network protocols
    I. Title
    004.6'2

    ISBN 1-58053-069-9

Cover design by Igor Valdman

© **2000 Philippe Byrnes**

International Standard Book Number: 1-58053-069-9
Library of Congress Catalog Card Number: 99-052313

10 9 8 7 6 5 4 3 2 1

To my parents, Anne-Marie and the late Dr. Kendal C. Byrnes,
in whose home learning was of more value than lucre,
with gratitude for all their support and encouragement.

Contents

Preface

This book presents a novel management framework, which I refer to as the *Me*asurement-*E*stimation-*S*cheduling-*A*ctuation (or MESA) model. This is derived from the control and queueing theory; and applies it to various aspects of computer communications, from modulation and coding theory to routing protocols and topology reconstruction.

Why a new approach to the subject? After all, the Internet and its underlying technologies and protocols are doing quite nicely, thank you. Likewise there is no shortage of well-written, current books on networking and communications. The short answer is that this is the book that I wish were available when I first started working for IBM on network management architecture in 1984. I believe it clearly details the tasks executed in computer networks, the large majority of which are in some sense management tasks. It derives from first principles a model of management in general and of management in computer networks in particular that is comprehensive, that is not tied to any particular protocol architecture, and that helps identify the basic elements of management that recur at different layers of the protocol stack.

This last point leads to a longer answer, namely, that in the course of researching what it means to manage the logical and physical resources in a computer network, I came to the conclusion that the effort was hampered by a naive reductionism that created artificial distinctions where none truly existed. A good example of this is what I would argue is the false distinction made between fault management per SNMP or RMON versus fault management executed by a protocol that uses retransmission or fault management as executed by dynamic routing. Decomposing these into the MESA management tasks reveals common denominators and unifies otherwise disparate areas of networking. In short, the MESA model was developed to help explore what we mean when we speak of management in computer networks and their protocols.

The results that came of these explorations are presented in this book, which is organized into four parts. The first part is devoted to management in the physical layer: with channels, signals, and modulation; with error control coding; and management at the physical layer interface. The second part moves up the protocol stack to the data link layer and examines the management tasks executed by data link protocols, including the principal serial and LAN protocols. Part III examines management in end-to-end protocols, focusing on the TCP/IP and SNA protocol suites. Finally, the fourth part of this book looks at the management tasks implicit in the various concatenation techniques used in computer networks: bridging, routing, and tunneling.

Acknowledgments

This book could not have been finished without the encouragement and support of many friends and colleagues. I would like to thank Mike Flynn, Wayne Jerdes, and Clyde Tahara of IBM; Christina Shaner of Incyte Bioinformatics; John Erlandsen, Michelle Swartz, and Greg Hobbs of Cisco Systems; Tom Maufer and Bob Downing of 3Com; Chuck Semeria of Juniper Networks; and others I am undoubtedly forgetting at the moment. Many of the strengths of the book are due to their comments.

I would also like to acknowledge my debt to my many teachers, and in particular to Richard Klein of the University of Illinois, Urbana-Champaign, and Arthur Bryson of Stanford University, for teaching me some of the many subtleties of control theory. To Klein in particular I am grateful for his tutelage in sundry matters of control theoretic and otherwise. Late afternoon discussions in his office over what he referred to as "some very good hooch" left me with memories that are still fresh over a decade later.

Philippe Byrnes
Palo Alto, California
December, 1999
Philippe.Byrnes@Computer.Org

Part I:
Introduction and Management in the Physical Layer

1

Introduction and Overview

1.1 The Internet, the MESA Model, and Control Systems

The Internet is arguably the greatest technological achievement since the microprocessor. A measure of its success can be seen in its explosive growth, which since 1969 has developed from an experimental testbed connecting a few dozen scientists into a commercial juggernaut, fueled by the explosion of the World Wide Web, connecting tens and even hundreds of millions of users everyday.

When a technology has been this successful, it may be unwise to suggest a new approach to the subject. Nonetheless, that is the purpose of this book: To propose a new management model, derived from control systems engineering, with which to analyze communications protocols and data networks, from simple LANs to the Internet itself. As we hope to make clear in the next 14 chapters, this is not a book about network management as the term is conventionally used but rather a book about management in networking.

1.1.1 Why a New Approach?

There are several reasons for advancing a new management framework. First, it is our belief that management is ubiquitous in modern communications networks and their protocols, at every layer and in ways not commonly appreciated. Indeed we contend that layering, central to every modern protocol architecture, is itself an instance of (embedded) management: Layering works by mapping from the abstracted layer interface to implementation details, and

3

this mapping is a management task. And yet, because of narrow and constricting definitions of what constitutes management, this fact is obscured.

Another example of this obscurity is the handling of faults. Most protocols include some mechanism(s) for the detection if not correction of communications faults that corrupt the transported data. Another type of fault recovery is provided by routing protocols and other methods of bypassing failed links and/or routers. Still another type of fault management is the collection of statistics for the purpose of off-line fault isolation. Our goal is to unify these, and for this we need a definition of management that treats all of these as instances of a common task.

Instead of such a common definition, today we are confronted by a reductionism that in the first place differentiates protocol management from network management. This book seeks to go beyond such distinctions by constructing from the first principles a very broad formulation of management and utilizing this to identify management tasks whether they occur in communications protocols, routing updates, or even network design. To borrow a notable phrase from another field, we seek to demonstrate the "unity in diversity" of management in data networking.

And the foundation of this construction is the concept of the manager as a control system. More precisely, from control engineering we derive a framework for task analysis called *MESA*, taking its name from principal task primitives: Measurement, Estimation, Scheduling, and Actuation. Corresponding to these four tasks are four classes of servers that make up closed-loop control systems: sensors, which measure the system being controlled (also known as the plant); estimators (also known as observers), which estimate those plant parameters that cannot be easily measured; schedulers (also known as regulators), which decide when and how to change the plant; and actuators, which carry the changes (Figure 1.1). Note that in Figure 1.1 the plant is a discrete event system, composed of a client and a server, along with storage for queued Requests for Service (RFSs). Everything in this book will revolve around this basic model, where management intervenes to control the rate at which work arrives from the client(s) and/or is executed by the server(s). The former we refer to as *workload management* and the latter we call *bandwidth* (or *service*) *management*.

1.1.1.1 Repairing Foundations

Another reason for a new approach is that networking suffers from cracks in its foundations. The theoretical framework on which networking at least nominally has been based for most of the last 20 years—epitomized by the OSI 7 layer model—has been rendered increasingly irrelevant by recent innovations and hybrids that have blurred once solid distinctions. For example, from the

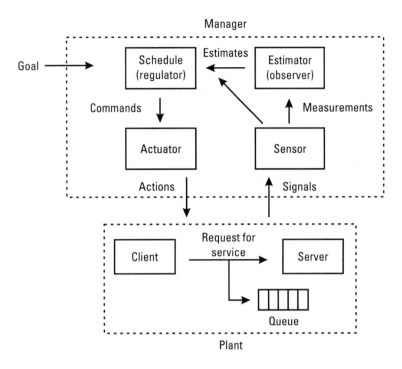

Figure 1.1 Basic control system: managing a discrete event system.

1960s onward, routing reigned supreme as the best—in fact the only—way to concatenate data links. When local-area networks (LANs) appeared, with their peer protocols and universally unique station addresses, it became possible to concatenate at layer 2. Bridging, as this became known, and its modern equivalent, Application-Specific Integrated Circuit (ASIC)-based incarnation known as (frame) switching, is simpler and often faster at forwarding data than conventional layer 3 routing.

Routing advocates counter that bridging/switching is considerably less efficient at using network resources and less robust at handling faults than routing. All this has led to heated arguments and uncertainty about how to design and implement large computer networks, as well as a slew of curious hybrids such as "layer 3 switches," "IP switches," and the like that may owe more to marketing hyperbole than any meaningful technical content. Separating the latter from the former requires, in part, a neutral vocabulary, and that is one by-product of our study of management. To repeat, our aim is to construct a unified theory of data networking, with unified definitions of tasks such as concatenating transporters, independent of the nature of the technology—bridges, routers, and so on—used in implementing networks.

1.1.1.2 Next-Generation Internets: New Management Models

Finally, the current model will not get us from where we are to where we want to go. The next generation of internetworks must be self-tuning as well as self-repairing. By monitoring traffic intensity (its constituents, workload arrival, and server bandwidth), throughput, and particularly response times, the managers of these networks will automatically adapt workload to changes in bandwidth, increasing the latter in reaction to growth in the former (for example, meeting temporary surges through buying bandwidth-on-demand much as electric utilities do with the power grid interconnect (Figure 1.2)).

Response time is particularly worth mentioning as a driving force in the next generation of internets: the visions of multimedia (voice and video) traffic flowing over these internets will only be realized if predictable response times can be assured. Otherwise, the effects of jitter (variable delays) will militate against multimedia usage. All of this points to the incorporation of the techniques and models of automatic control systems in the next generation of internetworking protocols.

1.1.2 Control Systems

This is precisely the tracking problem encountered in control engineering. Two tracking problems frequently used as examples in discussions of control systems are the home furnace/thermostat system, perhaps the simplest and certainly most common control system people encounter, and the airplane control system, perhaps the most complex. Someone who wishes to keep a home at a certain temperature sets this target via a thermostat, which responds to changes in the ambient temperature by turning a furnace on and off. Similarly, a pilot

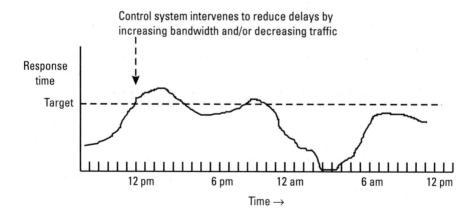

Figure 1.2 Managing response time.

who moves the throttle or positional controls (e.g., elevators, ailerons) is giving a goal to one or more control systems, which seek to match the target value. In both cases, as the goal changes the control system attempts to follow, hence, the term *tracking* problem.

A control system attempts to ensure satisfactory performance by the system being controlled, generally referred to as the *plant*, the fifth entity in Figure 1.1. The control system's scheduler receives a high-level goal or objective. Obviously, the goal is expressed in terms of the desired state value(s) for the plant. The scheduler seeks to attain this state by means of actuations that change the plant. Naturally enough, the actuations are executed by the control system's actuator(s). The roles of sensor and estimator are complementary: For those state variables that can be measured, the sensor executes this task; however, in many instances the plant may have state variables that cannot be measured and must instead be estimated—this estimation is the task of estimators. Finally, this information on the state of the plant, obtained either by measurement or estimation, is fed back to the scheduler, which uses it to schedule the next actuation(s).

1.1.2.1 Interacting With the Plant: Monitoring and Control

The obvious next question is "What about the plant are we measuring and/or actuating?" The plant is characterized by its *state*, that is, the set of variables and parameters that describe it. For example, the state of an aircraft is described by its motion, its position, its current fuel supply, and its currently attainable velocity; other variables and parameters may include its mass, its last servicing, the number of hours its engines have been ignited, and so on. The computer analogy of state is the Program Status Word (PSW), the set of state information that is saved when a process is suspended by an operating system and that is retrieved when execution is resumed. The state is the set of information needed to adequately characterize the plant for the purposes of monitoring and controlling it.

The actuator is the server in the control system that changes the plant's state. When an actuator executes, it changes one or more of the state variables of the plant; otherwise, if no variable has changed then by definition the execution of the actuator has been faulty or, put equivalently, the actuator has a fault. When the plant is an airplane, the actuators are its ailerons, its other wing surface controls (canards, variable geometry, and so on), its rudder, and its engines. By altering one or more of these, the airplane's position will change: open its throttle and the engine(s) will increase its acceleration and speed; pivot its rudder and the direction in which it is heading will correspondingly alter; change its wing surfaces and the amount of lift will alter, increasing or decreasing altitude. Note that not all of a plant's state variables may be directly actuatable; in this case those variables that cannot be directly actuated may be coupled by the plant's dynamics to variables that can be actuated, allowing indirect control.

The sensors in a control system provide information about the plant. Also, when it comes to sensors, a similar situation may arise: With many plants, it is impossible to measure all state variables for either technical or economic reasons. For this reason the literature on control and instrumentation introduces a distinction, which we will follow here, between state and output variables. Therefore, when we speak of a sensor it may be measuring a state variable or an output variable, depending on the plant involved.

An output variable does not describe the plant but rather its environment; in other words, an output variable represents the influence the plant has on its environment, and from which the state of the plant may be inferred. For example, there is no direct way to measure the mass of planetary bodies; however, by applying Newton's laws and measuring the forces these bodies exert, their respective masses can be estimated. This brings us to the role of the estimator in our control system model. Estimation is complementary to measurement. Using the state and/or output variables that can be measured, an estimator estimates those state variables that cannot be measured. Another term used by some authors for this is *reconstruction* (see, for example, [1]).

Finally we come to the scheduler. Although this statement may seem obvious, a scheduler *schedules*. For example, at the heart of a flight control system is a scheduler (regulator) that decides when the various actuators should execute so as to change some aspect of the plant, such as the airplane's position or momentum. The classic feedback control system operates by comparing the state of the plant, measured or estimated, with the target state: Any discrepancy, called the *return difference,* is used to determine the amount and timing of any actuations to drive the plant toward the target state, often called the setpoint. When the scheduler receives a target (goal), this is in fact an implicit request for the service. The simplest example of a feedback control system is perhaps the furnace–thermostat combination. The room's desired temperature is set and a thermometer measuring the room's actual temperature provides the feedback; when the difference between the temperature setpoint and the actual temperature is sufficiently great, the furnace is triggered, actuating the room's temperature to the setpoint.

1.1.2.2 Types of Control: Open Versus Closed Loop

Although every control system has a scheduler and actuator, not all control systems rely on estimators or even feedback. Some schedulers make their decisions without regard to the state of the plant. These are referred to as *open-loop* control systems. In other instances, the plant's state variables are all accessible to measurement; this is called *perfect information*, and no estimator is required to reconstruct them. The most complicated control system, of course, is the scheduler-estimator-sensor-actuator, which is required when one or more state variables are inaccessible to measurement (Figure 1.3).

1.1.3 The Plant and Its Model

It is worth briefly mentioning the role of the *model* in control systems. There must be some understanding of the plant and its dynamics if the scheduler is to create the optimum schedule or, in many instances, any schedule at all. In addition, the design of the instrumentation components (sensors and estimators) is based on the model chosen, otherwise it will not be known what data must be collected. The sensors and estimators change the state of the model, and the updated model is used by the scheduler (Figure 1.4).

A control system that uses feedback does so to correct discrepancies between the nominal model of the plant and the plant itself. Two common sources of these discrepancies are (exogenous) disturbances and uncertainties reflecting idealization in the mathematical models—the latter is introduced as a concession to the need for tractability. When a server suffers a fault, for example, this is an example of a disturbance.

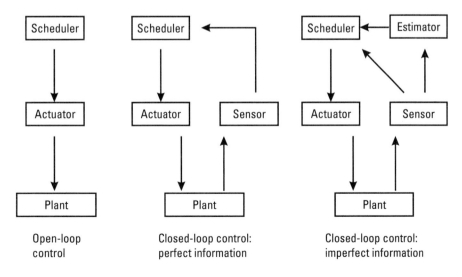

Figure 1.3 The three different types of control systems.

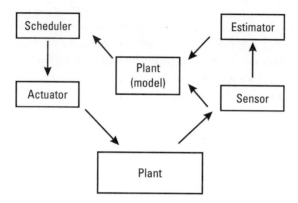

Figure 1.4 Scheduler and actuator changing reality, estimator and sensor changing model.

1.1.4 Control System as Manager

When it suits our purpose we would like to abstract the details of a given control system (open-loop, closed-loop with perfect information, closed-loop with imperfect information) and simply denote its presence as a manager, which in our vocabulary is the same thing as a control system.

1.1.5 Managing Discrete Event Plants

Let's move back from the plant being either a plane being flown or a room being heated to the plant being a discrete event system composed of a server and a client (Figure 1.5), where the client generates work for the server in the form of RFSs, which are stored in a queue for execution if there is sufficient space.

Reduced to essentials, we are interested in controlling the performance of a discrete event system (also known as a queueing system). Later we explain that the server is the computer network that is a collective of individual servers and the client is the collective set of computers seeking to exchange data; that is in

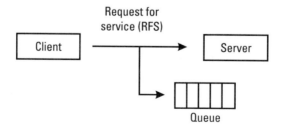

Figure 1.5 Plant client and server.

the case at hand, the plant is the communications network and its clients (i.e., digital computers), and the function of the control system is to respond to changes in the network and/or its workload by modifying one or both of these. For now, however, let's keep it simple.

The performance of a discrete event system can be parameterized by such measures as delay, throughput, utilization, reliability, availability, and so on (Figure 1.6). These are all determined by two (usually random) processes: the arrival process, which determines the rate at which workload arrives from the client; and the service process, which determines the rate at which the workload is executed by the server. This means that, for even the simplest discrete event (i.e., queueing) system, there are *two* degrees of freedom to the task of controlling the performance: actuating the arrival and service processes. Other ancillary factors include the maximum size of the queue(s), the number of servers, and details about how the queueing is implemented (input versus output versus central queueing), but the fundamental fact remains that overall performance is determined by the arrival and service processes.

1.1.5.1 Discrete Event System State Variables

The state of a discrete event plant is determined by the state of the client and the state of the server (leaving aside for now the question of queue capacity and associated storage costs) (Figure 1.7).

In the terminology of queueing theory, the client is characterized by the arrival process/rate. The arrival rate generally is shorthand for the mean

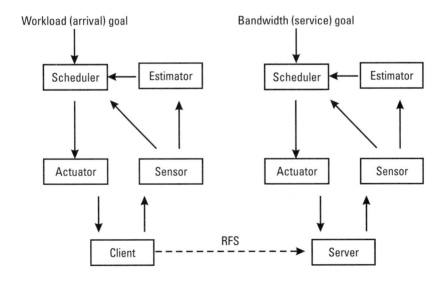

Figure 1.6 Separate control systems: workload and bandwidth.

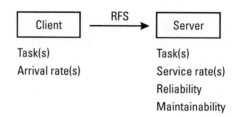

Figure 1.7 Client and server with parameters.

interarrival time, and is denoted by the Greek letter λ; more sophisticated statistical measures than simple means are sometimes used (Figure 1.8). Real-world clients generate requests for service (work) that can be described by various probabilistic distributions. For reasons of mathematical tractability the exponential distribution is most commonly used.

In addition, we must specify *what* the client requests. A client must, of necessity, request one or more types of tasks (this can include multiple instances of the same task). The task type(s) a client may request constitute its task set. Two clients with the same task set but with different arrival processes will be said to differ in degree. Two clients with different task sets will be said to differ in kind.

A server by our definition executes tasks. The types of tasks a server can execute constitute its task set. Take, for example, an Ethernet LAN with four clients: The LAN can execute a total of 12 tasks (move data from client 1 to client 2, move data from client 1 to client 3, and so on). Note that the task $A \rightarrow B$ is *not* equivalent to the task $B \rightarrow A$ so we count each separately. If we add a station (client) to the LAN then its task set grows to 20 tasks. If the server can execute only one task (type) then its client(s) need not specify the task. This is called an *implicit* RFS.

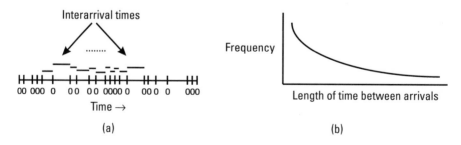

Figure 1.8 Interarrival times and exponential distribution: (case 1) arrival process and (case 2) exponential distribution.

As with clients, two servers with the same task set but with different arrival processes will be said to differ in degree. Two servers with different task sets will be said to differ in kind.

A server's task set is basically an atemporal or static characterization. To capture a server's dynamic behavior, we need to discuss its *tasking*—how many tasks it can be executing at one time. The obvious answer to this is one or many—the former is single tasking and the latter is multitasking. However, we must further differentiate multitasking between serial multitasking and concurrent multitasking. In serial multitasking, two or more tasks may overlap in execution (i.e., their start and stop times are not disjoint) but at any given time the server is executing only one task. Concurrent multitasking, on the other hand, requires a server that *can* have two or more tasks in execution at a given moment. Concurrent multitasking implies additional capacity for a server over serial multitasking. Figure 1.9 illustrates serial versus concurrent multitasking.

Obviously, if there is a mismatch between the task set of a server and the task set of its client(s), then there is a serious problem; any task that is requested that is not in the former will effectively be a fault. Corresponding to each task in a server's task set is a mean service rate, which we will refer to as the *bandwidth* of the server. It is denoted by the Greek letter μ. As with arrival rates, more sophisticated statistical measures than simple means are sometimes used. In all cases, however, one thing holds true: The service rate is always finite. This will, it turns out, have profound consequences on the type of workload manager that must be constructed.

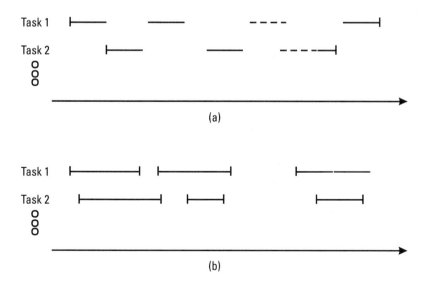

Figure 1.9 Multitasking: (a) serial versus, and (b) concurrent.

Now, there is an additional fact of life that complicates this picture: An actual server will have finite reliability. At times it will be unable to execute a task requested of it not because it has no additional bandwidth but because it has no bandwidth at all. There are several classes of faults that can cause problems in a server. A fatal fault, as its name indicates, "kills" a server; a server that has suffered a fatal fault cannot operate, even incorrectly. A partial fault will reduce a server's bandwidth and/or task set but neither totally.

We also must distinguish faults based on their duration. Some faults are persistent and disable a server until some maintenance action (repair or replacement) is undertaken by a manager. Other faults are transient: They occur, they disable or otherwise impair the operation of the server in question, and they pass. Exogenous disturbances are often the source of transient faults; when the disturbance ends the fault does, too. An immediate example is a communication channel that suffers a noise spike from an outside source such as lightning.

1.1.5.2 Bandwidth Management

Management of a server in a discrete event plant amounts to managing its bandwidth, that is, its service rate, and by extension of its task set. When we speak of bandwidth we mean its *effective* bandwidth (BW_e), which is the product of its nominal bandwidth (BW_n) and its availability A. Availability, in turn, is determined by the server's reliability R, typically measured by its Mean Time Between Failures (MTBF) and its maintainability M, typically measured by its Mean Time To Repair (MTTR). A "bandwidth manager"—arguably a more descriptive term than "server manager" or "service rate manager"—can therefore actuate the bandwidth of a server by actuating its nominal bandwidth, its reliability, or its maintainability.

Implementing the least-cost server entails making a set of trade-offs between these parameters. For example, a server with a high nominal bandwidth but low availability will have the same average effective bandwidth as a server with a low nominal bandwidth but high availability. Similarly, to attain a given level of average availability, a fundamental trade-off must be made between investing in reliability (MTBF) and maintainability (MTTR). A highly reliable server with poor maintainability (i.e., a server that seldom is down but when it *is* down is down for a long time) will have the same availability as a server that is less reliable but which has excellent maintainability (i.e., is frequently down but never for a long time). In both of these trade-off situations, very different servers can be implemented with the same averages, although it should be noted that the standard deviations will be very different. Figure 1.10 shows a MESA analysis of the process of making these trade-offs.

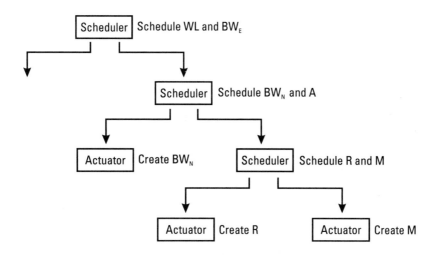

Figure 1.10 Bandwidth management: server parameter trade-offs.

When is a server's bandwidth (and/or other parameters) actuated? Figure 1.11 shows the occasions. The first is during its design and implementation. Implementation is an actuation of the server's nominal bandwidth from zero, which is what it is before it exists, to some positive value; and an actuation of its task set from null to nonempty. Up to this point, the server does not exist. Although it seems obvious to say, bandwidth management is open loop in the design phase since there is nothing to measure. Based on measurements and/or

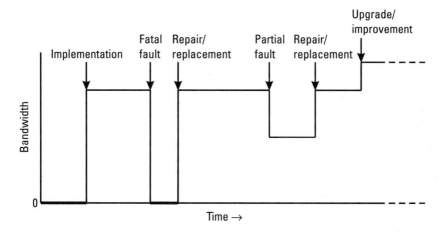

Figure 1.11 Events in server life cycle.

estimates of the client's demand, management will schedule the actuation of the server and its components.

This question has been studied considerably: Can a reliable server be constructed out of unreliable ones? The answer is yes, and the key is the use of redundancy. For this reason, implementing a reliable server is much easier when the plant is a digital plant that can be replicated at will, subject to limitations in cycles and storage. We will explore ways to do this in more detail in various chapters that follow, from coding theory to data link redundancy (bonding) to protocol retransmission to dynamic routing.

After this there *is* a server extant, and this means that bandwidth management may be, if desired, closed loop. The next instance of actuating a server's bandwidth generally occurs after a fault. As we remarked earlier, *all* servers have finite reliability. A server that is disabled by a fault has a reduced bandwidth. A partial fault may reduce the bandwidth but still leave a functioning server, whereas a fatal fault reduces the bandwidth to 0. Restoring some or all of the server's lost bandwidth is obviously an instance of bandwidth management.

This task is typically divided into three components: fault detection, isolation, and repair (or replacement). Of these, fault detection involves the measurement of state and/or output variables to detect anomalous conditions. For example, high noise levels in a communications line can indicate a variety of faults or vibrations at unusual frequencies or can mean mechanical system faults. Fault isolation generally requires estimators since it entails a process of inference to go from the "clues" that have been measured to identifying the failed component(s) of the server. The actuation of the server is effected in the last phase, repair or replacement. The reason this is bandwidth actuation is that after a successful repair the bandwidth of the server is restored to the *status quo ante.*

It might seem from the preceding discussion that a bandwidth manager must be closed loop to effect maintenance; and while feedback undoubtedly reduces the time from the occurrence of a fault to the server having its bandwidth restored, there are circumstances under which open-loop maintenance policies might be used instead. Such policies as age replacement and block replacement require the bandwidth manager to replace components of the server irrespective of their condition; such a policy will result in any failed components eventually being replaced, and many failures being prevented in the first place, albeit at the cost of discarding many components with useful lifetimes left. (For further discussion of various maintenance policies, see either Goldman and Slattery [2] or Barlow and Proschan [3].)

Typically, however, bandwidth managers responsible for maintaining servers are closed loop. Indeed, in the absence of sensors and estimators to infer the server's condition, the incidence of latent faults will only increase.

Therefore, a major part of most bandwidth managers is the instrumentation of the server to monitor its condition. In fact, because it can even be argued that the maintainability of a server is one measure of the service rate of a bandwidth manager that is responsible for fault detection, isolation, and recovery (repair or replacement), an investment in instrumentation that reduces downtime increases the bandwidth of the bandwidth manager.

Finally we come to deliberately upgrading or improving the server's bandwidth, as opposed to merely restoring it after a fault. The two basic degrees of freedom here are (1) the bandwidth of the server and (2) its task set. Consider first the instance where we have a server that can execute multiple types of tasks but we can change neither its total bandwidth nor its task set. By holding both of these constant, we can still change the bandwidth allocated to each task and this is still a meaningful change. An example would be to alter the amount of time allocated to servicing the respective queues of two or more competing types of tasks, such as different classes of service or system versus user applications. Because we are not changing the tasks the server can execute, the "before" and "after" servers differ only in degree, not kind. We will therefore refer to this as actuation of degree.

Another variant of actuation of degree is possible, namely, holding the server's task set constant but now changing the total bandwidth of the server. An example would be to replace a communications link with one of higher speed; for example, going from 10BaseT to 100BaseT, but not adding any additional stations. The task set would be unchanged but the bandwidth would be increased. We refer to this as actuation of degree as well, but to distinguish the two cases we will call this actuation of degree$_2$ and the first type actuation of degree$_1$. Of course, if a server can execute only one task, obviously, this collapses to a single choice, namely, the actuation of degree$_2$.

Changing a server's task set does transform it into a different type of server and we call this last type of change *actuation of kind*. Changing the task set of a server often entails significant alteration of its design and/or components. Of course, it can be as simple as adding a new station to a LAN. Generally, though, actuation of kind is the most complicated and extensive of the changes possible in bandwidth management. Figure 1.12 shows the continuum of change for bandwidth actuations.

Note that changing the nominal service rate and/or task set is not something undertaken easily or often. In some cases servers have two or more normal service rates that a bandwidth manager can actuate between, perhaps incurring higher costs or increased risk of faults as the price of the higher bandwidth. For example, increasing the signal levels in communications channels can improve the noise resistance but reduce the lifetime of the circuits due to increased heat. An example of a server that has several service rates is a modem

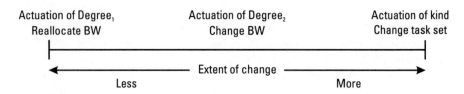

Figure 1.12 Bandwidth continuum of change: actuations of degree and kind.

that can operate at several speeds, depending on the noise of the communications channel.

We should also remark that for many servers, particularly those which are complex and composed of many component servers, the key parameters may not be known or known adequately and must be measured. In these cases, it may be up to bandwidth management to monitor (measure and/or estimate) the three key random variables/parameters: the service time process, the reliability process, and the maintainability process. If the server is composite, then the option exists to estimate these parameters for its components and, with knowledge of its composition, estimate the topology of the composite; or its internal details can be simply elided, lumping the components together and treating them as one entity.

1.1.5.3 Workload Management

Now we come to workload managers. The need for, indeed, the very existence of, workload management is a concession to the inescapable limits in any implementable server. This means, as we just discussed, accommodating a server's finite bandwidth and reliability. And, just as we identified three levels of actuation for changing the task set and/or service rate(s) of a server, so are there three levels of workload actuation.

The first level of workload management is access and flow control. A server with limited (i.e., finite) bandwidth cannot service an unlimited number of RFSs. (In addition, although we did not dwell on it in the state description given earlier, the limits on the queue size often constitute even greater constraints than the fact that bandwidth is necessarily finite.) A basic workload manager will actuate only the interarrival distribution, that is, the arrival process. Figure 1.13 illustrates an example of this called *traffic shaping*, with before and after graphs of the interarrival time distributions. We will refer to this as actuation of degree$_1$.

Various mechanisms can be used to actuate the arrival rate so as to allocate scarce resources (bandwidth, queue space, and so on). These mechanisms can be broadly divided into coercive and noncoercive. Coercive mechanisms include tokens, polling, and other involuntary controls. To change the

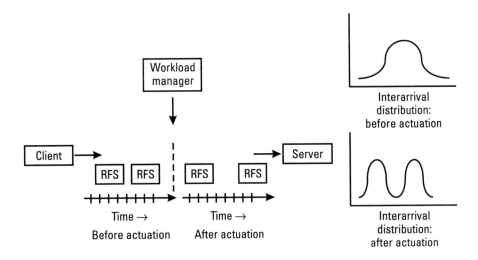

Figure 1.13 Traffic shaping: workload actuation of degree$_1$.

arrival rates of workload can also be done by buffering and/or discard, either in-bound or out-bound. Noncoercive mechanisms revolve around issues of pricing and cost: raising "prices" to slow down arrivals, lowering them to increase arrivals. Note that coercive and noncoercive mechanisms can be combined.

Examples of basic workload actuation (actuation of degree$_1$) abound in computer communications. The access control mechanisms in Ethernet require each station (i.e., client) to sense the status of the shared bus and, if it is not free, to stop itself from transmitting. SDLC uses a polling protocol with a single master that allocates the channel to competing clients. Token Bus and Token Ring systems (802.4 and 802.5) use token passing to limit access.

By altering the arrival rates at which the work (RFSs) arrives, workload management can avoid saturating limited queues, balance out the workload over a longer period, and avoid the possibility of a contention fault, when two or more competing clients prevent any from having their requests being successfully executed.

One of the most important types of workload actuation of degree extends beyond simply deferring or accelerating the rate of arrival of RFSs. If the original RFS is replicated into two or more RFSs, this is what we will call actuation of degree$_2$. The importance of this replication may not seem obvious but the whole concept of time slicing that lays behind packet switching is in fact actuation of degree$_2$ in which the plant is divided for piecemeal execution. (Packet switching works by dividing a message into smaller units called packets—see Part III for more on the technology and origins of the term.)

In the case of packet switching, this means sharing scarce communications bandwidth by means of dividing the bandwidth of a transporter into time slices, also called *quanta;* the result of this division is packets. That is to say, given a transporter of finite bandwidth, it follows that in a finite interval of time only a finite amount of data could be transported. Consider what it means to time slice a transporter of bandwidth M symbols per second (= Mk bits per second, where there are k bits/symbol). If we establish the time-slicing interval (quantum) as $1/J$ seconds then the maximum amount of data that can be transported in that interval is Mk/J bits. This is the maximum packet size (Figure 1.14). Equivalently, if we limit the maximum unit of plant (data) that can be transported with one RFS we effectively establish an upper bound on the quantum that can be allocated to an individual client.

The preemptive nature of packet switching is what distinguishes it from message switching, another technique for serially reusing (sharing) a transporter. With message switching, transporters are shared serially among two or more clients but the plant each client requests to be transported is sent intact, that is, without division. Consider, for example, message switching of a 9.6-Kbps transporter with two clients, one of which has file transfer that involves transporting 2 MB of data while the other needs to send a credit card verification of 265 bytes. If the file transfer occurred first and without interruption, then the transporter would be unavailable for almost 3.5 min; the credit card client would have no choice but to accept this delay. The advantage of packet switching's preemption is that such large delays and the associated "jitter" can be minimized if not eliminated.

Time slicing has another benefit, namely, fault management (Figure 1.15). Because the plant is now divided into components, these constitute units of recovery that are much smaller than the whole message. A fault that occurs during the transportation of the plant is unlikely to affect all of the

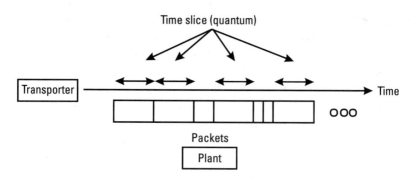

Figure 1.14 Transporter, time slices, and packets.

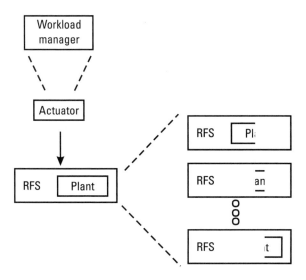

Figure 1.15 Workload actuation of degree$_2$: time slicing.

components, meaning that only the affected components need be retransmitted (if such reliability is necessary).

Such replication is a very powerful technique for managing transient faults in digital servers; by replicating the plant and RFS for either concurrent execution or serial reexecution, the effects of the transient faults can be mitigated. Examples of the former in digital communications include parallel transmission and forward error correction via error correction coding; examples of the latter include retransmission of data corrupted by faults on the channel. Table 1.1 compares workload and bandwidth management responses to transient and permanent faults. Note, though, that both of these stop short of actuating the type of the RFS. This brings us to workload actuation of kind: transforming one RFS into another of a different type. At first blush, workload actuation of kind is not a very useful tool for managing the dynamics of a discrete event plant. That would be true except for one vital application: When the server is a composite of other servers, then in effect RFSs to the composite server are aliases and workload management must "de-alias" the RFSs into RFSs for the component servers (Figure 1.16).

A workload manager actuating an RFS from one type of task to another occurs with layered implementations where the implementation details of a server are abstracted from a client. For example, when the server is a composite transporter, then this de-aliasing is precisely the concatenation task that we frequently call *routing* (although it could be bridging, switching, and so on). The

Table 1.1

Management Actions: Transient Versus Permanent Faults

Duration	Bandwidth	Workload
	Management Action	
Transient	Fix the cause of the transient faults although the server will recover by itself	Replicate the plant and RFS for reexecution/retransmission
Permanent	Fix the cause of the faults since the server will not recover by itself	No workload management action can help

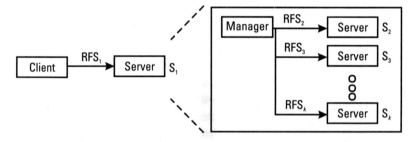

Figure 1.16 RFS mapping with composite servers: workload actuation of kind.

server S_1 in the figure could be an IP network, with S_2, S_3, ..., S_k as component networks and the workload manager is a router (bridge, switch, and so on). In other words, workload actuation of kind is precisely the relaying function. This brings us to one of our principal results, which we will explore in detail in the chapters of Part IV namely, that the composition of two or more servers is necessarily effected by a workload manager. In this instance we say that the workload manager is a *proxy client* for the actual client.

Figure 1.17 shows an example of an internetwork in which he relays R_1, R_2, R_3, and R_4 could be bridges, routers, switches, or some combination.

The transport task $N_1 \rightarrow N_2$ is realized by a set of component transport tasks between pairs of MAC addresses (and, internal to the relays, between LAN interfaces):

(1) $2001.3142.3181 \rightarrow 4000.1302.9611$

(1′) Across bus(ses) of R_1

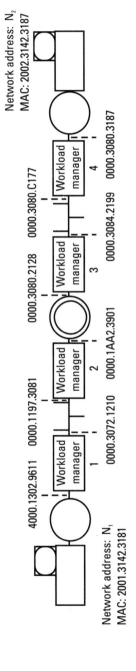

Figure 1.17 Mapping end-to-end transportation to component tasks.

(2) 0000.3072.1210 → 0000.1197.3081

(2′) Across bus(ses) of R_2

(3) 0000.1AA2.3901 → 0000.3080.2128

(3′) Across bus(ses) of R_3

(4) 0000.3084.2199 → 0000.3080.C177

(4′) Across bus(ses) of R_4

(5) 0000.3080.C178 → 0000.1118.3112

Looking over the range of workload actuations, as with bandwidth actuation, there is a clear continuum of change when it comes to workload actuation (Figure 1.18).

1.1.5.4 Coordinating Workload and Bandwidth Control

An issue remains to be considered, namely, the interaction of the bandwidth and workload managers. Of course, it is possible to have a monolithic manager responsible for both workload and bandwidth actuation. However, even in this case the question remains of which variable to actuate for controlling the performance variables of the discrete event system. A number of problems in management stem directly from the fact that the objectives of the respective control systems cannot be easily decoupled; the coupling is due to the presence of performance variables in any optimality criteria used to optimize the control of service rates and traffic, respectively. Because performance is a joint product of service and traffic rates, specifically the traffic intensity, the two indirectly influence each other.

In some instances, for example, fault recovery, it may be that both variables are actuated. For instance, beyond the case we discussed earlier where fault recovery is effected by retransmission, there are circumstances in which bandwidth management will attempt to restore the lost bandwidth while workload management attempts to reduce the arrival rate(s) until this happens. Table 1.2 lists various possible responses of the workload and bandwidth

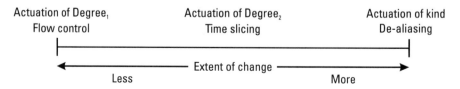

Figure 1.18 Workload continuum of change: actuations of degree and kind.

Table 1.2
Bandwidth Versus Workload: Responses

Event Responder	Workload Decline	Workload Increase	Bandwidth Decline	Bandwidth Increase
Bandwidth Scheduler	Reduce bandwidth (capacity)	Increase bandwidth (capacity)	Restore bandwidth (repair) if possible	Not applicable (unplanned capacity increases?)
Workload Scheduler	Stimulate demand Lower prices Alter priorities	Restrain demand Increase minimum Access priorities Raise prices	Restrain demand Increase minimum Access priorities Raise prices	Stimulate demand Lower prices Alter priorities

schedulers to various events in a discrete event system's life cycle. This means that the coupled schedulers must cooperate. One way to establish the rules of this cooperation is to define another scheduler that is a "master" of the schedulers of the two managers. This master scheduler receives the same state and parameter feedback from the respective monitoring servers (sensors and estimators) and using this information determines the targets that the workload and bandwidth managers will seek to attain with their respective plants.

The master scheduler can decide these targets economically. If the utility to the client of the service provided by the server is known and if the cost function of providing the service is likewise known then, using the well-known formula for profit maximization, MR = MC (from marginal economic analysis; see, for example, [4]). In other words, the master scheduler would set the bandwidth target such that the marginal revenue from client RFSs equals the marginal cost of providing the bandwidth; and the bandwidth scheduler would then seek to keep the server at that level. Difficulties in applying MR = MC include defining cost function and establishing price elasticity for the demand from the clients.

Figure 1.19 shows a master scheduler receiving feedback from the client and server monitoring servers and sending workload and bandwidth targets to the respective schedulers.

1.2 Managing Transport Clients and Servers

The previous section differentiated between servers insofar as they were components of managers (control systems)—indeed, the cornerstone of this analysis is

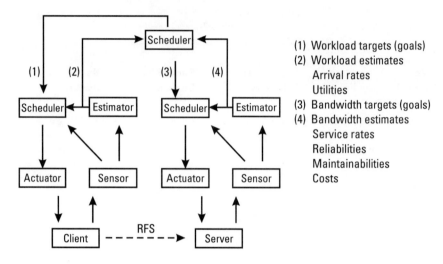

Figure 1.19 Coordinating workload and bandwidth managers.

the MESA model, with its schedulers, estimators, sensors, and actuators. However, no differentiation was made with regard to servers as plant to be managed. In this section we discuss just this: computer networks as servers to be managed.

Toward this end, we first ask a simple question: What is a computer network? One answer is to say that it is a special type of server; a subclass of actuator, what we will call a *transport actuator* or *transporter*, for short. Notice that we do not mention anything about its topology—whether it is built out of a single data link or thousands. Or the way it is put together—routers, switches, or other types of relays. That is the advantage of a functional description like "transporter," which is based on *what* is done, not *how*.

In this section we explore several points peculiar to managing transport actuators. One is the distributed nature of the plant, that is, the server and by extension the distributed nature of the workload. A particular consequence of this is the fact that a manager of a transport network has a locus of implementation—distributed or centralized. In addition, the distributed nature of the transport server means that, like the blind men and the elephant, it is possible to monitor and control the whole by having its parts report on their respective sums. This, of course, is the area of routing protocols, which we merely touch on here, but explore in detail in subsequent chapters.

1.2.1 Physical Transporters

So what does a transporter do? A transporter actuates the location of its plant, that is, it moves things. When the things are tangible items with shapes, weights, and so on, we speak of physical transporters such as trucks, trains, and planes. When the things to be moved are bits of information in files, media streams, and so on, we speak of information transporters, examples of which range from simple LANs to the Internet itself. (Of course, as Tanenbaum's example of a station wagon loaded with high-density computer tapes illustrates, there can be some overlap between the two types [5].) The plant is the data that is to be locationally actuated, that is, transported. The client is, in most cases, the source of the data to be transported, and in our case is assumed without any loss of generality[1] to be a digital computer. (Figure 1.20).

The physical layer provides the basic server of a computer communications network (internet). As we have indicated, such an entity is a transport actuator (transporter); it changes the location of information from the source to the destination. So what is the *actual* transporter here? It is a composite, the components of which are a communications channel plus the signal that the channel carries; the transmitter, an actuator that creates the signal; and the receiver, a sensor that measures it at the destination (Figure 1.21).

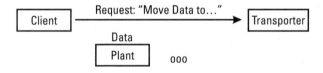

Figure 1.20 Client and transporter with plant (= data).

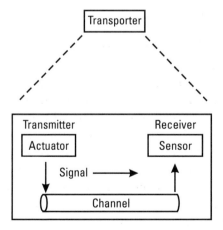

Figure 1.21 Transporter as transmitter, channel, signal, and receiver.

---

1. Human beings, telemetry equipment, and so on, can be "represented" by computers.

Few communications channels can carry the signals used within digital computers in an unmodified form. This is because of inherent limitations in the propagation of such signals over distances greater than a few feet. Because of this, the transmitter in most instances executes a transformation of the plant (i.e., the signal carrying the user data) from the form in which it is received from the client into a form that can propagate better over the channel. The generic term for this conversion is *modulation.* A corresponding demodulation then is executed at the receiving end to convert the plant back into the form expected by the receiving entity (typically, also a computer). The combination of a modulator and demodulator is termed a *modem.*

Along with modulation and demodulation, another topic that dominates digital communications is error control codes and coding theory. We referred earlier to such codes and their relationship to workload management. One result of Shannon's work [6] on the capacity of a channel was the search for error correcting codes that could utilize a higher fraction of the capacity than communications without such codes. The development in the past 15 years of higher speed modems such as 9600 and above is directly attributable to the success in finding such codes and their realization in inexpensive VLSI circuits.

These are the physical resources that move the data, and physical faults may affect them. Workload management is fundamentally the responsibility of the data link protocols. Bandwidth management basically amounts to monitoring the condition of the server, with instrumentation mainly measuring the state of the signal, the transmitter, the receiver, and the state of the channel to the extent that geographic distances allow this. Geography, however, is the key limitation to local management.

1.2.1.1 What Is Meant by Transportation?

The preceding definition of a transporter may seem unambiguous. For this reason, it may be surprising to find out that there has for many years been a deep philosophical debate in the computer communications community over what exactly constitutes communications. This is the debate over connections: Prior to the transportation of data must a connection be set up between the client, the destination, and the transporter, or can the client transmit without any prior connection being established? The most important benefit of a connection is, quite simply, that it ensures the destination will be there when the signal arrives. The disadvantage of connections is the overhead of and delay caused by the connection (and disconnection) process.

This issue of what is required for effective transportation mirrors the old paradox about the tree falling in the forest: If no one is about to hear it, does it make a sound? Clearly, the impact of the tree creates a physical "shock-wave" that propagates throughout the forest—is this what we mean by sound? If you say yes then you agree with the connectionless advocates. If you say no, there is no sound if there is no one to hear the impact, then you agree with the connection-oriented advocates. Similarly, if information via a signal is transmitted but there is no sensor to detect it, has the information been transported? Technically, the answer must be yes. Effectively, however, the answer is no.

The telephone network is a classic example of connection-oriented transportation of a signal. If the destination is unattainable, then the client will be unable to set up a connection (no one picks up the phone); there will be no uncertainty when an information-carrying signal is sent whether there was anyone to hear it. The connectionless model is used, for example, in mass communications and old-fashioned telegraphy. When a telegram is dispatched, the client who originates it does not know, unless some acknowledgment mechanism is used, whether the destination ever receives it.

Each school of thought, connection oriented and connectionless, has its advocates. The telephone world tends toward connection-oriented solutions, as can be seen in the connection-obsessive structure of broadband ISDN and ATM; the data communications community, on the other hand, tends toward connectionless models of communications, most notably in protocols such as IP. Neither is unambiguously the best approach, and we will see the debate between the two at almost every level of communications we will examine in this book.

1.2.2 Layering and Embedded Management

Let's now put back the layers of communications protocols that typically surround an actual transporter and in some way "virtualize" it. The layered model of communications is shown in Figure 1.22 using the nested (Chinese) box way of depicting this. Each layer encapsulates its predecessor, abstracting implementation details and allowing a client to request services without necessarily knowing any of the details of implementation.

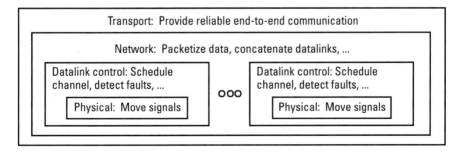

Figure 1.22 Nested boxes: layered communications.

The nature of a layered communications architecture is that intelligence is embedded within each layer to present an abstracted service to its immediately superior layer. The top layer, which provides the interface for the communicating applications, varies according to the different protocols used; but in the world of the Internet the top layer is generally TCP and/or UDP. Regardless of the top layer, as the client's request and accompanying data proceed down the protocol stack through the lower layers, the request and generally the data are both transformed into new RFSs and into multiple pieces, respectively. This is why, in the layered model of communications protocols such as the SNA, TCP/IP, or OSI, the $(n-1)$st layer can be a transporter to the nth layer, which is a proxy client for the real client, at the same time that the nth layer is a transporter to the $(n+1)$st layer, again a proxy client.

Although we have not stressed this fact, the definitions of server have all been "object oriented"; specifically, there is adherence to the principle of inheritance. In particular, the significance of this is that a composite server, such as a transporter plus a manager, may "inherit" the task(s) of the component server. In the case at hand, this means that a transporter plus a manager is a transporter (Figure 1.23). The service rates may be different, indeed the tasks may have been actuated, but they are nonetheless transportation tasks. This is the reason layering works. Because when all the layer logic (really, management logic as we have demonstrated and will show further in subsequent chapters) is stripped away, we are still left with a transporter that moves data from one location to another.

As we discussed in Section 1.1, this transformation is an actuation of the workload. Because one RFS is being transformed into a completely different RFS, this is actuation of kind. Figure 1.24 shows this process for several arbitrary layers in the OSI protocol stack, along with the workload managers that must be there to effect the transformation of one layer's Service Data Unit (SDU) into another (ICI refers to control information passed between layers).

1.2.3 Locus of Implementation: Centralized Versus Distributed Management

The Internet, and by extension modern data communications itself, comes from the marriage of two distinct technologies: the use of voice communications channels, made possible by inexpensive modems necessary for sending digital data over voice networks, to facilitate multiuser access to computing

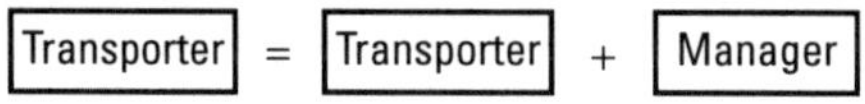

Transporter = Transporter + Manager

Figure 1.23 Inheritance: transporter = transporter + manager.

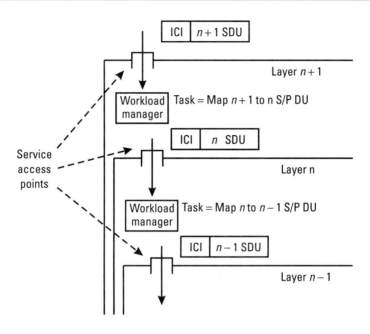

Figure 1.24 Workload managers with OSI layers: actuation of kind.

centers; and the development by Paul Baran et al. at the RAND Corporation of adaptive, fault-tolerant management protocols for highly available communications networks [7].

Baran's work was driven by the U.S. military's need for highly survivable communications networks, with the attendant requirement for no single point of failure. This precluded the use of management that relied on centralized monitoring and controlling. To get around this, Baran and his colleagues investigated a number of concatenation techniques, both open and closed loop, that did not rely on routing centers, and which had the vital property of adapting to the loss of communications channels and/or relays. Prior to this dynamically based/distributed routing approach, other networks used centralized route determination computers to calculate optimal routes, which were then disseminated to the individual routers.

1.2.3.1 Estimators and Routing Protocols

Perhaps the most complicated topic we will discuss in this book is that of routing protocols. The irresistible analogy for a routing protocol is the story of the blind men and the elephant. This fable tells of a certain number of sages, who might be more generously referred to as visually challenged, confronted by an unknown entity. Each touches the part nearest him and comes to the appropriate conclusion: The man who touches a leg thinks he has felt a tree, the man who touches the trunk thinks he has felt a snake, and so on. Only by sharing with each other their respective impressions do the sages realize they are feeling up a pachyderm. The pachyderm's reaction to all this is unrecorded, and the story ends before we find out.

The analogy with a routing protocol is apt. Assume each relay (workload manager) has a bandwidth manager as well, measuring the condition of the local channels and the relay itself. This local state information is then broadcast or otherwise disseminated to the other relays, each of which uses the sum of this local information to reconstruct the global topology of the network. Such reconstruction is clearly the task of an estimator, and this points out one part of a routing protocol: the presence of a bandwidth estimator to put together the "big picture." (Seen another way, a routing protocol's collection of topology information is a type of configuration management; this tells us something about the functional areas of network management, a topic that we return to in the next section.)

The reason for collecting local topology (state) information and reconstructing the global topology is to efficiently use the components of the network—the data links and relays. Recall that the challenge of "computer networking" is knitting together multiple transport actuators. When it comes to this concatenation, the important question is "What is the path?" This brings us to the major second part of a routing protocol: scheduling. Determining the best next stage in a multistage transporter can be done in several ways, such as distance vector or link state (more on these later in the book), but in any case this is the responsibility of a workload scheduler.

We see that a routing protocol can be decomposed into a workload scheduler and a bandwidth estimator (at a minimum, other management components are usually present). We explore this in much greater detail in later chapters.

1.2.4   Costs of Implementing Management

This discussion of routing protocols gives us the opportunity to touch on something often ignored: the cost of management. Management is not free. There is a significant expense to the various management servers necessary to monitor and control the network. The question arises "How much should be invested in management servers and how much instead should be devoted to additional servers for the network itself?" An answer to this depends on the relative contributions to performance obtained from added management bandwidth versus added transport bandwidth.

These costs to managing a network come in two varieties: the fixed costs, generally from the implementation of the management servers, and the variable costs that are incurred when these servers execute their respective management tasks. For example, monitoring a communications network requires instrumenting it, that is, implementing sensors to measure the network's state, as least as it is reflected in the output (that is, measurable) variables. The "upfront" cost of implementing the sensors is fixed. Whether these servers are monitoring/measuring/executing or not, they still must be "paid" for.

On the other hand, certain costs accrue from operating the sensors: Power is consumed, memory and CPU cycles may be used that might otherwise be employed in executing communications and other tasks, and, not least, the measurements made by these sensors are usually sent to management decision servers (estimators and/or schedulers) located elsewhere in the network. This last cost can be particularly significant because management traffic (consisting of such things as these measurements from sensors, estimates from estimators, and commands from schedulers) either competes with the user traffic for the available bandwidth of the network or must flow over its own dedicated communications network.[2]

1.2.4.1 The Cost of Management: Routing Overhead

An example of this is the routing protocol's topology data exchange, which uses network bandwidth that is consequently unavailable for transporting end-user data. And, unfortunately, as the size of the internetwork grows, the amount of topology data exchanged grows even faster, in fact as $O(n^2)$. To reduce this quadratic scaling, various techniques have been devised that, generally speaking, involve aggregation of routing information, with a concomitant loss of granularity. This modularization, in fact, results precisely in the hybrid locally central, globally distributed decision structure we referred to earlier (Figure 1.25).

One design objective, therefore, when planning the monitoring network is to reduce the volume and frequency of measurements that must be sent over the network. Two types of sampling can reduce this volume: spatial sampling and time sampling. Sampling in time brings us to the sampled data control system. Spatial sampling is considerably more complex and is applicable to distributed parameter systems such as computer networks.

---

2. The former is often referred to as *mainstream management*, whereas the latter is called *sidestream management*.

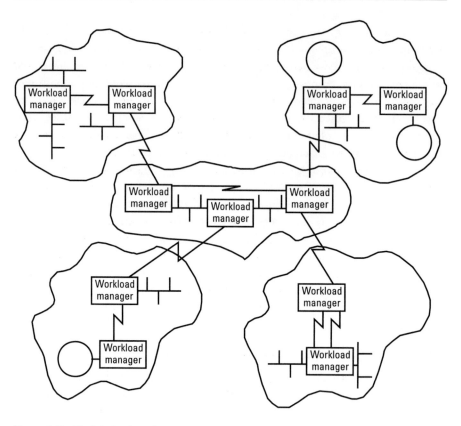

Figuro 1.26 Modulorizotion of routing information.

1.3 False Dichotomies, Reconsidered

As noted earlier, a principal catalyst for developing the MESA model of management—bandwidth versus workload, actuation of kind and actuation of degree (both types)—was to eliminate from network theory certain false dichotomies, that is, artificial distinctions based on practice and precedence, perhaps, but not well founded in analysis. Conspicuous among these are the prevailing wisdom that protocol issues are distinct from management issues, and that the events of a network's life cycle constitute distinct and dissimilar phases.

1.3.1 Management Versus Protocol

In the development of computer networks and their associated protocols, at an early point tasks were divided into two categories: those that were part of enabling computers to send and receive data among themselves and those that were part of managing the enabling entities, that is, managing the communications networks themselves. The first of these we have come to call *computer networking,* the second *management.* Yet, we will demonstrate in this book that, apart from the simple transportation of the plant as it was received from the client, all other tasks executed in a computer network are management tasks.

Take, for example, two computers sending and receiving data from each other. Computer A is sending data to computer B. If a fault occurs that causes the link to corrupt the data but it still arrives at computer B (what we call a *latent fault*) or if the fault causes the link to fail to transport anything at all (what we call a *fatal fault*) then B sends feedback to A informing it that, by inference from the condition of the received data (or its failure to arrive at all), a fault has occurred in the transmission.

Reliable transport means that, at some level in the protocol stack, there is a mechanism for fault detection and recovery by, in the first instance, resending the data that failed to arrive at the destination correctly. Such retransmission is predicated on the fact that many faults are transient and, in effect, the server "repairs" itself when they occur. Most connection-oriented data link protocols (for example, SDLC or LAP-B) specify that the sender will retransmit a finite number of times in case the fault proves to be transient. All fault detection, isolation, and recovery is automated, implemented in a combination of hardware and software.

Contrast this, however, with the situation that arises if the communications link(s) suffer a more persistent fault, in which retransmission will be of no avail. If the network's topology has been laid out such that alternative routes are available linking client and destination *and* if the network possesses sufficient intelligence to reroute traffic, then automatic recovery is still possible. Whether such is the case or not, however, full recovery typically will require sending a service technician to examine and repair the physical links in question. This may involve pulling configuration records from a computer database (or file cabinet), setting up line probes and analyzers to identify the faulty component(s), replacing or repairing those components thus isolated, testing the repaired communications link(s) to check the repairs, and so on.

In computer networks even most of these "management" tasks are executed manually. On the other hand, the protocol management tasks are not ever executed manually, that is to say, by people examining each frame, deciding if it is bad, scheduling retransmission, and so on. Any way we look at it, the prospect is absurd. Yet to distinguish between identical servers because one is implemented digitally and the other manually is no less absurd—all we care about in terms of servers is their nominal bandwidth, their reliability, their maintainability, and their task set. The difference between communications protocols and network management comes down to *how* the tasks are executed (i.e., manual versus mechanized). In fact, what we have are two bandwidth managers that are components of a multiechelon maintenance server (bandwidth manager).

1.3.2   Unified Life Cycle Management

In the section on bandwidth management, we discussed the occasions when a server's bandwidth was actuated, namely, when it is initially designed and implemented, during fault recovery, and at the time of upgrades or other improvements (Figure 1.11). This integrated approach is consistent with the experience of software developers, in which the life cycle concept has grown in popularity in recent years as the best way to address the design, implementation, and maintenance of software. The core lesson is that all phases of a project's existence should be considered from the beginning when making trade-offs in design and implementation.

A side benefit to considering a new approach is that it allows us to step back and look at the whole network life cycle and see if we can integrate or unify its activities, in other words to develop a new approach to managing the life cycle of the computer network, from its preimplementation design to fault management to growth and decline as traffic ebbs and flows. What is needed is a unified approach encompassing four areas of computer networking traditionally addressed separately: network design, tuning, management, and operations (including routing).

Toward this end, we can identify certain well-defined events that punctuate the life cycle of a computer network. First of all, there is the initial design and implementation; this is unique because there is no existing network to consider. The next event to typically occur is a fault; thus fault detection/isolation/recovery must be considered and the servers implemented to execute the requisite tasks. Finally, as the traffic workload changes, the network itself must adapt: A workload increase requires capacity expansion, whereas a workload decrease may necessitate some capacity reduction. Although the latter may seem unlikely in this time of explosive Internet growth, even the normal ebb and flow of daily commerce can impact the design of the transport infrastructure—especially provisioning communications bandwidth (discussed later).

1.3.2.1   Timescales of Network Dynamics

We can arrange the various tasks executed in a computer network according to the frequency with which they happen. As a matter of fact, a convenient way to look at these is by their "interarrival times" (Figure 1.26).

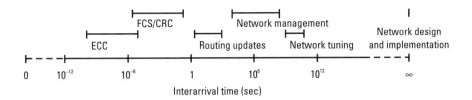

Figure 1.26 Time constants of network events.

- The least frequent is, naturally, the design of the network; this can happen only once, because all subsequent changes to the network will have a functioning network to start with.

- The next most frequent activity is network tuning/redesign. This occurs approximately every 3 months to every 3 years; that is to say, every 10^7 to 10^8 seconds.

- Likewise, the events of network management occur approximately every 10^3 to 10^6 seconds (20 minutes to 11 days). If such events occur more frequently, then that means the network is suffering faults more frequently than normal operations would allow.

- Finally, network operations represent the shortest timescales of all. Consider a 1-Gbps channel, more or less the upper limit on communications today. Assuming that error correction coding (ECC) is employed then a decision must be made approximately 100,000,000 times per second as to whether a fault has occurred. Likewise, a frame with a Frame Check Sequence to be calculated will arrive from 100 to 1,000,000 times per second.

1.4 Summary

We said at the outset that this book was predicated on two assertions. First, that management pervades computer communications networks, their architectures, and their protocols. Second, that management is best understood using the concepts and vocabulary of control engineering. Toward this latter end, this chapter has introduced a new conceptual framework, borrowed from control engineering, which we have called the Measure-Estimate-Schedule-Actuate or MESA model.

The advantages of the MESA/control systems model can be summarized as follows:

1. Results in less unplanned, uncoordinated redundancy in the implementation;

2. Encompasses the life cycle of an internet within one unifying framework; and

3. Defines a new vocabulary—consistent and (almost) complete.

Taking this last point first, terminology bedevils any technology discussion, computer communications more than most. This chapter has pointed out several key areas where the current approach to internetworking has created artificial distinctions and unnecessary complexity—our "false dichotomies." What has been needed is a new vocabulary for discussing the relevant concepts free from the "baggage" of current definitions.

This brings us to our next point: More than just a vocabulary, however, this chapter has shown that the MESA model gives us a "toolbox" with which we can constructively define what it means to manage a discrete event system in general, and an internet specifically. First, we showed this amounts to actuating workload and bandwidth. Second, with respect to each of these, we showed the range of possible actuations: from actuation of $degree_1$ (no task transformation, simply changing the arrival rate) to actuation of $degree_2$ (no task transformation, but with task replication = time slicing) to finally actuation of kind (with transformation of the task).

Likewise, we demonstrated that layering and encapsulation, concatenation, and even "traditional" network management issues such as security and accounting all can be best explained with these basic concepts. The task(s) executed by computer networks involve the transportation of data from one location to another. The actual transportation of data is effected using one or more communications channels (and their respective signals)—the *sine qua non* component(s). However, that is only the beginning of the process. The vast majority of tasks executed are management tasks. Indeed, as we showed in Section 1.1 both workload and bandwidth management tasks are present everywhere:

- Scheduling mechanisms such as token passing, carrier sensing, and polling are all examples of workload actuations of $degree_1$.

- Packetizing data for time-division multiplexing of communications channels and relays is workload actuation of $degree_2$.

- Finally, layered protocol models with their emphasis on abstracting a layer's implementation details from its client(s) in upper layer(s) involve workload actuation of kind.

This chapter has discussed other advantages to approaching networking and network management from an automatic control theory perspective. Control engineering is concerned with ensuring the stability and optimality of controlled systems, called the *plant*; here the plant is the transport network being managed. The control system does this by mathematically modeling the dynamics of the plant and by applying various optimization techniques to determine the best management policies. Part of the challenge of this is that simple models do not work when it comes to controlling computer networks and their traffic flows because of their spatially varying nature. At its heart is a set of optimization decisions concerning revenue and costs, traffic flows and service capacities, and the related parameters that are relevant to network operations. The concept, at least, is simple:

1. Establish an objective to be attained, for example, a range of acceptable response times, server utilizations and traffic intensities, or other.

2. Monitor the arrival of traffic and its servicing by the network's links.

3. Based on these measurements and/or estimates, determine the necessary intervention (if any) to optimize the network's performance.

4. Finally, effect the change using the available management actuators—workload and/or bandwidth.

References

[1] Kwakernaak, H., and R. Sivan, *Linear Optimal Control Systems,* New York: John Wiley & Sons, 1972.

[2] Goldman, A., and T. Slattery, *Maintainability: A Major Element of System Effectiveness,* New York: John Wiley & Sons, 1964.

[3] Barlow, R., and F. Proschan, *Mathematical Theory of Reliability,* New York: John Wiley & Sons, 1965.

[4] Nicholson, W., *Microeconomic Analysis,* 3rd ed., Dryden Press, 1985.

[5] Tanenbaum, A., *Computer Networks,* 2nd ed., Upper Saddle River, NJ: Prentice Hall, 1988, p. 57.

[6] Shannon, C. E., "A Mathematical Theory of Communication," *Bell Sytems Technical Journal,* Vol. 27: pp. 329, 623 (1048), reprinted in C. E. Shannon and W. Weaver, *A Mathematical Theory of Communication,* University of Illinois Press, 1949.

[7] Baran, P., On Distributed Communications: XI. Summary Overview, Mem. RM-3767-PR, The Rand Corporation (August, 1964). For a popular treatment of Baran's books, see K. Hafner and M. Lyon, *Where Wizards Stay Up Late,* New York: Simon and Schuster, 1996.

2

Management of the Basic Transporter: Channels, Signals, Modulation, and Demodulation

2.1 Introduction

This chapter presents the MESA analysis of the mechanisms involved in transporting signals across communications channels and especially the modulation and demodulation required to do so. While the intricacies of modulation, transmission, and demodulation extend beyond the scope of this book, we can nonetheless derive meaningful insights into the tasks involved without delving into all of the minute details. The purpose here and in subsequent chapters is to focus on placing the tasks in the context of our control theoretic model of workload and bandwidth management; and identifying where the respective tasks of estimation, scheduling, actuation, and measurement are executed in different types of implementations.

As we saw in the introductory chapter, workload and bandwidth management can occur at many levels, particularly when the server is composite. That is the case here: The transporter is composed of a channel and associated signals, as well as the modulator that created the signals and the demodulator that recovered the signals after they had been transported across the channel and subjected to its noise and other faults. This combination of modulator–channel–demodulator is, in communication and information theory, referred to as a discrete memoryless channel (DMC) for reasons we explain later.

Putting things into the context of workload/bandwidth management, this chapter demonstrates that the modulator is an actuator that creates the signals. In fact, the transmitter is a composite: a scheduler that selects which signal waveform to transmit and the actuator that creates the waveform selected. This constitutes both bandwidth actuation, regarding the channel/signal as the server, and, in a different frame of reference, workload actuation of kind since it can be looked on as transforming the RFS to the DMC into RFS to the channel within the DMC; that is to say, the client of the DMC issues an RFS to the DMC, but conceptually this must be mapped or transformed into an RFS to the actual transport actuator, namely, the channel that is contained within the DMC. Returning to modulators, the chapter examines the various types of actuation, that is to say modulation, and the corresponding signals they create. The "big picture" can be stated very simply: The modulator actuates the plant into a form that matches the channel's characteristics and that, subject to constraints on cost and realizability, optimizes or otherwise improves the chances of successful understanding (estimation) at the receiver.

Complementing this (bandwidth) actuation, the chapter then proceeds to examine the mechanisms of demodulation, that is, the measurement and estimation of the server (the channel/signal composite). Such instrumentation is at the heart of designing the receiver; in particular, this chapter discusses the design of demodulators to reconstruct the transmitted waveform. The chapter also explores the efficacy of demodulation and its intimate connection to the types of signals and types of modulation used.

2.2 Channels

2.2.1 What Is a Channel?

It is appropriate to begin the chapter with the channel. One definition in the literature is as follows:

> **channel** (1) That part of a communications system that connects the message source with the message sink. In information theory in the sense of Shannon the channel can be characterized by the set of conditional probabilities of occurrence of all the messages possible received at the message sink when a given message emanates from the message source. (2) A path along which signals can be sent.... [1]

The definition proposed by Shannon himself provides some examples of channels:

The *channel* is merely the medium used to transmit the signal from transmitter to receiver. It may be a pair of wires, a coaxial cable, a band of radio frequencies, a beam of light, etc. [2]

Although we hesitate to take issue with Shannon, we would alter his definition with concern to the radio band and the beam of light: Both electromagnetic waves are signals, and the channel is the space through which it propagates. In other words, channels are not always wires or other tangible entities. Wireless communication, for example, with microwave links and satellite channels relies merely on space for the propagation of its signals. Figure 2.1 shows a wireless communication scenario built around two ground stations, each a pairing of transmitter and receiver, and a geosynchronous satellite, itself comprised of two pairings of receivers and (re)transmitters. The channel is the volume of space through which the radio signals propagate between the two earth stations and the satellite. Likewise, the channel used with an infrared communications link is just the space transited by the infrared beams.

2.2.2 Channels: Actual Versus Ideal

An ideal channel is quite simple to define, if not to implement: It would transport a signal without changing it. Actual channels, however, are neither that

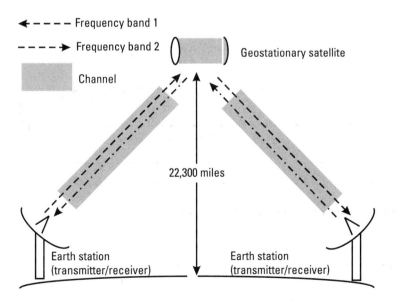

Figure 2.1 Satellite channels.

simple nor that benign. Signals are attenuated, delayed, and distorted by the addition of noise from the channel itself, from amplifiers and other ancillary components, and from exogenous sources outside the communications system. Complicating this picture even further, many of these deleterious effects vary according to the frequency of the signals.

Attenuation, for example, is typically uneven across the frequency spectrum, with the consequence that channels are often modeled mathematically as linear bandpass filters. The dimensions of this passband, commonly measured from the high and low frequencies that have been attenuated by 50% from the nominal response, are referred to as the bandwidth of the channel (and corresponding filter). Converting this 50% loss into the decibel (dB) measurements frequently used in communications engineering, the passband is the band of frequencies for which the attenuation is less than 3 dB off the ideal or nominal response (Figure 2.2).[1] Note that the bandwidth of the channel is *not* the same thing as what we've been calling the bandwidth, that is, the service rate, of the channel/signal composite. The former is a factor in determining the latter, which in information and communications texts is called the *channel capacity* (more on this subject later).

The situation is similar with delay. Figure 2.3 shows the delay that might be experienced with a given channel over a certain distance. Few, if any, channels propagate signals of different frequencies uniformly, that is, at the same rate. Indeed, one widely held misconception of Einstein's theory of relativity

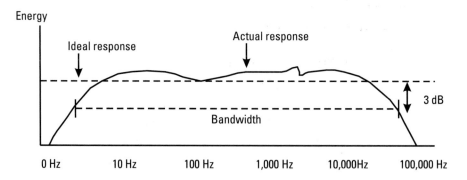

Figure 2.2 Bandwidth: attenuation versus frequency.

---

1. A decibel is a measure of gain, that is, amplification (or attenuation). Decibels are expressed logarithmically. If A is the amplitude of a variable before it has been subject to gain and B is its amplitude after the gain, then the value of the gain G in decibels is given by the equation $G = 20 \log_{10}(B/A)$.

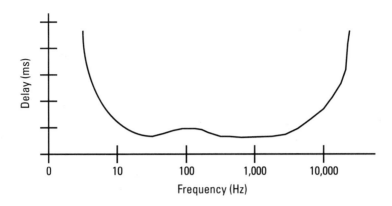

Figure 2.3 Channel delay versus frequency.

concerns the speed of light: Although electromagnetic radiation propagates at 300,000 km/s in a vacuum, through nonvacuous media different frequencies propagate at different speeds. This, in fact, is the basis of the phenomenon of refraction. Different frequencies of light propagating at different speeds produce the refractive, spreading effect. The same is true of communications channels, in which certain frequencies are retarded relative to others, inducing relative phase skews that enormously complicate the problems of demodulation (see later discussion).

2.2.3 Noise

Beyond its bandwidth, a channel is characterized by its noise—roughly speaking, its reliability. Noise is a sequence of latent faults that corrupts the signal propagating over the channel. There are different types of noise—notably the thermal (Gaussian) noise inherent in any physical system, and the impulse noise associated with individual events such as relays closing, accidental contacts due to careless maintenance, and even lightning strikes. Models of noise must balance capturing these real-world details versus mathematical tractability. The particular "type" of noise used in most models of channels is additive white Gaussian noise (AWGN); this results in the so-called AWGN channel encountered frequently in discussions of digital communications, modulation, and coding theory. As with attenuation and delay, noise is seldom uniform across all frequencies.

Imperfections and faults in the channel introduce noise. A partial list includes linear distortion, nonlinear distortion, frequency offset, phase jitter, and impulse noise.

2.2.4 Channel Topologies

To borrow a phrase, no channel is an island. It exists within the context of a communications system that consists of the channel; a transmitter, which contains an actuator that creates the signal; and a receiver, which contains a sensor that measures the received signal; and if necessary an estimator that applies estimation techniques to reconstruct the original signal (Figure 2.4).

We can interpret the relationship of transmitter to channel in two ways. In both interpretations the transmitter is an actuator—the difference between the two interpretations being what we regard as the plant. In the first instance, the transmitter creates (actuates into existence) a signal that the channel transports to the destination; that is, the plant of the transmitter is the signal, and the channel is just the actual transport actuator of the signal. The channel "picks up" the signal and carries it to one or more destinations. The second interpretation regards the transmitter as an actuator that changes the state of the channel, that is, the channel is the plant of the transmitter. In other words, whether the channel is a pair of electrical conductors, a fiber optic cable, or, in the case of wireless communication, just plain space, the transmitter changes the state of the channel. For example, if the channel is a pair of wires then the transmitter changes the state of the pair by introducing a current/voltage waveform.

Such is also true in terms of the relationship of channel to receiver. At the destination, a sensor is present that measures one or more state and/or output variables of the channel to detect the transported signal. One can regard the channel as the plant of the sensor or one can regard the signal as the plant. It essentially amounts to whether we regard the channel as distinct from the signal or inextricably bound with it; arguments can be made either way, and for our purposes it is not important where we draw the boundary.

Leaving philosophy, there is the question of channel topology. The topology of a channel—more broadly, the topology of the transporter—can be characterized by the relationship of transmitter to channel to receiver. The simplest, clearly, is a one-to-one-to-one relationship: The topology is a point-to-point one, connecting a single transmitter to a single receiver. This is the simplest

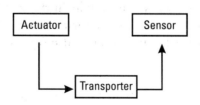

Figure 2.4 Communications channel with transmitter and receiver.

possible transport actuator. The task set of the transport actuator contains only the single source → destination transport actuation task.

Point-to-multipoint (one-to-one-to-many) topology allows a single transmitter to transmit over a channel and multiple sensors to receive the signal without creating untenable reception problems for each other (Figure 2.5). This is obviously the mass communications model, in which a single transmitter broadcasts either line of sight or via such intermediary mechanisms as satellites or tropospheric scattering, and then distributes its signals to millions of receivers—television and radio stations, for example. However, it is also the model of a communications bus—Ethernet LANs, for example. In both instances, the signal is sent simultaneously to multiple destinations.

Finally, we come to a many-to-one-to-many topology. A channel that supports multiple transmitters is clearly the most complicated to effect, and it requires workload management mechanisms to allocate the channel bandwidth. We defer a discussion of workload management, including frequency- and time-division multiplexing, until later chapters.

Note that the type of channel and signal frequently dictate limitations on the channel topology. Many channels are inherently broadcast in nature: wireless channels, for example, and even some electrical channels such as busses. Others, for instance waveguides, tend to be limited to point-to-point configurations. Such limitations can, generally speaking, be overcome by use of various relays and repeaters. For example, Token-Ring LANs are actually a set of point-to-point channels, each channel having a single transmitter and a single receiver, joined at the physical layer.

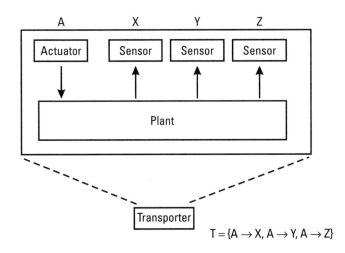

Figure 2.5 Broadcast transporter with actuator, plant, and three sensors.

2.2.5 Discrete Memoryless Channels

Within the physical layer a key abstraction is that of the discrete memoryless channel (DMC; Figure 2.6). The essence of digital communications is that information is discretized. The analogy can be made to sampled data or digital control systems, in which the measurements and the actuations are quantized to a finite set and occur at periodic intervals. A discrete channel is one that receives a finite set of input messages $\{X_i,\ i = 1,\ \ldots,\ m\}$ from a source and, after transporting the corresponding signals, outputs a finite set of messages $\{Y_j,\ j = 1,\ \ldots,\ n\}$ to a sink. By abstracting the entire communication system—transmitter, channel, and receiver—a discrete channel results.

As for memorylessness, recall the first part of the definition of channel offered earlier: "the channel can be characterized by the set of conditional probabilities of occurrence of all the messages possible received at the message sink when a given message emanates from the message source." With a memoryless channel, the conditional probabilities associated with a given output message depend only on the input message. A DMC refers to one in which the sets of input and output messages are both finite, and in which the conditional probabilities of incorrect execution are independent of the previous messages sent (memoryless). It is frequently convenient to talk about discrete memoryless channels as abstractions, and indeed when we discuss coding theory in the next chapter we will do so.

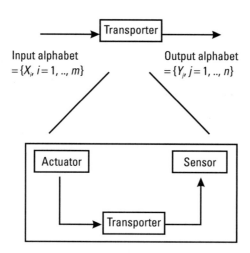

Figure 2.6 Discrete memoryless channel.

2.3 Signals

2.3.1 What Is a Signal?

After channels, the obvious next thing to explore is signals. Indeed, as we tried to explain in the previous section, a channel without a signal is hardly a channel at all—at the very least, it is not able to function as a transport actuator to carry information. This is why we stress the fact that the composite of the channel and the signal is the working unit—each is necessary, neither is sufficient.

So what *is* a signal? One standard reference work reads:

> **data transmission** (1) A visual, audible, or other indication used to convey information. (2) The intelligence, message or effect to be conveyed over a communications system. (3) A signal wave; the physical embodiment of a message. [3]

A second definition offers a more "active" view of signals: "In communications, a designed or intentional disturbance in a communications system" [1]. Looking at these two definitions, several things are clear. First, signals have transitions. A constant, undisturbed wave, for example, does not convey anything because it never changes. In terms of information content, a constant signal has a probability of one and an information content of zero; to see this, recall that information content is measured by the negative of the logarithm of the probability, and the logarithm of 1 (the probability) is 0. (Just think of the blaring car alarms in urban neighborhoods—few if any listeners pay any attention because there is always an alarm going, forming an acoustic drone not unlike water torture.)

The second thing that is clear is that signals are, essentially, energy. Indeed, in this respect at least, signals and noise have much in common—they are both forms or states of energy. When energy has a nonrandom (intentional) content it carries information; the random component is called noise. As we saw when discussing actual versus ideal channels, all physically realizable mechanisms of transporting information that use signals/energy suffer from degradation, in which the noise increases and signal decreases. The crucial parameter here is the signal-to-noise ratio (S/N): the energy contained in the signal versus the energy contained in the noise on the channel.

Note that we can measure the respective energy or power in the signal and the noise in many different ways. Some of these include Root Mean Square (RMS) and peak signal-to-peak noise methods. Whichever method is used, the higher the S/N, the less likely a receiver is to mistake the incoming message for another. In conventional telephone networks, for example, it is typical for noise to be 20 to 30 dB below the signal level.

2.3.2 Capacity of the Channel/Signal Composite

Now that we have touched on signals and the S/N, we can elaborate on the whole question of bandwidth—the bandwidth of the channel versus what we have called the bandwidth (i.e., service rate) of the transporter of which it is part. Channel or communications bandwidth—a term we have had cause to reconsider or at least nuance in the course of the chapter—is but one factor in determining the bandwidth (as we use the term) of the transporter. The other factor is the power of the signal it carries relative to the strength of the channel's noise, that is, the S/N.

This is the substance of Shannon's Channel Capacity Theorem, which established, using information theoretic arguments, upper bounds on the attainable capacity (i.e., what we refer to as the bandwidth of a server) of a channel with a given noise level carrying a signal of a given power level. Given a transmitter that creates (= actuates) a signal with power S and a channel that has communications bandwidth BW and noise N, what in information theory is called the *maximum capacity* (service rate) of the combination is given by the equation:

$$\text{Capacity} = BW \times \log_2\left(1 + \frac{S}{N}\right)$$

This is Shannon's result and is frequently called the *fundamental theorem of information theory*. Notice that if S/N is zero then so is the capacity. If S/N = 1 then the capacity is equal to the bandwidth of the channel. Although it might seem possible to increase channel capacity arbitrarily simply by increasing the signal power, the realities of implementing the mechanisms involved—transmitters, channels, and receivers—constrain signal levels because higher signal levels require components that are both more expensive and more unreliable. This is part of the system designer's cost structure when budgeting for a communications system.

2.3.3 Signals and Signal Rates

Now that we know the maximum possible bandwidth of a transporter, we can address the question of how much data we wish to send across the channel and how best to accomplish this. There are essentially two types of signals:

broadband and baseband. The difference between the two types comes down to the presence (broadband) or absence (baseband) of a carrier signal that is modulated to encode the incoming messages. It is worth noting that the signals used within computers are, with few if any exceptions, always baseband (digital signals without any modulated carrier). Baseband signals, by their very nature, are always electrical, whereas broadband signals can be either electrical or wireless (electromagnetic) (Figure 2.7).

A note on terminology: Recent years have seen a new meaning assigned to the term *broadband*, as in Broadband-ISDN (B-ISDN), generally as a synonym for Asynchronous Transfer Mode (ATM) and the very high bandwidth communications links of the Synchronous Digital Hierarchy. The term *broadband* was originally used in telephony to refer to refer to "a bandwidth greater than a voice-grade channel (4 Khz) and therefore capable of higher-speed data transmission" [1]. While ATM over fiber optic channels does use a form of carrier-based communications (on–off keying; see later discussion), ATM over electrical channels such as DS-3s uses baseband signaling. Consequently, we will try to be careful to specify *which* broadband is meant when we use the term.

Back to signals. A signal $s(t)$ can be mathematically represented by the function

$$s(t) = a(t)\cos\left[2\pi f_c t + \theta(t)\right]$$

where $a(t)$ is the amplitude of the signal, cos is the sinusoidal carrier with the frequency f_c, and $\theta(t)$ is the phase of the signal [4]. From this equation we can see that three basic degrees of freedom are required when modulating a carrier

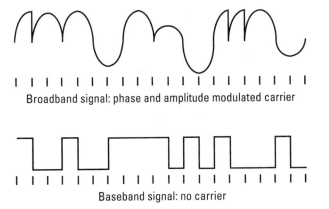

Broadband signal: phase and amplitude modulated carrier

Baseband signal: no carrier

Figure 2.7 Broadband and baseband signals.

so that it can carry information: amplitude, frequency, and phase. These correspond, respectively, to amplitude, frequency, and phase modulation, three common techniques of modulation (think of AM and FM radio, for example).

If the signals within computers are baseband, then why bother to use broadband signals for digital communications? Quite simply, because of distance. Typically, baseband signals have broad spectra with significant high-frequency content. A Fourier analysis of the signal reveals the reason for this: the sharp transitions or "edges" in the signal. And, as noted earlier when we discussed bandwidth-limited channels, their attenuation of frequencies is uneven, with high-frequency signals being more affected than most. This means that recovering the original baseband signal becomes problematic once the signal has propagated more than a limited distance. For this reason, baseband signals are typically limited to LANs (both 802.3 and 802.5 use baseband signals) or to very specialized applications within the telephone network where digital repeaters reform the signal frequently to overcome the attenuation effects.

Broadband signals, on the other hand, are especially well suited to traversing long distances over channels with minimal attenuation and delay. This is partly because, with some care exercised in their design, carrier-modulated signals have spectra that are limited in breadth and consequently do not suffer from the extreme gradients of deterioration that characterize long-distance channels at the extremes of their passband. Of course, many channels can support either type of signal so it is a designer's choice (see Table 2.1).

2.3.4 Waveforms

Recall that we said a discrete channel, memoryless or not, has a finite alphabet of input and output symbols. That is to say, unlike analog communications systems (for example, radio and television), which must transport signals with potentially unlimited variations, digital communications systems must

Table 2.1
Types of Channels: Baseband Versus Broadband

Channel	Baseband	Broadband
Twisted pair	Yes (e.g., 802.5)	Yes (e.g., modems)
Coaxial	Yes (e.g., 802.3)	Yes (e.g., 802.4)
Fiber optic	No	Yes
Infrared	No	Yes
Radio	No	Yes

accommodate only a finite number of input variations—two at a minimum since 1 and 0 define the minimum "symbol alphabet" that a discrete memoryless channel can be presented—in which case it is called a binary channel.

As a consequence, the modulator in a digital communications system need only create a finite set of signals corresponding to the number of input "symbols" that the system accepts. These modulated signals are called *waveforms* and constitute the basic unit of communication across the channel. Such waveforms are, from an information theoretic perspective, symbols with respect to the channel, and much of waveform design is concerned with minimizing and mitigating what is called *intersymbol interference* as the waveforms traverse the channel.

Just as there are baseband and broadband channels, so are there corresponding waveforms: digital waveforms for baseband channels and analog (carrier-modulated) waveforms for broadband channels. Unfortunately, the communications field has two meanings for the term *waveform* (yes, Virginia, we come to another instance of inconsistent terminology and overloaded definitions). The first, more restrictive meaning, limits *waveform* to analog carriers. In the context of this more restrictive definition, the term *waveform channel* is used to denote a channel that propagates analog (broadband) signals. For example, if phase modulation is used two instances of the carrier might have phases of 45° and 135°; with frequency modulation, one waveform might have a frequency of 88.100002 MHz while another has 88.099998 MHz.

The second, broader meaning of *waveform* includes not just analog signals but digital as well. Obviously, a digital waveform does not use a carrier. Digital waveforms exist in various shapes, but the key thing is that they use a finite number of levels and a finite number of transitions between these. The propagation characteristics of digital electric signals are very different than the propagation characteristics of analog electric signals. Digital electric signals suffer greater losses due to impedance than analog signals, and they are less impervious to certain types of noise. On the other hand, digital signals can be regenerated by a combination of sensor, estimator, scheduler, and actuator. (The longer distance baseband channels in the telephone network have digital repeaters spaced as frequently as every 1.2 km.)

We now examine analog and digital waveforms in more detail.

2.3.4.1 Analog Waveforms

As was just noted, the term *waveform* in some texts means carrier-modulated (i.e., analog) signals, and this is why when we speak of modems (modulators/demodulators), we are specifically referring to analog waveform signaling. A carrier can be modulated in three basic ways: amplitude, frequency, and phase. More complicated waveforms involve a combination of two or more of the amplitude, frequency, and phase. Among the issues involved in designing analog waveforms are the carrier and its sidebands: Is there a single sideband used in the modulated signal or dual sidebands on either side of the carrier, is the carrier itself present or suppressed in the transmitted signal, and so on? The interested reader should consult one of the standard texts on digital communications, such as [4], [5], or [6].

Amplitude and frequency waveforms are relatively straightforward to envision, since they are basically similar to the familiar radio modulation techniques. Figures 2.8 and 2.9 show two hypothetical sets of four waveforms, one amplitude modulated and the other frequency modulated (with the frequency differences between waveforms greatly exaggerated).

Phase modulation, on the other hand, has no radio equivalent, but does have advantages that make it particularly suited for higher speed signaling (for the mathematics, consult one of the previously mentioned texts [4–6]). One variant of phase modulation is *differential* phase modulation, where the information is encoded in the difference in the phase from the previous waveform. Figure 2.10 shows four phase-modulated waveforms, which may be differential or not depending on whether there is a fixed association of symbols to waveforms or if the association of symbols is to the transitions. For example, a transition of 45° might correspond to the input symbol 00, a transition of 135° to the input symbol 01, and so on.

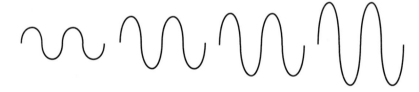

Figure 2.8 Four AM waveforms.

Figure 2.9 Four FM waveforms.

Figure 2.10 Four PM waveforms.

Different modulation techniques have been employed, particularly as technologies such as integrated circuits have evolved and speed requirements have increased. Early modems used amplitude or frequency modulation, and even today's less-expensive low-speed modems, for example V.21 modems, still employ these. Phase modulation is used at greater speeds, as for example in V.27. Among the more sophisticated modems, one popular technique is called *quadrature amplitude modulation* (QAM). QAM combines amplitude and phase modulation to create a set of waveforms with improved noise resistance and bandwidth utilization (see later discussion). The V.29 and V.32 technologies use QAM. Figure 2.11 shows a QAM constellation diagram, in which the radial angle of a point corresponds to the phase (or phase differential) of the waveform and its distance from the origin indicates the waveform's amplitude. In very high speed modems it is not uncommon to have constellation diagrams with 64 or more points.

Finally, optical systems use signaling that might, in some lights, be considered the limiting case of carrier amplitude modulation. On–off keying (also called amplitude-modulation keying; Figure 2.12) uses two waveforms, one of the carrier at nominal power and one of the carrier at little or even no power;

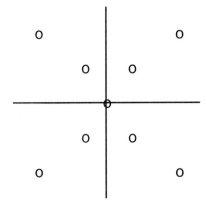

Figure 2.11 QAM scatter plot.

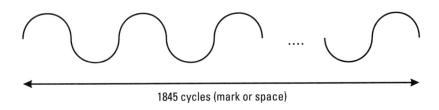

1845 cycles (mark or space)

Figure 2.12 On–off keying.

the latter can be considered a "null" waveform. As an example, consider an FDDI or similar system using a diode laser that emits light at 1300 nm to carry a signal that changes at 125 MHz—then a unit of signal would consist of 1845 wavelengths of light or the equivalent "time slice" of no light.

So, with on–off keying, a space or mark is indicated by a signal, whereas a mark or space is indicated by the absence of a signal. The absence can be regarded as a signal of zero amplitude. Hence, we can regard on–off keying as an instance of amplitude modulation.

2.3.4.2 Digital Waveforms

Digital waveforms are used in LANs and in the digital links that comprise the telephone backbone in the world. Digital waveforms can be broadly divided into return-to-zero (RZ) and non-return-to-zero (NRZ) signals. An NRZ encoding has constant value for the entire bit interval. An RZ encoding, in contrast, takes part of the bit interval to encode a value and then it changes back to the previous level. NRZ-Inverted (NRZI) is a non-return-to-zero code that changes levels to indicate the occurrence of a one in the incoming signal stream. Return-to-zero codes are important principally because of the issue of clock recovery—the extra transition helps ensure that the estimation mechanisms in the demodulator can keep synchronization.

The importance of synchronization can be seen from the fact that the digital waveforms in both 802.3 and 802.5 LANs (although each is theoretically media independent and hence allows for broadband realizations, in practice both use baseband signals) and alternate mark inversion signaling all use some form of return-to-zero encoding.

In the case of 802.3, its symbols are Manchester encoded, which splits the bit period into two equal parts and enforces a transition midway to ensure recovery of timing information. The two symbols, called Clocked Data 0 and Clocked Data 1, are complementary: CD0 starts out high and transitions low, whereas CD1 starts out low and transitions high (Figure 2.13). In addition to the CD0 and CD1 symbols, the 802.3 standard specifies two control signal

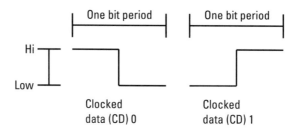

Figure 2.13 Data waveforms for 802.3 LANs: Manchester encoding.

waveforms, CS0 and CS1, that do not use Manchester encoding and are only used between the DTE and its local multistations access unit (MAU), that is, the control signals do not appear on the LAN's channel.

The 802.5 standard also uses Manchester encoding but it uses *differential* Manchester coding, which avoids fixed assignment of waveforms to bits. A binary 0 is encoded by issuing a symbol that has opposite polarity to that of the trailing segment of the previous symbol, whereas a binary 1 is encoded by issuing a symbol that has the same polarity to that of the trailing segment of the previous symbol. In addition, the 802.5 standard specifies two other nondata symbols, called J and K, which lack the midway polarity transitions but which otherwise follow the same coding rule. The J symbol has the same polarity to that of the trailing segment of the previous symbol, and the K symbol has opposite polarity to that of the trailing segment of the previous symbol (Figure 2.14).

Finally, the digital waveforms used over telephone channels most often use a variant of bipolar signaling based on alternate mark inversion (AMI) with modifications designed to ensure that long streams of 0's do not cause loss of synchronization at the receiver [7]. AMI uses alternating polarity transitions whenever 1's occur in the data signal (Figure 2.15). Examples of these variants include high-density bipolar codes (HDBn), where *n* refers to the maximum number of consecutive zeros allowed; and bipolar with *n* zero substitution (BnZS), where again *n* is the maximum number of consecutive zeros allowed.

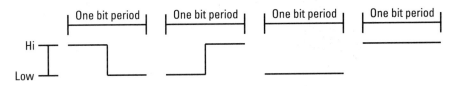

Figure 2.14 Waveforms for 802.5 LANs: differential Manchester encoding.

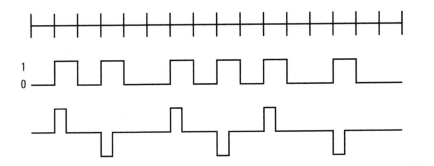

Figure 2.15 Bipolar waveforms with alternate mark inversion.

(The difference between HDBn and BnZS codes concerns how the coder modifies the bit stream when more than the allowed number of consecutive zeros occurs.)

Also used are block codes that group multiple bits together and use coding with, for example, three different digital waveforms. Codes used in practice include Four Binary, Three Ternary (4B3T), which takes 4 bits and represents the 16 possible combinations by three ternary symbols, that is, $3^3 = 27$ possible combinations. Also used is the Six Binary, Four Ternary (6B4T) code, which encompasses 64 four-bit combinations by 81 ternary combinations.

2.3.4.3 The Waveform Set

It should have been clear from the preceding discussion that waveforms do not exist in isolation. That is to say, just as we saw that a signal without transitions has zero information content, so a single waveform does not carry any information. It is the *transition* between waveforms, the sequence of waveforms sent, that carries the information. Therefore, it is more appropriate to speak of the waveform set $\{s_m(t)\}$, $m = 1, 2, \ldots, M$ than of individual waveforms. Digital communications systems and their properties are, to a considerable degree, dependent on the set of waveforms that the designer chooses to use.

To some extent, the choice is constrained by the channel to be used and its associated noise levels. Just distance or medium (e.g., wireless) can force the issue of baseband versus broadband signaling, so some waveforms work better on noisy channels than other waveforms, generally at the expense of increased bandwidth requirements. Conversely, some waveforms are bandwidth efficient, that is, less bandwidth is required per bit transmitted, but they require "cleaner" channels. In the remainder of this section we touch on some of the principal issues involved with selecting a set of waveforms.

Size of the Waveform Set

The number of waveforms required for signaling digital information is determined by the number of bits of information each waveform is to carry. The minimum waveform set contains two elements. This would correspond to 1 bit per waveform. As the size of the set of waveforms increases, more bits can be encoded per waveform: If there are M waveforms, then each waveform can encode $k = \log_2 M$ bits, that is, $M = 2^k$.

The size of the waveform set is critical. When just two waveforms are used, this is called binary signaling. When more than two waveforms are used, it is called *m*-ary signaling. Binary versus *m*-ary signaling touches on one of the many minor points of confusion that plague the discussion of communications: *bit* rate versus *Baud* rate. These are frequently—and incorrectly—used as synonyms. The Baud[2] rate is the rate at which the signal changes. The bit rate is the rate of binary information conveyed by the signal. By definition, when binary signaling is used the bit and Baud rates are the same; otherwise, the bit rate is a multiple of the Baud rate.

The basic trade-off that exists between bit and Baud rates is that the greater the size of the waveform set, the lower the Baud rate and the larger the *signaling interval*, which is the reciprocal of the Baud rate. Obviously, the higher the Baud rate, the smaller the signaling interval. As we will see in this section, this is a very important parameter. For example, with frequency-modulated signals, the smaller the signaling interval, the greater the frequency spacing required for waveform orthogonality (see later discussion). Another advantage of larger waveform sets is that, for a given probability of error in demodulating the waveforms, there is a corresponding reduction in the signal-to-noise ratio per bit.

Waveform Energy

A waveform $s_m(t)$, like any signal, has associated with it an energy:

$$E_m = \int_0^T s_m^2(t)\, dt$$

where E_m is the energy of the waveform and T is the signaling interval determined by the rate of the symbols (the Baud rate).

To ease demodulation (see later discussion), the waveforms in the waveform set should have equal energy; this simplifies implementing estimators for recovering many of the lost waveform parameters such as phase, carrier, and clocking. The reason is that if all waveforms have equal energy, then their attenuation by the channel will be identical and this factor (attenuation) can be left out of the complex task of estimating the missing pieces of information about the waveform.

Cross-Correlation Between Waveforms

Another very important design point is the desired and/or attainable cross-correlation between different waveforms in the set. Two waveforms $s_m(t)$ and $s_j(t)$ have a (complex-valued) cross-correlation coefficient given by

$$\rho_{jm} = \frac{1}{2\sqrt{E_j E_m}} \int_0^T u_m(t)\, u_j(t)\, dt$$

---

2. Named after Emile Baudot, a nineteenth-century French communications engineer.

where $u_i(t)$ is the low-pass equivalent of the waveform $s_i(t)$. This can be expressed directly in terms of the waveforms as

$$\text{RE}\left[\rho_{jm}\right] = \frac{1}{\sqrt{E_j E_m}} \int_0^T s_m(t) s_j(t)\, dt$$

A set of waveforms is said to be orthogonal if the cross-correlation of any two waveforms is equal to zero. More generally, a set of waveforms is said to be equicorrelated if the cross-correlation of any two waveforms is equal. In other words, orthogonal waveforms are a subset of equicorrelated waveforms, where the correlation is identically equal to zero. Orthogonal waveforms are advantageous in demodulation because the noise terms can be more easily filtered out by the demodulator (more on this topic later).

However, zero cross-correlation is not necessarily optimal. For example, with binary signaling, antipodal waveforms are optimal: $s_1(t) = -s_2(t)$. In this case, the correlation coefficient is -1. Extending the antipodal concept brings us to the *simplex* set of waveforms. These are equicorrelated with the cross-correlation coefficient $_r$ given by the relation

$$\rho_r = \frac{-1}{M-1}$$

where M is the size of the waveform set. Notice that if $M = 2$ (binary signaling) then $\rho_r = -1$, which is the result for antipodal waveforms; that is, antipodal waveforms are simplex with $M = 2$.

Waveform Efficiency

Several different methods are available for measuring the efficiency of a digital modulation system. One metric is the ratio of information rate to the required bandwidth in hertz. A basic trade-off in waveforms is between bandwidth efficiency, in particular as measured by the increase in bandwidth required to accommodate an increase in the number of waveforms, versus the signal-to-noise ratio required per bit to achieve a specified level of performance. That is to say, as the size of a waveform set increases, a corresponding increase occurs in either the bandwidth of the channel or in the signal-to-noise ratio. In addition, the various types of modulation and waveforms have their own efficiency characteristics. For example, efficiency is a downside to orthogonal waveforms: Their requirement for transport bandwidth—and corresponding receiver complexity—increases exponentially with the size of the waveform set. Likewise, waveforms that are created by modulating carrier phase and/or amplitude are more efficient users of channel bandwidth than orthogonal waveforms formed, for example, by frequency modulation of the carrier.

Waveform Spectra and Pulse Shape

The spectral characteristics of an information signal are very important, particularly when it comes to controlling the effects of intersymbol (symbol here meaning waveform) interference. With bandwidth-limited channels, such interference can be the principal impediment to waveform recovery/estimation. To quote from a text on digital communications, "A system designer can control the spectral characteristics of the digitally modulated signal by proper selection of the characteristics of the information sequence to be corrected." (In Chapter 3 we discuss the question of selecting the information sequences by encoding.) Exploring the spectral characteristics of various pulse shapes would take us too far from our focus, but the interested reader is advised to consult the previously referenced works on digital communications.

2.4 Modulators

2.4.1 What Is a Modulator?

We started the two previous sections by asking the questions "What is a channel?" and "What is a signal?" It seems almost unnecessary to pose this question about modulators, because most of their aspects were more or less covered in our discussion of waveforms. That is, when designing a waveform set, in terms of its size, the energy and cross-correlation of the waveforms, their shape, and so on, one is fundamentally designing the corresponding modulator.

Nonetheless, we should provide a definition of what we mean—we have been using the terms *modulate* and *modulator* quite freely but have not provided a rigorous definition. Unfortunately, just as we discovered with waveform, *modulate* is an overloaded term. One popular definition of modulate is:

> 1. *v.t.* Regulate, adjust; moderate. 2. adjust or vary tone or pitch of (speaking voice); alter amplitude or frequency or phase of (wave) by wave of a lower frequency to convey a signal. [8]

Even a more technical dictionary gives the same, analog-flavored definition:

> **data transmission** (1) (Carrier) (A) The process by which some characteristic of a carrier is varied in accordance with a modulating wave.... (2) (Signal Transmission System) (A) A process whereby certain characteristics

of a wave, often called the carrier, are varied or selected in accordance with a modulating function. [3]

Clearly, these definitions focus on analog waveforms and modulation. Our meaning is broader: to modulate is to create a signal bearing information, whether that signal is carrier-based or digital.

What, in terms of the MESA model, is modulation? It is actuation. Remember from the introduction to the MESA model that, if something changes, some actor or agent changed it, and that agent we called an actuator. Within a communications system the actuator that creates and changes the signal is, as we indicated before, the modulator. Baseband modulation is the actuation of the voltage (or, in some cases, the current) of the signal. Carrier modulation (amplitude, frequency, and/or phase) is the actuation (= change) of one or more parameters of the carrier. The type of modulation, obviously, depends on the set of waveforms used on the channel. Some of the types we have already discussed include:

- Pulse Amplitude Modulation (PAM): actuate the amplitude of the signal;
- Frequency Shift Keying (FSK): actuate the frequency of the signal;
- Phase Shift Keying (PSK): actuate the phase of the signal; and
- Quadrature Amplitude Modulation (QAM): actuate the phase and amplitude of the signal.

Figure 2.16 summarizes the various possible actuations that a modulator (actuator) can execute according to the input and waveforms. Notice that all four possible combinations are typically encountered in communications engineering, although analog-to-analog modulation is not often employed in digital communications systems except as an intermediate stage between digital-to-analog and analog-to-digital modulations, for example, with microwave line-of-sight or satellite links using carrier-modulated waveforms to transport end-to-end digital signals.

We should say a word about analog-to-digital conversion and specifically pulse code modulation (PCM). When the source is analog, the plant (its signal) is a continuous time, continuous state. A digital channel, in contrast, requires a plant that is discrete time, discrete state. Recall that we characterize a signal by its state and temporal characteristics and that a signal that can only assume a finite number of states is said to be discrete; otherwise, it is continuous. Likewise, a signal that changes only at finite instances in time is said to be discrete

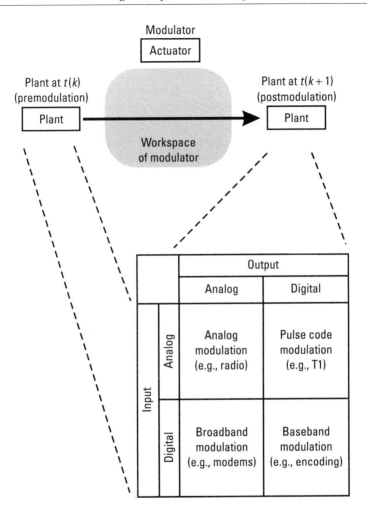

Figure 2.16 Input versus output: analog and digital signals.

time; otherwise, it is continuous. When a signal is in a finite state then we can speak of its alphabet $X = \{X_i, i = 0, \dots, q - 1\}$.

Unlike the discrete modulators we implicitly considered when discussing discrete waveform sets, a PCM converts a continuous or denumerably infinite input alphabet of symbols into a finite input alphabet. It does this by measuring the signal level at a fixed interval, called the *sampling frequency*, and converting the measurement into a digital "sample." There are two variables here: the sampling frequency and the number of levels of quantification, generally speaking, the number of bits in each sample.

This brings us to another fundamental result of communications theory, namely, the Nyquist theorem on sampling, which states that a continuous signal can be reconstructed from a sample sequence if the sampling frequency is at least twice the highest frequency in the continuous signal. By using filters to shape the input signal, the telephone network uses a sampling frequency of 8000 Hz, which is clearly adequate to handle voice-band signals. The number of bits in a sample determines the granularity of quantification and also the signal-to-noise ratio. The rule of thumb is that each additional bit adds 6 dB.

2.4.2 Modulation Management

In this limited definition of the modulator as actuator (the "brawn"), some sort of decision-making entity (the "brains") must exist to instruct the modulator regarding which actuation to execute, that is, which waveform to create. To quote from a standard text on digital communications systems, "The modulator performs the function of mapping the digital sequence into waveforms that are appropriate for the waveform channel" [4]. This mapping is scheduling. Therefore components of the transmitter then include a waveform actuator, the task set of which is the waveform set of the signals it can create, and a scheduler that manages the actuator. This scheduling is workload actuation of kind: mapping the RFS to the discrete channel composite (modulator plus channel plus demodulator) to the RFS to the modulator/waveform actuator itself (Figure 2.17).

We have just demonstrated that modulation is necessarily a composite task. Recall that a modulation is an actuation: The plant is changed. There are two possibilities: If the plant of the modulator is never changed, then the

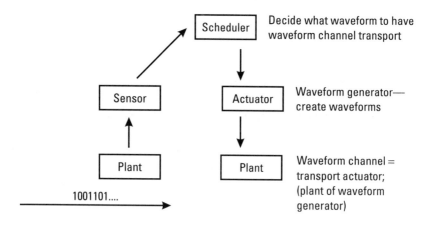

Figure 2.17 MESA decomposition of modulation of waveforms.

modulator is vacuous and can be replaced by the null server (this is equivalent to being discarded). Otherwise, the plant of the modulator is changed at least once. We must consider three possibilities: The plant is always changed, and each time the same way, or the plant is not always changed, or the plant is not always changed the same way. Recall the reason (*raison d'etre*) of the modulation: to convey information. If the modulation always produces the same change, then the probability of this event is equal to one and its information content is zero.

Therefore, we can rule out the first case. This leaves the two remaining possibilities, namely, that the modulation either is not constant in time or not constant in kind/degree. In the first of these, the unmodulated signal (= plant) carries one "piece" of information, whereas the modulated plant carries the second; it is possible to convey information with these two states. Likewise, assume that the modulator always changes the plant but *not* always in the same way. The plant may be modulated *a, b,* and so on. In either of these instances, a scheduling is required. The client may schedule the modulation tasks but this merely moves part of the composite workload to the client.

2.4.3 Modulation as Channel Adaptation

Finally, we can look at modulation as an adaptation mechanism between two or even three different types of channels (and signals). As we just discussed, different types of channels are suited to carrying different types of signals. The interconnection of computers typically involves distances that preclude the use of signals similar to those within the computers. Hence, we experience the need to mix different types of channels and to adapt signals between them.

Consider a discrete memoryless channel which, as we discussed earlier, is composed of a channel, a modulator, and a demodulator. The modulator and demodulator "encapsulate" the channel, isolating its implementation details and in effect abstracting it. In this way we can view modulation as signal adaptation, between the waveforms used by end systems and those used by the communications channel that interconnects them (Figure 2.18). The channel used may, according to needs and availabilities, be baseband (DS0, DS1) or broadband, meaning analog waveforms are carried.

2.5 Demodulation

We began this chapter by first exploring channels, then signals, and last modulation and modulators, fitting these into our MESA analysis. Now we complete the picture with demodulation and demodulators, the last of the four basic

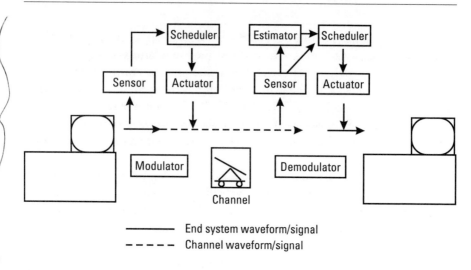

Figure 2.18 Modulation as channel adaptation.

building blocks of the physical layer. Where does demodulation fit into the larger context of workload and bandwidth management? Recall that together the signal and the channel that carries it make up the server we are managing, namely, our information transporter; and that modulation is, as an actuation of the server, part of bandwidth management. It follows, therefore, that demodulation, with its reconstruction or estimation of the waveforms received over the channel, is also part of bandwidth management, namely, bandwidth (or server) monitoring.

The task of the demodulator would be relatively simple if the channel propagated the signal without any deleterious or destructive result. Undo the effect of the modulator by taking the received waveform as measured by the sensor, look up the corresponding bit(s), and output these for transport to the destination (Figure 2.19). But what we saw in the previous sections is precisely that a channel corrupts a signal it propagates—both endogenously, by attenuation and delay, as well as exogenously, by the noise coming from various "fault" processes. Indeed, as we just discussed in the previous section, much of the analysis involved in designing/selecting the modulator's waveforms is to find the trade-off between signal level, channel bandwidth, pulse shape, and so on, that optimizes the performance of the composite transporter with respect to such metrics as the probability of incorrect reception.

As a prerequisite to recovering or estimating the bits as they first appeared to the transmitter, the demodulator relies on the receiver's sensor(s) to first measure the received signal. The sensor that executes this signal measurement is part of a bandwidth manager. Measuring the signal (and, of course, at the same

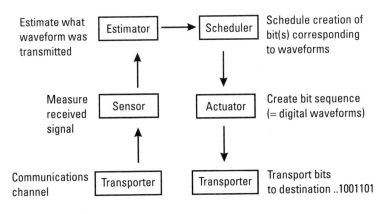

Figure 2.19 MESA decomposition of receiver with demodulator.

time measuring the interference from the channel's noise) is measuring the state of the transporter. Therefore, the first-level bandwidth instrumentation of the channel/signal composite is the receiver's sensor. We will pass over the details of the sensor(s), simply noting that they are appropriate to whatever channel/signal combination is used: volt meters for baseband electrical signals, tuned circuits for wireless signals, and so on.

2.5.1 Demodulation and Signal Uncertainty

The complexity in designing a demodulator is to find the optimal estimation mechanisms to mitigate the effects of the channel on the received waveform, to "undo" the corruption, and to decide what the original waveform was as transmitted by the modulator. This recovery is only possible because of one advantage that the demodulator (and demodulator designer) has: Notwithstanding uncertainties in the received signal due to the random nature of noise, the effects of intersymbol interference on bandwidth-limited channels, and so on, the waveform set used by the modulator *is* known.

In fact, the nature of the demodulator depends greatly on the nature of the modulation used and specifically the set of waveforms employed. First of all, the waveforms may be baseband or broadband—digital or analog signals. A demodulator that is to estimate baseband waveforms has a much simpler task to execute than its broadband counterpart; the baseband demodulator must concern itself principally with two issues, namely, synchronization and recovering from intersymbol interference on bandwidth-limited channels. A broadband demodulator, in contrast, must deal with both of these *and* with the additional complexities that come with carrier-modulated signals: uncertainties in amplitude, phase, and frequency introduced by the channel and/or by the modulator itself creating signals (waveforms) that deviate from the nominal. Another source of uncertainty concerns the time characteristics of the channel, notably delay. Delay in the channel, and possibly at the modulator, can produce clock/synchronization errors for the demodulator.

Due to these uncertainties, the received signal can differ from the transmitted signal in one or more ways. Mathematically the received signal is represented with greatest generality by the equation

$$r(t) = \operatorname{Re}\left\{\left[\alpha(t)u_m(t - t_0)e^{-j2\pi f_c t_0} + z(t)\right]e^{j2\pi f_c t}\right\}$$

where t_0 is the time delay of the channel (which, as we noted before, is actually variable over different frequencies); $\alpha(t)$ is the attenuation introduced by the channel (also, as we noted before, it is actually variable over different frequencies); and $z(t)$ is the additive noise, a zero mean Gaussian stationary random process in the case of AWGN channels.

Note that the carrier phase ϕ of the received signal is itself dependent on the time delay $\phi = 2\pi f_c t_0$, and since the carrier frequency f_c used in most digital communications signals tends to be in megahertz and higher, a small uncertainty in its value can result in large uncertainty in the phase ϕ. Beyond frequency instability in the oscillators in the modulator and/or demodulator, other reasons for uncertainty in phase include phase instability in the oscillators and/ or variable or random time delay in the channel. Depending on whether the demodulator does or does not know the phase, the demodulation is referred to as coherent or noncoherent (more on this later).

The rest of this section addresses the recovery of clocking and phase information, its use in (coherent) demodulation of FSK, PSK, and PAM signals, and, finally, noncoherent demodulation.

2.5.2 Synchronization and Clock Recovery

Digital demodulators receive signals, as measured by the receiver's sensors, that change at discrete intervals. With such demodulation, therefore, the question of synchronization is both more important and, as we will see, more complicated than might first appear. The essence of digital communications is that information is discretized—the analogy can be made with sampled data, also known as digital control systems. In both instances, timing is, if not everything, then at least crucial.

The idea behind synchronization can be illustrated by recalling a common sight to viewers of World War I movies, namely, a pilot firing his machine gun through the moving propeller of his own plane. The secret that made this work was a synchronization of the machine gun to the drive shaft that powered (actuated, to be precise) the propeller. This coordination (scheduling) prevented the ignominious spectacle of dashing airmen shooting themselves out of the skies.

It is putting things mildly to say that a considerable simplification in the demodulation task occurs when the receiver is synchronized to the transmitter. For starters, if the receiver is synchronized to the transmitter (and also the phase is known) then the delay t_0 is known and the consequent reduction in uncertainty means that the received signal can be represented by

$$r(t) = \alpha e^{-j\phi} u_m(t) + z(t)$$

There is, in addition, an impact on the probability of correct estimation from any loss of synchronization between transmitter and receiver. Figure 2.20 shows an actual versus a nominal digital waveform, where the deviation from nominal is due to such effects as capacitance and inductive loading of the electrical circuits, and finite rise times. The result of these imperfections is that measuring and/or estimating the waveforms toward the beginning or the end of the signaling interval increases the probability of error. That is to say, the question of *when* the signal is assessed becomes paramount: too early or too late and the probability of correct measurement and/or estimation diminishes markedly.

The timing of these changes originates at the modulator, since it creates the signal in the first place. And, just as the demodulator is "in the dark" about many other aspects of the transmitted (as opposed to received) signal, so too with the exact timing used by the modulator. For this reason most if not all digital demodulators include, in the absence of an explicit signal from the

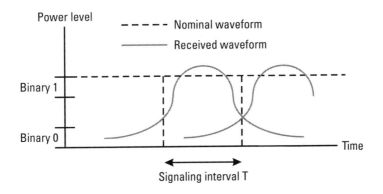

Figure 2.20 Nominal versus actual waveforms.

transmitter, a mechanism called a *synchronizer* to extract clock information from the received signal. It does this by tracking the timing of the changes in the signal.

If we examine this for a moment, it is clear that the synchronizer is an estimator; after all, the signal is measured but the clock is not, it is reconstructed or otherwise estimated (Figure 2.21). By recovering the transmitter's clock, the synchronizer is creating the schedule to be used by the demodulator to measure and/or make estimation decisions about the incoming signal. In this respect, it could be considered a scheduler but it is actually a scheduler *manque,* a proxy. The actual scheduler is at the transmitter, and the synchronizer at the receiver is recovering the timing (scheduling) information.

Why not just transmit a separate clock signal along with the data signal? That is, one simple solution to the synchronization problem is to have the transmitter send a clock signal along with the data-carrying waveforms. There are two reasons why this is seldom used. The first is that this would require part of the transmitter's power to be devoted to nondata purposes. The second is that, to ensure adequate receiver separation for demodulation, the clock and data signals would together ensure inefficient use of the channel bandwidth.

The alternative, as we noted in the section on line encodings, is to embed the clock into the data-carrying signal in such a way that it can be extracted at the receiving end by the aforementioned synchronizer. Most of the codes we discussed, including AMI, Manchester, and NRZI are self-clocking, and their respective demodulators include estimators (synchronizers) that recover the transmitter's clock from the transitions in the data signal. A requirement to recovering the clock of the transmitter from transitions in the signal is that

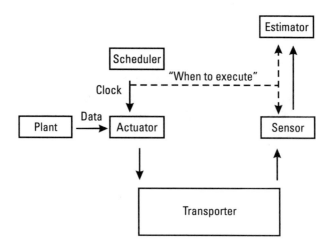

Figure 2.21 Transmitter clock scheduling receiver measurements.

there be sufficient transitions in the signal to allow the estimator to keep in sync. When users' computer equipment interfaces directly with digital telephone channels, the service provider frequently requires that a certain "density" of transitions be assured by the interfacing CSU/DSU.

Several different mechanisms can be used for this recovery of clocking signals from a signal. Two of the most common mechanisms for executing clock recovery are the early–late gate synchronizer and the phase-locked loop (PLL). Because we are going to explore PLLs with respect to recovering or estimating the carrier itself in broadband demodulation, we concentrate here on the first of these.

An early–late gate synchronizer, like demodulators (see later discussion) themselves, can be realized using a pair of either matched filters or cross-correlators. The signal $r(t)$ is fed into the filters or correlators, along with a reference waveform; the exact waveform chosen as reference is not material as long as the same one is sent to both the early and the late estimators. Whether filter or correlator is used, the pair produce estimates of the autocorrelation of the signal $r(t)$. If one estimator executes some time before and the other an equal time after the peak of the autocorrelation, then the estimates will be equal (autocorrelation is an even function, hence symmetric about the peak). On the other hand, if the transmitter's clock is not synchronized with the clock encoded in the signal's transitions, then the two estimators will not execute equally before and after the peak; one will be closer to the peak and the other will be further away from it. Using the autocorrelation function of a rectangular pulse, Figure 2.22 shows two examples, one in which the transmitter and receiver clocks are synchronized and one in which they are not. Clearly, in the first example the difference of the two estimates will be zero while in the second it will not.

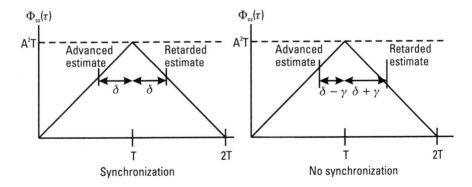

Figure 2.22 Early–late estimates of autocorrelation.

Subtracting one estimate from the other yields an estimate of the error between the transmitter's clock and the receiver's clock. The synchronizer is essentially a feedback controller that actuates the receiver's clock, advancing or retarding it to match the clock used by the transmitter, based on the return error difference between the two clocks. The scheduler then instructs an actuator, which actuates the receiver's clock until the return error difference is driven to zero (Figure 2.23).

Other realizations of early–late synchronizers are possible, including ones based on square-law devices. For more details, see [5].

2.5.3 PLLs and Carrier Recovery

After the synchronizer, the next estimator we want to discuss is the carrier recovery mechanism. Just as it is possible to recover a clock signal from a "well-behaved" data signal (i.e., one with adequate signal transitions), so it is also possible to actually reconstruct the carrier, including its phase, from a signal that is not too noisy. The mechanism that executes this reconstruction is called a *phase-locked loop*, as mentioned in the preceding section.

Like early–late synchronizers, PLLs use feedback to modify estimates derived from the incoming signal. One common implementation of a PLL

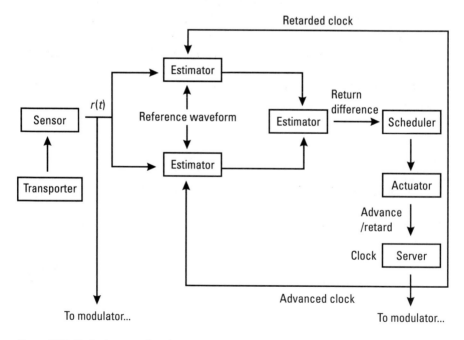

Figure 2.23 Early–late synchronizer.

passes the incoming signal through a squared-law device and then a bandpass filter tuned to the carrier frequency f_c. This produces a double-frequency signal $\cos(4\pi f_c t + 2\phi)$, and it is this that the PLL actually tracks by actuating a voltage-controlled oscillator (VCO) that produces an estimate signal $\sin(4\pi f_c t + 2\phi*)$, where the term $\phi*$ is the estimate of the phase. The product (call it the *return product*) of these two signals is then fed into a loop filter, which actuates the VCO to drive the error as measured by the product to zero. The VCO converges to the true value of ϕ (actually, 2ϕ) if the original signal $r(t)$ is not too noisy. Finally, the output of the VCO is fed into a frequency divider, which produces $\sin(2\pi f_c t + \phi*)$, and it is from this that the demodulator gets the estimate of the incoming signal's phase (Figure 2.24).

2.5.4 Optimum Demodulators

Now that we have seen how the clock and the carrier can be provided, it is time to focus on the demodulation process itself. The challenge of a demodulator is that every signaling interval T it receives a waveform that has been mangled and garbled by the noise on the channel, by interference from preceding waveforms (symbols), and which in fact may have started out with some problems when it was created by the modulator due to degradations and/or faults in the latter. All the demodulators we will consider execute their respective reconstruction of the waveforms by comparing the measured waveform with instances of the waveforms in the waveform set. Confining ourselves to broadband demodulators, we see these can be divided into those that make this comparison using cross-correlators and those that use matched filters.

Of course, the exact type of demodulator will depend on the nature of the modulation involved. Frequency shift keying and related orthogonal waveform

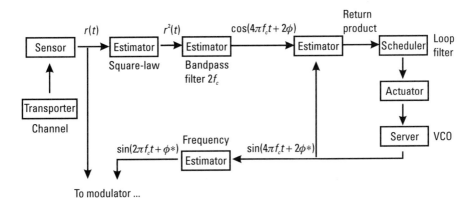

Figure 2.24 Phase-locked loop.

modulation techniques require that the demodulator perform M comparisons, where M is the size of the waveform set. That is to say, the demodulator for these waveforms has M cross-correlators or M matched filters. In the former case, the received waveform is fed into each cross-correlator along with the complex conjugate of one of the waveforms in the waveform set; the output of each cross-correlator is an estimate of the "match" of the received waveform and the candidate waveform. In the latter case, the received waveform is fed into M filters, each matched to the equivalent low-pass impulse response corresponding to one of the waveforms in the waveform set; as with the cross-correlators, the outputs of these filters are estimates of the "match" between received and candidate waveforms.

Whether they have been derived by cross-correlation or from matched filters, these estimates of the closeness are used to compute M decision variables by multiplying each estimate by $e^{j\phi}$, taking the real part of the product, and then subtracting the bias terms corresponding to the attenuated energies αE_m for each of the waveforms in the waveform set. An advantage of equal-energy waveforms is that the bias terms for all the waveforms are the same and they can be ignored rather than calculated; this includes the attenuation factor α, which would otherwise have to be estimated.

The resulting M decision variables are fed into a comparator, which selects the largest of these as representing the best "match" between the received waveform and the set of all possible waveforms. This is called *maximum a posteriori* (MAP) probability decision making and the comparator is, in our analysis, an estimator. To see this, note that the plant is the channel/signal composite and in particular the received waveform, which is being "processed" to mitigate the effects of the channel transit (noise, delay, attenuation); the effect of this is to reconstruct or estimate the original waveform. The two-tiered estimator depicted in Figure 2.25 could correspond to either the cross-correlator or matched filter implementation.

A somewhat different demodulator structure is used if the waveform set is constructed by phase and/or amplitude shift keying. In the case of PSK, the phase modulated waveforms $s_m(t)$ are constructed from one basic waveform by phase shifting:

$$s_m(t) = \mathrm{Re}\left\{ u(t)e^{j\left(2\pi f_c t + \frac{2\pi}{M}(m-1)+\lambda\right)} \right\} \quad \text{for } m = 1, 2, \ldots, M, \quad 0 \le t \le T$$

where $u(t)$ is the equivalent low-pass waveform. The same is true of various forms of ASK such as PAM. In the case of PAM, the waveforms $s_m(t)$ are constructed from one basic waveform by actuating the signal amplitude A_m

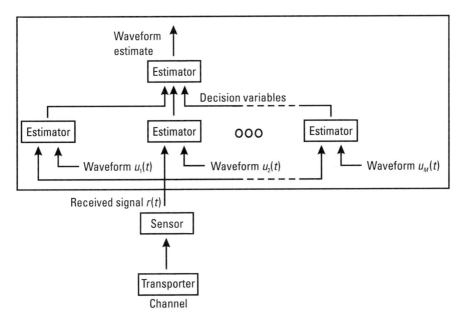

Figure 2.25 Structure of a demodulator.

$$s_m(t) = A_m \, \mathrm{Re}\!\left[u(t) e^{j2\pi f_c t} \right] \quad \text{for } m = 1, 2, \ldots, M, \quad 0 \le t \le T$$

where $A_m = 2m - 1 - M$, $m = 1, 2, \ldots, M$.

In both cases, because essentially one waveform is involved in constructing the waveform set, demodulation no longer requires M comparisons of the received waveform with the M waveforms $s_m(t)$. This means that demodulators for PSK or PAM (or QAM) systems need only one cross-correlator or matched filter, that is, only one estimation.

PSK demodulators nonetheless still require M decision variables, which are created by taking the inner product of this estimate and M unit vectors corresponding to the phase shifts. These M variables are then sent to a comparator and the largest is selected, just as with FSK demodulation. Differential PSK is similar but uses a delay to allow comparison between successive received waveforms.

PAM demodulators differ from both of these in several significant ways. First of all, the decision to be made centers on the amplitude or level of the received waveform, meaning only the levels of the waveforms need be compared; however, such a decision is necessarily predicated on knowing (or estimating) how the channel has attenuated the arriving waveforms. Complicating this estimation, the waveforms used in PAM do not have equal energy, meaning that the bias terms that we were able to ignore in other demodulators cannot be ignored here. The solution is to incorporate an estimator/compensa-tor called automatic gain control (AGC) to mitigate the attenuation (more on AGC in the next section).

2.5.4.1 Noncoherent Demodulation

In the discussion of demodulators up until now, we have implicitly assumed that the demodulator knew or had estimates, via carrier recovery, of the phase. If the demodulator contains no mechanism (PLL or otherwise) for recovering or estimating the phase of the received carrier, then it is still possible to demodulate the signal and estimate the waveform that was transmitted, albeit with worse performance than if the phase were known or estimated and used in the demodulation decision making. By assuming that the phase of the incom-ing signal $r(t)$ is a random variable uniformly distributed over the interval $[0, 2\pi]$, a worst-case analysis can be performed on the probability of erroneous demodulation estimates.

Noncoherent demodulators have a multitiered structure similar to that of coherent demodulators. In both instances, the first tier usually consists of estimators based on either matched filters or cross-correlators that estimate the "closeness" of the incoming waveform to the various waveforms in the wave-form set. However, the absence of phase information means that another stage of estimators is needed to produce a set of decision variables that will be used to decide which waveform was received. This second stage consists of envelope detectors or square-law detectors, both of which produce magnitude estimates that are independent of the phase. These decision variables are then fed to the third and last stage estimator which, as with the demodulators we discussed above, uses MAP probabilities.

2.6 Summary

The objective of this chapter was twofold: to outline the elements of digital communications systems and to place these in the MESA framework we devel-oped in Chapter 1. Toward this end we saw in this chapter that there are four components to a communications system: (1) a transmitter, (2) the signal, (3) a communications channel, and (4) a receiver (a sensor that measures the received signal).

A transmitter is an actuator—it creates the signal. We also discussed how bandwidth in the context of actual communications channels has a slightly different meaning than in Chapter 1. We saw that there are two basic types of signals, namely, baseband signals, which are digital, and broadband signals, which are analog.

The chapter also explored why, because the long-distance channel has different characteristics than the channel(s) inside a digital computer, the task of modulation is to convert the signals used inside the computer to signals better suited for transport over the long-distance channel.

This exploration of modulation brought us, naturally, to demodulation. On the surface, the task of demodulation is quite simple to define, namely, to reverse the effects of the modulation process. We saw that demodulators belong to one of two families, those that use matched filters and those that use cross-correlators. Both types of demodulators have a two-tiered structure. The first tier of estimators produces a set of decision variables that is then sent to the second-tier estimator where an estimate of the waveform is produced.

However, we saw that in the real world of imperfect communications systems, the demodulator may lack other information concerning the received signal, information that must either be estimated or done without in the waveform estimation process. Synchronization in digital communications involves the receiver scheduling its measurements relative to when the sender (actuator) changed the signal. We explored the extraction of timing information and the corresponding implementation of synchronization mechanisms.

The chapter concluded with an examination of adaptive equalization. The reason for this is twofold. First, adaptive equalization is an unequivocal example of a bandwidth manager employing explicit feedback control to actuate the channel, in this instance, for the purpose of changing its transmission characteristics. Second, adaptive equalization is worth studying in its own right. Along with error-correcting encoders and decoders (covered in Chapter 3), it is principally responsible for the enormous increase in the speeds and price/performance of modems thaty we have witnessed in the past two decades.

References

[1]   Sippl, C. J., *Data Communications Dictionary,* New York: Van Nostrand Reinhold, 1976.

[2]   Shannon, C. E., "A Mathematical Theory of Communication," *Bell Sytems Technical Journal,* Vol. 27: pp. 329, 623 (1048), reprinted in C. E. Shannon and W. Weaver, *A Mathematical Theory of Communication,* University of Illinois Press, 1949.

[3]   *Standard Dictionary of Electrical and Electronics Terms,* 4th ed., New York: IEEE Press, 1988, p. 896.

[4]   Proakis, J., *Digital Communications,* New York: McGraw-Hill, 1983.

[5]   Bennnet, W.R. and J. R. Davey, *Data Transmission,* New York: McGraw-Hill, 1968.

[6]   Lucky, R. W., J. Saltz, and E. J. Weldon, Jr., *Principles of Data Communications*, New York: McGraw-Hill, 1968.

[7]   Owen, F. E., *PCM and Digital Transmission Systems*, New York: McGraw-Hill, 1982, p. 221.

[8]   *The Concise Oxford Dictionary*, 7th ed.

3

Managing Channel Faults via Error Control Coding

3.1 Introduction

Recall from Chapter 1 that we said the need for management in discrete event plants is due in part to the finite reliability of servers. We saw in Chapter 2 that, although a DMC's bit error rate can be made arbitrarily small by using high enough signal power and suitably "distant" waveforms, realizing error-free DMCs are not practical because of the associated inefficiencies and costs such as those due to high-power levels. Consequently, the question becomes this: Can we substitute fault recovery for fault prevention with the DMC? In other words, can we repair or replace the faulty data the DMC transported? The answer is yes and the solution, as is so often the case with high availability systems, is to employ redundancy. In modern communications systems, this redundancy is in the form of error control coding of the client data.

The chapter begins by reviewing what it means to detect and repair faults when the server is a discrete memoryless channel; the focus is on error control coding as DMC maintenance. It then proceeds to outline the three basic elements to error control coding systems: the code, the encoder, and the decoder. Just as much of Chapter 2 was taken up with discussing signals and waveforms, this chapter is organized around the different families of codes for error detection and correction. There are two very different ways to generate error correction codes, namely, block and convolutional encoding, and this chapter explores the corresponding differences in the respective encoders and decoders. An example of a block code is the parity code used in many simple communications links; an example of a convolutional code is trellis code modulation (TCM) used in such modem protocols as V.32 and V.33. Both these and other related codes are discussed. As in Chapter 2, our focus is not on presenting a comprehensive survey of the topic of coding but rather on the MESA analysis.

As we will see, from a MESA perspective there are marked similarities between the tasks of encoder and modulator. Just as a modulator executes a mapping of data to be transported into waveforms better suited to the exigencies of channel transit, so the encoder executes a mapping of data into coded forms that are designed to better withstand faults. Both instances involved scheduling and actuation tasks. Such is also the case between the tasks of decoder and demodulator: Each executes a reconstruction that reverses the effects of its partner, all with the ultimate purpose of more reliable transportation of data. As with demodulators and modulators, decoders are much more sophisticated than encoders because of the greater amount of uncertainty they must accommodate in making their decisions—uncertainty introduced by the channel.

3.2 Error Control Coding

3.2.1 DMC Maintenance: Fault Detection, Isolation, and Repair

Traditional reliability theory decomposes maintenance into fault detection, isolation, and recovery. Further, detection and isolation are frequently aggregated into a single task, while recovery may mean repair or replacement. In data communications the equivalent terms are error detection and error correction. To detect that a fault in transmission has introduced an error into the received data, we need to receive an unambiguous indicator that a fault has occurred. However, to repair the received data, that is to say, to correct the error, an unambiguous indicator of *which* fault has occurred is required. If it can be determined which bits have been changed, then the original data can be reconstructed.

What is a DMC fault? Quite simply, a fault occurs if the symbol a DMC delivers to the destination(s) is not the symbol the DMC received from the client. And notwithstanding the best efforts of the demodulator, a DMC is inevitably going to suffer from faults—the destination will get an incorrect symbol due to the demodulator's inability to accurately reconstruct the transmitted waveform.

We can characterize DMC faults according to their severity and duration. The severity of the fault is determined by the DMC's ability to transport any signal, however erroneous the result may be. If no symbol is delivered then the fault was fatal; otherwise, it was a latent fault—the DMC appears to be functioning correctly but it is in fact introducing errors into the transported data. Obviously, latent faults are more difficult to detect than fatal faults and such detection represents a central challenge to managing DMCs.

In terms of duration, DMC faults may be transient or persistent. Persistent faults, as the term implies, do not go away. A persistent fault is generally caused by equipment failure due to factors ranging from overheating transformers to backhoe operators digging up telephone channels (fiber optic or copper cables). Transient faults, on the other hand, will resolve themselves (hence use of the term *transient*). The vast majority of transient faults are caused by noise on the communications channel.

Table 3.1 lists the four possible combinations corresponding to various types of faults for any server, including the discrete memoryless channel, and the necessary management intervention, if any.

Management of the DMC's faults varies widely depending on the nature of the fault in question. Persistent faults require the intervention (actuation) of a bandwidth manager to execute maintenance, either repair or replacement, on the components of the DMC (channel, modulator, and/or demodulator); otherwise, the server (DMC) will not recover and will remain in the faulty condition. Transient faults, on the other hand, do not require management intervention for the server to recover, because they come and go with the highly stochastic channel noise. If transient faults are too frequent, indicating for example an excessive noise level, bandwidth management may be indicated; otherwise, bandwidth managers are not necessary to recover from transient faults—by definition, they are self-limiting.

Table 3.1
Faults: Severity Versus Duration

	Severity	
Duration	Fatal	Latent
Transient	The server fails completely but recovers by itself	The server fails but continues operating; recovers by itself
Persistent	The server fails completely and will not recover without outside maintenance (repair)	The server fails partly and will not recover without outside maintenance (repair)

However, the effects of the DMC's faults, even transient faults, on the plant (the data being transported) do not go away so easily. A fault can have two consequences: for the server and for its "output"—depending on the particular server, fault, and plant. If the server is an actuator or scheduler, the plant may be affected; if the server is a sensor or estimator, the information about the plant may be affected. A DMC is a transport actuator, and as such its faults can corrupt the data it is moving, with potentially disastrous consequences for the client.

So how do we repair the damage of a DMC fault? We touched briefly on the solution in Chapter 1 when we remarked that, with a digital server, redundancy effectively "repairs" the consequences of transient faults such as those that are caused by noise on a communications channel. We distinguish between redundancy as being either concurrent or serial. In the former, the redundant information is sent at the same time the original data is sent, while with the latter the redundant information is sent following the original data.

Concurrent redundancy is ordinarily accompanied by replication of the server to gain the greatest degree of immunity to faults, notably persistent faults; in this case the replicated servers may still fail but the composite does not as long as a sufficient number of component servers are unaffected by the fault(s). An example would be bonding k DMCs together and sending the same data down each. However, because the focus here is on *transient* faults, this extra degree of protection is generally unnecessary. We nonetheless still use the term *concurrent redundancy* to refer to any scenario in which client data is transported along with the redundant data all at once; this technique is also called *forward error control* (FEC).

Serial redundancy involves retransmission: Keep sending the message until the destination indicates the message has arrived without error. This is also called *backward error control* (BEC) and, of course, implies that there is at the destination a means to detect that errors have occurred. The advantage of forward error control is that the destination will receive the redundant data at the same time it receives the client data, eliminating the delay attendant on retransmission; the advantage of backward error control is that redundant data is transported only if necessary.

In both FEC and BEC, transient faults are being remedied by workload management. Recall that a workload manager is necessary to actuate the Requests for Service (RFSs) arriving from the client. With BEC, what had been a single RFS "Task = Transport (client data)" from the client is mapped into k RFSs "Task = Transport (client data)" by the workload manager, where the number k is determined by the number of times the transporter (the DMC) fails to transport the data without fault. With FEC, the RFS is transformed from "Task = Transport (client data)" to "Task = Transport (client data +

redundant data)." In addition, of course, a workload scheduler (not necessarily the same one) is at least conceptually present to map the RFS "Task = Transport (client data)" into the three RFSs "Task = Encode," "Task = Create Waveform," and "Task = Transport (client data)"—this mapping is required because the transporter is now a composite server.

Concurrent redundancy is an open-loop maintenance policy; serial redundancy, on the other hand, is a closed-loop maintenance policy. Recall from Chapter 1 that in open-loop maintenance, such as age or block replacement, the component is replaced without reference to its actual condition; with concurrent replication k copies of the RFS and the plant are created irrespective of the server's condition. Closed-loop maintenance, in contrast, takes maintenance action only in response to feedback about the system being maintained; similarly, serial replication only occurs when it has been determined that the server has suffered a fault and the task must be reexecuted.

3.2.2 Forward Error Control and Error Control Coding

A simple implementation of forward error control can be terribly inefficient. Consider a message of k symbols, two copies of which are transported. This in effect is 100% redundancy overhead. Each copy of the message will arrive with from 0 to k symbols changed by faults in the DMC. By comparing the two copies symbol by symbol, it is possible to detect many of the faults—although not all of them, since a comparison of two symbols affected identically by faults will not reveal anything. Another shortcoming of this simple twofold replication is that it is impossible to *correct* any of the faults thus detected. Of course, a threefold replication allows correction of any single error in the message, but at the cost of 200% redundancy overhead.

The question therefore arises: Can we combine the advantages of serial and concurrent redundancy, that is, transport the redundancy information along with the client data so as to eliminate the delay of serial redundancy but utilize a form of redundancy more sophisticated than simple repetition to minimize the inescapable overhead of concurrent redundancy? The answer is yes. Use of an error control coding based on advanced mathematical concepts, such as the theory of finite fields, yields the benefits of concurrent redundancy—recovery from faults on the channel—with much less overhead than simple replication. This is the subject of coding theory, and we explore different types of codes in subsequent sections.

As with so much else in communications theory, the origins of error control coding can be traced to Shannon's landmark paper "The Mathematical Theory of Communication" [1]. Beyond defining channel capacity and its relationship to signal and noise levels (see Chapter 2), in this paper Shannon proved that channel capacity was attainable with some error correction coding system:

> Shannon's basic coding theorem ... states that with a sufficiently sophisticated channel encoder and decoder, we can transmit digital information over the channel at a rate up to the channel capacity with arbitrarily small probability of error. [2]

What Shannon's coding theorem says is that we can exchange information "gain," in the form of encoding and decoding, for information loss due to the occurrence of faults on the channel. In other words, faults in the channel "lower" the amount of (useful) information in the signal; coding mitigates the effects/consequences of the information loss due to the introduction of noise.

Just as there is a fundamental trade-off with any server between reliability and maintainability, so is there a trade-off here between the investment in the DMC to make it more reliable and the investment in the coding servers (encoder and decoder) to make the composite transporter more maintainable. The design trade-offs we discussed for a DMC—signal and noise levels, waveform set, and demodulator sophistication—are reliability concerns. Error control coding addresses the maintainability of the transporter. Fundamentally, Shannon's result is a restatement of the reliability/maintainability trade-off as applied to transporter design. Indeed, as one author remarked,

> Shannon's theory of information tells us that it is wasteful to build a channel that is too good; it is more economical to make use of a code. [3]

3.2.3 The Components of a Coding System: Code, Encoder, and Decoder

As we said earlier, the three components of an error control coding system are the code, the encoder, and the decoder. What must come first is the code, for its nature determines much of the structure of the encoder and decoder. Although Shannon's proof gave impetus to the search for error control codes, it did not say how to construct such codes and it was more than a decade before codes were discovered that began to approach the performance Shannon promised. Today, however, many different types of codes are used in commercial and government communications systems; powerful error control codes have been developed for applications as diverse as deep space probes and low-cost modems. Indeed, beyond communications systems proper it is increasingly common for highly available computer systems to use error control coding in their memory subsystems; the data is stored in encoded form and when it is retrieved the error control coding (ECC) information is used to ensure that no corrupted data are sent to the CPU.

3.2.3.1 Types of Coding Systems: Block Versus Convolutional

Error control codes can be divided into block and convolutional. As the name implies, block coding operates by encoding client data in units called blocks of k symbols in length. Each k symbol block is mapped into an n symbol code word, where obviously $n > k$. Convolutional coding, on the other hand, is more analogous to a "continuous flow" process, in which there are no clearly defined units of client data encoded independently.

In the case of block codes, as we see later, the aim is to construct code words of maximum dissimilarity. This is analogous to modulation, where by selecting channel waveforms for maximum dissimilarity (for example, orthogonal waveforms) subject to constraints such as waveform efficiency, the greatest degree of immunity to noise can be attained. The set of code words that define a block code is generally chosen to exploit certain algebraic properties. Convolutional codes, on the other hand, do not have code words or algebraic structures to exploit; rather, they are based on probabilistic properties useful in decoding.

All codes share some common elements. To quote a standard text on the coding theory:

> Although individual coding schemes take on many different forms ... they all have two common ingredients. One is redundancy. Coded digital messages always contain extra or redundant symbols. The second ingredient is noise averaging. This averaging effect is obtained by making the redundant symbols depend on a span of several information symbols. [4]

The mechanism that introduces the redundant information is an encoder. An encoder is a scheduler/actuator composite server. The scheduler receives the input symbols and selects the appropriate output symbols; it then schedules the appropriate actuation to "create" this new symbol. There is a fundamental isomorphism between the tasks of modulation and encoding. In both cases there is a mapping of input to output; in the former scheduling channel waveforms and in the latter scheduling code words, which are then actuated into existence.

> The purpose of the discrete channel encoder is to introduce redundancy in the information sequence which aids in the decoding of the information sequence at the receiver. [2]

An encoder's actuation of the input symbols can be regarded as either actuating by modifying an existing bit pattern or actuating by creating a new bit pattern.

The parallel, again, is with modulation where we saw that the actuator either created or modified waveforms.

Returning to the differences between block and convolutional codes, a major distinction centers on the use of state information in the encoding and decoding processes. A block encoder makes the same coding decision irrespective of the past data encoded: A block of k symbols will be encoded into the same code word irrespective of the symbol content of the preceding block(s). In contrast, a convolutional encoder's scheduler includes the past data in its decision as to which code will be created. The distinction is identical to that which separates combinational and sequential circuits.

Whether the encoding is block or convolutional, the redundancy introduced by the encoder is used to "repair" the plant, following its transportation, by the decoder (Figure 3.1). The decoder is an estimator that reconstructs, from information passed to it by the demodulator, the data as it was originally passed to the encoder by the client: The demodulator is a device which estimates which of the possible symbols was transmitted [4].

To do this reconstruction, the decoder relies on its knowledge of the rules by which the encoder selected which code word to use for a given set of bits. Again, the similarity is with demodulation: Whereas a demodulator uses its a priori knowledge of the waveform set used by the modulator to arrive at the most likely estimate of which waveform was sent over the channel, a decoder uses its a priori knowledge of the code used by the encoder to arrive at the most likely estimate of the coded "information sequence" that was sent over the channel.

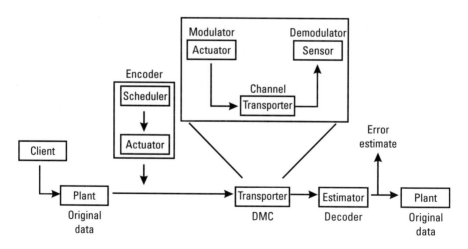

Figure 3.1 Encoder and decoder encapsulating DMC.

With both block and convolutional codes, decoding involves identifying syndromes corresponding to any errors that the channel (in the case of soft decoding) or the DMC (in the case of hard decoding) has introduced. A conventional definition of a syndrome is a set of "concurrent symptoms in disease" [5]. More precisely, a syndrome is a set of state and/or output variables that characterizes a given fault. For example, when the plant is an electrical motor that has suffered a burnt-out armature, the failure syndrome might include infinite resistance across the coil due to the open circuit. Similarly, if the plant is a person the failure syndrome might be his or her temperature, blood pressure, and so on. It is on the basis of the syndrome's values that an estimate of the fault can be made. The choice of code words is specifically designed to optimize the code estimator.

This illustrates the isomorphism between the tasks of demodulation and decoding. In both cases a reconstruction is executed at the destination that effectively undoes the mapping effected at the transmitter. Because of the noise and resulting errors introduced by the channel, the complexity of reconstructing the transmitted waveform (in the case of demodulation) and the transmitted code word (in the case of decoding) is greater than the complexity of scheduling the waveform or code word in the first place.

Neither the demodulator nor the decoder will be perfect. An error in received data means either that one or more waveforms has been incorrectly demodulated into information symbols different than those transmitted or that the coded sequence has been incorrectly decoded into data different than that which the client sent to be transported. Regardless of the sophistication of the demodulator used, such faults can occur. In this respect, the decoder provides a second echelon of maintenance for the automatic detection of such faults and, if possible, their correction.

Berlekamp [6] makes an important distinction between decoding failures, in which the decoder is unable to decide which code was transmitted, and decoding errors, in which the decoder mistakenly decides that code word A was transmitted when in fact it was code word B. The decoding failure will at least indicate the presence of an error in the received code word and leave it to other management entities to decide what to do (e.g., retransmission at a higher layer in the protocol stack). The danger of a decoding error is that the upper layers may or may not learn that an error occurred, and in any case will assume that the decoder has taken care of the problem.

3.2.3.2 Hard-Decision Versus Soft-Decision Decoding

A demodulator provides estimates used by the decoder. There are several ways in which this demodulator/decoder interaction can happen. If the demodulator sends estimates of the symbols transmitted, then this is called *hard-decision decoding.* On the other hand, if the demodulator sends to the decoder estimates that have yet to be quantified to symbols then this is called *soft-decision decoding.*

3.3 Block Codes

We now focus on block codes and their corresponding encoders and decoders. Block codes are some of the principal codes that are in use today primarily because of their power and the rigorous foundations provided by the mathematics of finite fields, linear spaces, and groups. For simplicity, we'll focus on binary block codes, but it should be noted that block codes based on nonbinary symbols are in common use, for example Reed-Solomon codes.

3.3.1 Block Code Characteristics

As we said earlier, the heart of (block) coding for error detection and error correction is the mapping of blocks of the data to be transported (the plant) into code words specifically chosen for their resistance to channel faults. Two tasks are involved in this depending on the particular block of data that arrives at the encoder: (1) a scheduling of which code word is to be created (actuated into existence) and (2) the actuation itself. Of course, prior to this is the determination of the code words to be used. Many different codes are in use, each representing a particular design trade-off between various costs and benefits (more on these later).

We can illustrate how the selection of code words can minimize the effects of propagating over a channel with a nondigital example. Consider a primitive community that has settled a large valley. The community wants to be able to communicate by shouting a relatively small number of important messages, such as "attack," "fire," "meeting," "dinner," and so on. The communications channel, namely, the valley and the air in it, will distort and attenuate the signals (the words) as they propagate. They would probably devise a vocabulary of code words that have a reasonable chance of being correctly understood after propagating across the valley. The individual code words would also be chosen for dissimilarity from each other to the maximum extent possible so as to minimize the chances that one might be confused with the other.

The simplest example of a block code is the parity code that is often used on asynchronous (start–stop) links. The parity bit is chosen so as to make the sum of the character's bits even or odd, depending on the type of parity used. The idea behind this is simple: Take a block of data, generally a 7-bit ASCII character, and add up its bits. For instance, the bits in the ASCII character 1001101 (the letter *M*) add up to 4, which obviously is even. If even parity is being used, then the parity bit for this character is set to 0; if odd parity is being used, then the parity bit is set to 1. This is equivalent to Modulo 2 addition. The encoder appends the parity bit to the character and passes it to the DMC for transportation to the destination, where it is processed by the decoder and an estimate of the channel condition (i.e., did a fault occur?) is made based on calculating the parity of the received character and comparing it to the parity expected (odd or even).

A parity encoder maps a 7-bit block into an 8-bit code word, meaning that out of 256 possible code words only 128 are used (Figure 3.2). If a single bit, or in fact any odd number of bits (3, 5, or 7 bits), is changed by the effects of a fault, then the decoder will detect a parity error since this will transform any of the 128 code words into one of the 128 invalid combinations. On the other hand, the well-known shortcoming of parity coding—that it cannot detect a fault that results in an *even* number of bits being flipped—is because the resulting combination *is* a valid code word.

Consider the creation (actuation *ex nihilo*) of the code word by a block encoder. The plant is the block of *k* information bits. These are "measured" and a code word is created. This code word is substituted for the original plant, and it is this that is transported across the channel. The received code word, possibly corrupted by fault(s), is "measured" and an attempt is made to reconstruct the original code word.

3.3.1.1 Block Code Parameters

So what is a block code? It is a set of 2^k code words of *n* bits each in length; obviously, $n > k$, otherwise some of the code words would be duplicated. Some

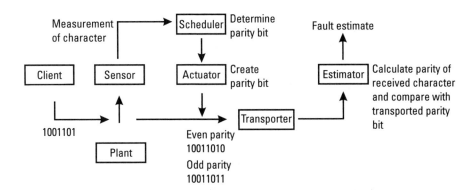

Figure 3.2 Parity encoding and decoding.

of the parameters of a given block code include those discussed in the following subsections.

Block Size

As we indicated earlier, block codes are characterized by the fact that the data being encoded are discretized into units called *blocks* of k bits, each of which is encoded separately, rather than being encoded in a continuous process a la convolutional coding. The resulting encoded blocks of n bits, $n > k$, constitute an (n,k) code with $n - k$ redundant bits of information per block of k bits of data. The greater the lengths of k and n (or, equivalently, $n - k$) the greater the noise averaging and redundancy, respectively. For an (n,k) code, there will be 2^k code words out of a total 2^n possible.

We can make an analogy between the size of a waveform set and the number of code words in an ECC implementation. In both instances, the complexity and overhead scale as the numbers increase. As we will discuss, some block codes, notably the cyclic codes, possess properties that reduce the overhead.

Code Rate

A closely related concept is the *code rate*, the ratio of the size of the original unit of data to the size of the code words created by the encoder. This is often denoted by R_c. A simple parity code such as we used in the example where the original block of data (module) is 7 bits and the output is 8 bits has a code rate $R_c = 7/8$.

Hamming Distances

Recall that the problem with parity codes is that the code words created by the concatenation of the character plus the parity bit are too "close" to each other. This concept of closeness has been rigorously defined in coding theory and is called the *Hamming distance*. Given two code words of n bits (or symbols) in length, the Hamming distance between them is the number of positions in which they differ.

A crucial parameter of any code is the number of erroneous bits that the decoder can correct. For a linear code the number of errors that can be corrected is determined by d_{min}, the minimum Hamming distance between any two code words in a code. This is directly related to the minimum Hamming distance d_{min} according to the formula

$$t = \left[\frac{d_{min} - 1}{2} \right]$$

where $[x]$ returns the largest integer less than or equal to x.

Given the similarity between encoding and modulation, it is not surprising that there is a connection between the Hamming distance of two code words and the cross-correlation of the corresponding waveforms. The determination of the 2^k *n*-bit code words is an equivalent task to determining the set of waveforms used by the modulator.

Linearity

A code is said to be *linear* if any linear combination of code words is also a code word. A code that is not linear is said to be *nonlinear*. By implication, a linear code must contain the all-zero code word. Most of the codes in use are linear codes.

Code Word Weight

The *weight* of a binary code word is the number of its nonzero elements. The weights of the code words in a code constitute the *weight distribution* of the code. If the code words all have the same weight then this is called a *fixed-weight* or *constant-weight* code. This means that no equal-weight code can be linear since the all-zero code word must be included in a linear code and its weight is zero, while the remaining code words must have nonzero weights. Nonetheless, for a linear code the weight distribution is important because from it we can determine the distance properties of the code.

Cyclicity

A linear code is cyclic if code words can be generated from each other by a rotation of the bits in any one of the code words. Cyclic codes are a subclass of linear codes. Many of the most important linear block codes are cyclic: Golay, Bose-Chaudhuri-Hocquenghem (BCH), Reed-Solomon (a subset of BCH codes), and maximum-length shift register codes are all cyclic; in addition, Hamming codes are equivalent (see later discussion) to cyclic codes. The principal reason cyclic block codes are so powerful is that their cyclicity can be exploited to allow very efficient realizations of encoders and decoders, enabling larger block size codes than would otherwise be possible.

3.3.1.2 Examples of Codes

We have already looked at parity codes, the simplest of block codes. Next we summarize the parameters of some of the more important linear block codes.

Hamming Codes

These are $(2^m - 1, 2^m - 1 - m)$ codes where *m* is a positive integer; if $m = 3$, for example, then we obtain a (7,4) Hamming code. For any value of *m*, the corresponding Hamming code has $d_{min} = 3$. By the formula given earlier, such codes can correct one erroneous bit. For this reason Hamming codes are known as single-error correcting, double-error detecting (SECDED) codes. For the (7,4) Hamming code, there are $2^7 = 128$ possible code words but of these only $2^4 = 16$ are required. These are listed in Table 3.2. The Hamming distance $d_{min} = 3$ of the code words is clear from inspecting them.

Golay Code

This is a (23,12) cyclic code with $d_{min} = 7$. The Golay code can correct up to three errors in a code word. Whereas the (7,4) Hamming code uses 16 out of 128 possible code words, the binary (23,12) Golay code uses $2^{12} = 4096$ out of $2^{23} \approx 8 \times 10^6$ possible code words. Since the Golay code is cyclic, it can be generated using the polynomial

$$g(p) = p^{11} + p^9 + p^7 + p^6 + p^5 + p + 1$$

This gives the basis code word $[00000000000101011100011]^T$. By the cyclicity property it is clear, for example, that $[00000000010101110000110]^T$ is another code word. We discuss generating cyclic codes using generator polynomials more in the next section.

The extended Golay code is created by adding a parity bit to the Golay code, creating a (24,12) code where $d_{min} = 8$. Note that this extension does not increase the number of errors the extended code can correct: the largest integer less than or equal to $(8 - 1)/2$ is 3, same as for $(7 - 1)/2$.

Bose-Chaudhuri-Hocquenghem (BCH) Codes

One of the most powerful families of codes is the BCH codes codiscovered by Bose-Chaudhuri and Hocquenghem. These are cyclic $(2^m - 1, k)$ codes with $d_{min} = 2t + 1$, where t is determined by the inequality

Table 3.2
The (7,4) Hamming Code: Data Block and Code Words

Data Block	Code Word	Data Block	Code Word	Data Block	Code Word	Data Block	Code Word
0000	0000000	0100	0100111	1000	1000101	1100	1100010
0001	0001011	0101	0101100	1001	1001110	1101	1101001
0010	0010110	0110	0110001	1010	1010011	1110	1110100
0011	0011101	0111	0111010	1011	1011000	1111	1111111

$$n - k \leq mt$$

or equivalently,

$$\frac{2^m - k - 1}{m} \leq t$$

BCH codes were first constructed that with $t = 2$, which would enable a decoder to correct two errors in a code word (the next level of error recovery after the single-error correction of Hamming codes). However, it was immediately apparent that the BCH technique was extensible to provide any level of error correction, unlike the Golay and Hamming codes, which are limited to three and one errors, respectively. The power of BCH decoding comes from what is called the *error locator polynomial*. We discuss this in the section on decoding.

Maximum-Length Shift Register Codes

These are cyclic $(2^m - 1, m)$ codes with $d_{\min} = 2^{m-1}$. These codes are attractive in part because they are equidistant: The distance between every pair of code words is the same. Equidistance of code words makes analysis of the code easier and allows tighter bounds on the code's distance properties. Another family of codes that possesses this property is the family of Hadamard codes (for more on Hadamard and other equidistant codes, see [2] and [6]). Maximum-length shift register codes are also attractive because of the simple mechanism required for the encoder, namely an m-stage shift register with feedback (more on this in the next section). Table 3.3 lists the $m = 3$ (7,3) maximum-length shift register code words and the corresponding data blocks.

3.3.2 Encoding Block Codes

As we said earlier, an encoder can be regarded as a scheduler/actuator composite. The scheduler receives the input symbol and selects the appropriate output symbol; that is to say, it schedules the appropriate actuation to create this new

Table 3.3
The (7,3) Maximum-Length Shift Register Code: Data Block and Code Words

Data Block	Code Word	Data Block	Code Word	Data Block	Code Word	Data Block	Code Word
000	0000000	010	0100111	100	1001110	110	1101001
001	0011101	011	0111010	101	1010011	111	1110100

symbol, which can be seen equally either as actuating by modifying an existing bit pattern or actuating by creating a new bit pattern.

The implementation of an encoder for linear block codes can take many forms. We can distinguish between different types of linear block encoders on the basis of how they schedule these actuations. With this in mind, we will consider four broad classes of encoders: (1) table lookup, (2) generator polynomial using Galois field methods, (3) generator matrix using vector space methods, and (4) maximum-length shift registers.

The most straightforward implementation is table lookup: The k bits of the data block to be encoded are the "key" to search the table, and the code word found is the output. The problem with this tabular approach is that, as the size k of the block increases, the size of the table increases exponentially. If $k = 30$, for example, the table has 2^{30} or more than 1 billion entries. An encoder with gigabytes of memory is not practical, even with today's low memory prices, and the memory requirements of table lookup implementations were even more unacceptable back in the 1950s and 1960s when much of the foundational work on coding was being developed.

This led to the search for alternative implementations that did not have such high storage requirements. This necessarily involves the encoder calculating code words based on the input block of data. In contrast to the extensional implementation of table lookup, such an intentional or algorithmic implementation of the encoder effectively "substitutes" arithmetic operations for storage. The algorithm is, in fact, the schedule of the actuation tasks involved in creating the code words.

Most linear block encoders are implemented using algorithms based on associations: the association of the block of data to be encoded, and the corresponding code word, with either polynomials over finite fields; or alternatively with vectors and linear vector spaces. The cornerstones of these two approaches are, respectively, the generator polynomial and the generator matrix. A third type of implementation, the maximum-length shift register, offers a different way of scheduling not based on either of these. We next examine these three techniques.

The generator polynomial approach is grounded in abstract algebra. Recall that we said the prominence of cyclic codes was due in large measure to their mathematical tractability and what this means for the implementation of encoders (and decoders). As one author comments,

> Cyclic codes are important because their underlying Galois-field description leads to encoding and decoding procedures that are algorithmic and computationally efficient. Algorithmic techniques have important practical applications, in contrast to the tabular decoding techniques that are necessary for arbitrary linear codes. [3]

We won't go into Galois field theory here; the interested reader can consult references on abstract algebra (such as [7] or [8]). Suffice it to say that modern coding theory is in large part built on Galoisian foundations.

To generate the code words of a cyclic (n,k) code, we first start with the generator polynomial $g(p)$ where $g(p)$ is a factor of the polynomial $p^n + 1$ and is of degree $n - k$. The coefficients of the polynomial are drawn from elements of the field—in communications applications, the symbols that make up the data block, here being 1 or 0 because we said we are confining ourselves to binary codes. The key to this is the association of each n bit code word C_k with a polynomial $C(p)$ of degree $n - 1$ or less. Likewise, a block of data to be encoded is associated with a polynomial $X(p)$ where the k coefficients x_i of the polynomial are the symbols of the block of data. There are 2^k possible polynomials of degree less than or equal to $n - 1$ which are divisible by $g(p)$. These are the code words that the code can generate.

Given a generator polynomial, the task of creating a code word for a given block of data becomes one of polynomial multiplication over the field. That is,

$$C(p) = g(p)X(p)$$

For example, we saw that (23,12) Golay codes are generated with the polynomial

$$g(p) = p^{11} + p^9 + p^7 + p^6 + p^5 + p + 1$$

Notice that this polynomial is of degree 11, which indeed is $n - k$. To generate the code word for a block of 12 bits of data, the encoder performs a multiplication of the generator polynomial $g(p)$ and the polynomial $X(p)$, of degree less than or equal to 12, that corresponds to the data to be encoded. The result is the desired code word. If the block of data is, for example, [1100000011000]—an arbitrary choice to illustrate the procedure—then the result is

$$C(p) = \left(p^{11} + p^9 + p^7 + p^6 + p^5 + p + 1\right)\left(p^{12} + p^{11} + p^4 + p^3\right) =$$
$$\left(p^{23} + p^{22} + p^{21} + p^{20} + p^{19} + p^{16} + p^{15} + p^{14} + p^{12} + p^8 + p^5 + p^3\right)$$

Translating back into bits, the code word is [111110011101000100101000]. It should now be clear why the degree of the generator polynomial must be $n - k$: Since the "data" polynomial $X(p)$ may be of degree up to k, if the generator polynomial were greater than $n - k$, then the code word produced could be greater than n symbols long.

Many codes not generated polynomially are nonetheless cyclic. Looking at the (7,4) Hamming code words, it is clear that the 16 code words can be divided into four: the all-zero code word [0000000], the all-one code word [1111111], and the rotations (seven each) of the code words [0001011] and [0011101]. The (7,4) Hamming code, in other words, is cyclic. For more details on cyclic codes, consult [3, 6, 9].

In contrast to generator polynomials and their roots in abstract algebra, the generator matrix/vector space approach to coding theory and implementation rests in large part on linear algebra. We start by regarding the code words as $n \times 1$ vectors, that is, vectors with n elements. A set of n linearly independent vectors each of which has n elements defines (or spans) an n-dimensional linear space S. If the set has $k < n$ linearly independent vectors (or code words) then it spans a subspace S_c of the space S. Therefore, an (n,k) code defines a subspace S_c of an n-dimensional space.

A generator matrix G, which has k rows and n columns, spans the subspace S_c. A generator matrix is a very concise way to represent the schedule (i.e., the algorithm) of these operations. G is called a generator matrix because it has the property that, when multiplied by the vector X corresponding to the k bit information block, the result generated is an n bit code word C:

$$\mathbf{C}_{1 \times n} = \mathbf{X}_{1 \times k} \mathbf{G}_{k \times n}$$

Note that the arithmetic operations are defined relative to the field whose elements are the symbols used in the information to be encoded. For example, if the code is binary the elements are 1 and 0 and addition, for instance, is performed modulo 2. Technically speaking, this is the Galois Field GF(2).

Given that there are 2^k possible information blocks the result is that 2^k possible code words can be generated in this way. The question is this: What are the elements of the generator matrix G? Consider the (7,4) Hamming code in. It is clear that the block of 4 bits to be encoded is transferred directly to form the four leftmost bits of the corresponding code words. The remaining three bits, the parity bits, are generated by the following equations:

$$\mathbf{C}_{m5} = x_{m1} \oplus x_{m2} \oplus x_{m3}$$
$$\mathbf{C}_{m6} = x_{m2} \oplus x_{m3} \oplus x_{m4}$$
$$\mathbf{C}_{m7} = x_{m1} \oplus x_{m2} \oplus x_{m4}$$

From these equations we, in fact, can derive a generator matrix for the (7,4) Hamming code:

$$G = \begin{bmatrix} 1 & 0 & 0 & 0 & 1 & 0 & 1 \\ 0 & 1 & 0 & 0 & 1 & 1 & 1 \\ 0 & 0 & 1 & 0 & 1 & 1 & 0 \\ 0 & 0 & 0 & 1 & 0 & 1 & 1 \end{bmatrix}$$

Note that the rows g_i of G are code words of the (7,4) Hamming code. In other words, for the (7,4) Hamming code it turns out that the rows g_i of G are code words of the code we seek to generate.

$$G = \begin{bmatrix} g_1 \\ \vdots \\ g_k \end{bmatrix}$$

This is true for all linear codes, not just Hamming codes. The rows of a generator matrix are code words, and because any linear combination of code words is a code word, so is the product XG.

Because G is made of k code words, out of the total 2^k code words in the code, there is no one-to-one correspondence between a linear block code and a generator matrix. In particular, there are

$$\binom{2^k}{k} = \frac{2^k !}{k!(2^k - k)!}$$

possible generator matrices for a code that has k information bits. As k increases in size (increasing the noise-averaging effect of the code), the number of possible generator matrices grows rapidly.

Of the generator matrices corresponding to a given code, one stands out. A generator matrix in "systematic form," also called standard or echelon canonical form, has a particular structure consisting of a $k \times k$ identity matrix adjoined to a $k \times (n - k)$ matrix that specifies how the redundant bits are encoded.

$$\mathbf{G} = \begin{bmatrix} 1 & 0 & 0 & \cdots & 0 & p_{11} & \cdots & p_{1n-k} \\ 0 & 1 & 0 & & & & \ddots & \\ 0 & & 1 & & & & & \\ \vdots & & & \ddots & & & & \\ 0 & & & & 1 & p_{k1} & & p_{kn-k} \end{bmatrix}$$

or

$$\mathbf{G} = \begin{bmatrix} i_{k \times k} \vdots P_{k \times n-k} \end{bmatrix}$$

A systematic code has the advantage that the structure of its generator matrix reduces the number of arithmetic operations required to generate a code word since the first k bits are simply identical to the k bits of the block of client data being encoded. Using a generator matrix for a nonsystematic code requires nk arithmetic operations, whereas using a generator matrix for a systematic code requires $(n - k)k$ arithmetic operations, a savings of k^2 operations. For large block size codes this can amount to thousands of operations saved per code word generated. Any linear block code can be reduced to systematic form by suitable matrix operations.

Not all code words are generated using a generator polynomial or generator matrix. The maximum-length shift register codes are encoded using shift registers with feedback. The code word generation begins by loading the m-stage register with the block of m symbols (bits, if the code is binary). The register is then left shifted $2^m - 1$ times, each shift producing 1 bit of the code word and, with the feedback, the result of some modulo 2 addition of two or more of the stages being fed back into the shift register.

3.3.3 Decoding Block Codes

Just as there is more than one way to implement linear block encoders, so, too, with decoders are there multiple ways of achieving the same end, namely, the reconstruction (estimation) of the code word as it was originally, prior to transportation. What all these estimation techniques have in common, however, is that they exploit the properties of the code being used, such as the distance between code words. Much as a demodulator relies on its a priori knowledge of the waveform set used by the modulator, so the decoder knows which code words could have been created by the encoder and which must be the result of faults in the DMC. For example, the decoder may maintain a table of all valid code words and use pattern matching with a "closeness" metric to decide which is the best estimate of the original code word.

Tabular decoding, however, suffers from the same problem as tabular encoding: the exponential growth in the size of the table of code words as the length of the code's code word increases. And, as was the case with encoding, the alternative is to use algorithmic methods to reconstruct the transmitted code word either on the basis of the received code word (in the case of hard-decision decoding) or the decision variables estimated by the demodulator (in the case of soft-decision decoding). Not surprisingly, these methods have a strong connection to the techniques used in linear block encoding. It turns out, in particular, that at the heart of this decoding there is an intimate connection between the generator matrix for a linear block code and the generator matrix of what is called its dual code.

Every linear (n,k) block code has a dual $(n,n-k)$ linear block code. Recall that the code words of a linear block code, regarded as linear vectors, span a k-dimensional subspace S_c of an n-dimensional space S. The $(n-k)$ dimensional subspace that is not spanned by the code words of the (n,k) code is the null space of S_c. The code words of this dual $(n,n-k)$ code span the null space of the (n,k) code. The two sets of code words are orthogonal, and the generator matrix H for this dual code possesses the special property that the product of any code word C_j of the (n,k) code and H is the null vector 0.

What this means is that with H we can determine whether or not a received code word is in fact part of the (n,k) code. H is called the parity check matrix for the original (n,k) code and it is the heart of the decoding estimator. As with the generator matrix, the parity check matrix specifies a schedule for a set of field arithmetic operations, out of which comes the estimate of the transmitted code word. Finally, if the generator matrix G of the (n,k) code is in systematic form, then the generator matrix of its dual code likewise has a special structure related to the structure of G as follows:

$$H = \left[-P' \vdots I_{n-k} \right]$$

where P' is the transpose of the matrix P that is part of the matrix G. As was the case with G in systematic form, H in systematic form is computationally advantageous since $(n-k)^2$ multiplications can be avoided.

Recall the generator matrix G for the $(7,4)$ Hamming code we presented earlier. Because this matrix was listed in systematic form, the generator matrix for the dual $(7,3)$ code is also in systematic form. That is, H is

$$\mathbf{H} = \begin{bmatrix} 1 & 1 & 1 & 0 & 1 & 0 & 0 \\ 0 & 1 & 1 & 1 & 0 & 1 & 0 \\ 1 & 1 & 0 & 1 & 0 & 0 & 1 \end{bmatrix}$$

Now, consider a code word $\mathbf{C}_j$ delivered to the decoder by the demodulator (as we said earlier, for the sake of simplicity our discussion of decoding will be confined to hard-decision decoding). The term $\mathbf{C}_j$ is an $n \times 1$ vector. $\mathbf{H}$ is an $(n - k) \times n$ matrix. The result of multiplying the two is an $(n - k) \times 1$ vector, which is the null vector $\mathbf{0}$ if $\mathbf{C}_j$ has not been corrupted by a DMC fault; or rather, $\mathbf{C}_j$ has not been corrupted into another valid code word $\mathbf{C}_k$. This is why the distance property of a code is so important. If $\mathbf{C}_j$ has been corrupted (but not into another valid code word) then the product $\mathbf{S} = \mathbf{HC}_j$ will be a nonzero $(n - k) \times 1$ vector. This vector $\mathbf{S}$ is called the syndrome of the error pattern resulting from the fault, and it is used to both detect the fault and "repair" the consequent error by restoring the code word to its original form.

A syndrome $\mathbf{S}$, however, is intrinsically limited in that it is an $(n - k) \times 1$ vector, meaning there are only 2^{n-k} syndromes. Given that an encoder has created a code word for an (n,k) code, the problem that the decoder faces is that there are $2^n - 1$ possible erroneous code words, of which $2^k - 1$ are also valid code words. An obvious difficulty is that the correspondence between errors and syndromes is not one to one; the syndrome may unambiguously determine if an error has occurred but not which error.

To decide which error caused a particular syndrome (fault isolation) requires another estimation of the most likely error pattern corresponding to the syndrome. This is, in effect, a two-tiered estimator (Figure 3.3). Depending on the code used, this second estimation may be implemented by table lookup even though this was impractical for the first estimation considered earlier. That is because the syndrome table, as it is called, has only 2^{n-k} entries. If there

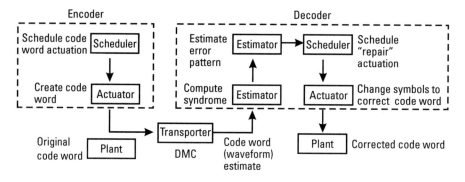

Figure 3.3 Two-tiered block decoder: syndrome and error estimators.

are relatively few parity bits in the code (i.e., $n - k$ is small relative to k) then this can be efficient.

Table 3.4 lists the syndromes and most likely (minimum distance) error patterns for the (7,4) Hamming code we have used in our examples. Note that all the error patterns in the table involve a single symbol (bit). Given that the code is single-error correcting, this is not surprising. Nonetheless, it also illustrates how a double error will cause an incorrect error pattern to be estimated. For example, if the original code word **C** was [0100111], corresponding to the block of data [0100], but the received code word was [1000111] then the syndrome **S** = **HC** will be [010], which the table says indicates the error pattern [0000010]. Correcting **C** to reflect this estimate, the recovered code word would be [1000101]—which is doubly unfortunate, because this is a valid code word corresponding to the block of data [1000]. This is, as we defined earlier, a decoding error.

Given this limitation of syndromes, the question arises: Beyond parity check matrices and syndrome tables, is there an algorithmic way to derive the error patterns that is more powerful? The answer is yes, and it is called an *error locator polynomial*. Error locator polynomials are the basis of the powerful BCH codes, or rather their decoders.

The process of decoding BCH code words begins by finding syndromes, as just done, but with BCH decoding the calculation proceeds very differently. For starters, a BCH syndrome **S** is a $2t \times 1$ vector, where t is the number of errors the code can correct. As with BCH encoding, the procedure associates a code word with a polynomial. **S** is then found by solving over the field of the code $2t$ simultaneous equations for its $2t$ components S_i evaluated at the $2t$ roots of the code's generator polynomial.

Finding the syndrome **S** is just the first step. From these S_i the error locator polynomial is derived, and from its roots are calculated the locations of the symbols (bits) in the code words which are in error. For more details beyond this very abbreviated account, consult [3], [6], or [10].

Table 3.4
Syndromes and Error Patterns for (7,4) Hamming Code

Syndrome	Error Pattern	Syndrome	Error Pattern	Syndrome	Error Pattern	Syndrome	Error Pattern
000	0000000	010	0000010	011	0001000	110	0100000
001	0000001	100	0000100	101	0010000	111	1000000

3.4   Convolutional Coding

We now come to *convolutional codes*. Other terms used to designate the family of codes and coding systems (encoder/decoder pair) that does not rely on a formal set of code words are *sequential* and *tree*. Such codes are the focus of this section and, while we limit ourselves to binary codes as we did in examining block codes, it is important to note what we discuss here generalizes to nonbinary symbols and codes.

3.4.1   Convolutional Codes and Encoders

Unlike block coding systems, it is difficult to discuss convolutional codes separately from convolutional encoders. To describe (n,k) block code one can just list the 2^k n-bit code words without discussing generator polynomials, matrices, and so on. There is no similarly simple way to describe a convolutional code. Instead, it is necessary to discuss a convolutional code and the encoder together.

    The basic mechanism of a convolutional encoder is a shift register. This shift register, however, is generally not monolithic but rather is composed of L stages, where each stage can hold k symbols (bits if the code is binary). The data is shifted into a convolutional encoder k bits (or symbols) at a time, so that with each shift the k symbols move from the first stage to the second until finally it is shifted from the Lth stage. (Figure 3.4 shows such a convolution encoder.) A

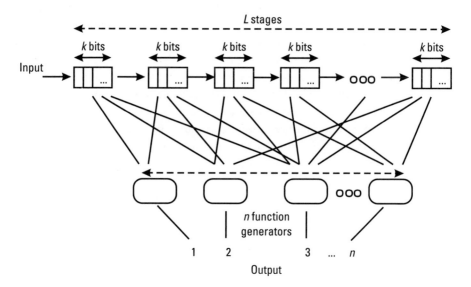

Figure 3.4   Convolution encoder.

good analogy is between convolutional coding and continuous flow manufacturing, whereas block coding is similar to batch manufacturing.

It is from the k symbols held in each of these L stages that the encoded output is created. The encoder puts out n bits generated by n linear algebraic circuits called *function* or *arithmetic generators,* generally adders. The client data is fed into the shift register and moves through it, in the process creating the convolutional code by means of connections between stages in the shift register and adders, which then create the encoded output. The code rate of a convolutional code has the same meaning as in the case of block codes, namely, the ratio of the k symbols input to the n symbols output.

Of course, not every one of the k positions in each of the L stages is necessarily connected to every function generator; the topology of the connections is, in fact, one way to characterize a given code and encoder. The connections for each function generator can be listed by a vector with Lk bits; a zero if there is no connection, and a one if there is one. When the function generators are adders (over the field of the code's symbols—for example, modulo 2 adders for binary codes) these vectors offer a concise description of the code—the closest thing to a generator matrix, in fact, and for this reason the n vectors are referred to as the code's generators.

For a given encoder, this product of the number k of symbols shifted per time and the number L of its stages is called its *constraint length.* The constraint length is important because, in part, it determines the "protecting" power of a convolutional code. Recall that with block encoding, each k symbol block of data is encoded into an n symbol code word. With convolutional encoding, each k symbol block is, along with the k symbol blocks in the other $L-1$ stages, encoded in up to Ln symbol units of encoded data (*not* code words). The span of noise averaging of block codes is one k symbol block of data. For convolutional coding, the constraint length kL represents the "memory" of the code, that is, the span of its noise averaging.

At least two difficulties, however, come with increasing constraint length. First, the complexity of the decoder increases and so does its storage requirements. Second, the latency introduced by the encoder and decoder grows with the constraint length. Choosing the parameters k, L, and n such that the constraint length is not too great is part of the challenge of designing convolutional codes and encoders.

The first question in designing convolutional codes is: How many symbols (bits if the code is binary) are shifted into the encoder at each time interval? The number ranges from a minimum of one to an arbitrarily large number. Next comes the question of how many output symbols, n, created for each k symbol, are shifted into the encoder. Finally, there is the question of how many stages the encoder has, where each stage is a shift register capable of holding k symbols. Note that if $L = 1$ then effectively the convolutional encoder has become a block encoder: k bits are shifted into the encoder and n bits are shifted out; there is no extra noise averaging from symbols in other stages, and such an encoder would be effectively "stateless."

This, however, would violate what we said earlier was one of the distinguishing characteristics of convolutional encoders: that the output of the encoder depends not just on the data being encoded but on the state of the encoder, and that the state of the encoder at a given moment is determined by the $L - 1$ recent data inputs. From simple multiplication we can see that the dimension of the state is $k(L - 1)$. That is to say, in addition to the k symbols shifted into the first stage of the encoder, the state of the encoder used in making the decision of which n symbols to output is the k symbols in each of the remaining $L - 1$ stages. If for example $k = 1$, $L = 2$ then the state of the encoder is the one symbol in the second stage—if the symbols are binary then there are two possible states. On the other hand, if $k = 3$, $L = 6$ then the state is given by the 15 symbols in the remaining five stages and, assuming binary symbols again, there are 32,768 states for the encoder.

Much of the power of convolutional coding systems comes from this adaptability. With block coding, the decisions of the encoder's regulator are fixed with the selection of the code: The same k bit block will result in the same n bit code word being created every time, irrespective of the past blocks that were encoded. In contrast, the regulator in a convolutional encoder is "closed loop" in that its encoding decisions (schedules) adapt to the patterns of the data being encoded, meaning that when the same k bit block is shifted into the encoder, the n bits output will vary according to the previous $L - 1$ k bit blocks (Figure 3.5).

3.4.1.1 Generating Convolutional Codes

Even though an encoder can be described by the number of stages it has, the number of symbols per stage, and the number of symbols created per input symbol, it is still necessary to specify *how* the encoder generates its outputs when it receives an input sequence of data. One way to do this, as we said earlier, is with the n generators of the encoder. Technically speaking, it is even possible to construct a generator matrix for a convolutional code, but this is not practical because its dimension would, because the input sequence is unbounded and the dimension of the generator matrix corresponds to the length of the input sequence, be equally unbounded if not infinite. Instead, the generation of a convolutional code is generally described by one of three means: a tree diagram, a trellis diagram, or a state diagram. Each of these is an equivalent representation of the scheduling executed by the encoder.

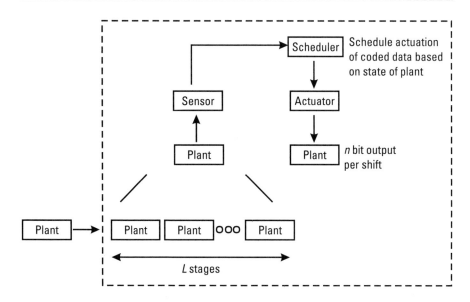

Figure 3.5 Convolutional encoder with closed-loop decision making.

For reasons of brevity, we examine only trellis diagrams. We start with the encoder's $m^{k(L-1)}$ states—where m is the number of input symbols and $k(L-1)$ is the state dimension of the encoder. These are listed in columns. The process begins by convention with the encoder in the all-zero state. As with tree diagrams, from each state there are m^k branches to m^k other states. Also as with tree diagrams, along these branches indicating the transitions between states are listed encoding decisions of the encoder. The result is a trellis diagram, from which the term *trellis code modulation* is derived. The advantage of trellis diagrams over tree diagrams is that they are bounded, and they show more concisely the state-dependent nature of the convolution encoder's scheduling. Figure 3.6 shows a trellis diagram with states and transition branches.

3.4.2 Convolutional Decoders

Much of the popularity of convolutional codes is due to the availability of relatively simple yet powerful decoders. As with block code decoders, both hard-decision and soft-decision decoding techniques are employed. In either case, the objective is the same: to estimate (or reconstruct) which k symbols were the input to the convolutional encoder from the client for each n symbol that arrives over the channel.

The most important convolutional encoder is the so-called Viterbi decoder, which can be implemented either with hard-decision decoding

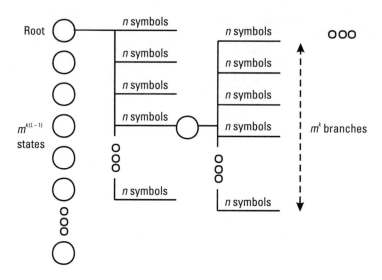

Figure 3.6 Trellis diagram for *k, L, n* convolutional encoder.

(the decoder receives symbols) or soft-decision decoding (the decoder receives unquantified decision variables). The only difference in the decoder is the metric used to discriminate between possible decodings: Hamming distance for hard-decision decoding and Euclidean distance for soft-decision decoding. In either case, the algorithm at the heart of the Viterbi decoder attempts to retrace the path taken by the encoder through its states as it encoded the client data. If the decoder chooses the correct path out of *B* possible paths, then the original data will be correctly recovered. Otherwise, a decoding failure will occur.

The Viterbi decoder does this by calculating branch metrics, based on the conditional probability of a sequence of symbols *Y* arriving given a sequence *C* was transmitted; and from these branch metrics are calculated path metrics corresponding to a sequence of branches taken by the encoder (in response, obviously, to the input symbols). The Viterbi algorithm compares possible paths and, as subsequent symbols arrive to be decoded, chooses the path that has the highest metric. At each stage, a survivor path is chosen for further comparisons. This allows the reduction of the number of decoding comparisons necessary from growing exponentially as the length of the sequence to be decoded grows.

The Viterbi algorithm does, nonetheless, suffer from scaling with the constraint length. As the state of the encoder grows according to $m^{k(L-1)}$ the decoder must consider $m^{k(L-1)}$ survivor paths and $m^{k(L-1)}$ metrics. For this reason, the Viterbi decoder is typically used with encoders that have small constraint lengths.

3.5 Summary

In this chapter we discussed error control coding (also known as forward error control) of data for transport across a noisy channel. We placed the discussion in the context of server maintenance: how to employ redundancy to optimally repair the consequences of transporter faults. If a fault in a transporter results in the corruption of the data being transported, then a repair has been effected when the data are finally transported correctly. To ease fault detection and to utilize channel bandwidth more efficiently, we saw that an error control encoder maps input symbol(s) to different symbol(s), these being transported over the channel; at the receiving end, after symbol reconstruction has been executed by an estimator in the demodulator, a corresponding decoder reconstructs the original symbol(s) as received by the encoder.

In addition, just as the modulator and demodulator encapsulated the channel to create an abstraction (i.e., the discrete memoryless channel), so the encoder and decoder encapsulate the discrete memoryless channel and create an abstraction, namely, a channel with much reduced bit error rate (BER) and hence improved reliability. Similarly, just as we saw that a modulator was an actuator the plant of which is the signal the client requests to be transported, transforming it from the waveforms used by the client (a computer) to the waveforms optimized for the channel, so it is with the task of error correction coding: It is an actuation of the plant, transforming it from its original form into a new form that will better resist the noise introduced by the channel.

We discussed the two types of codes: block codes and convolution codes. Notwithstanding their differences, we saw that certain basic facts about error control codes are true whether we are talking about block or convolutional codes. First of all, error control codes add redundant information. The redundancy introduced by the encoder is used to "repair" the plant, following its transportation, by the decoder. With decoders we are reminded of the real reason for codes and encoders: to increase the robustness of the transporter by facilitating the reconstruction of transported code words.

References

[1]   Shannon, C. E., "A Mathematical Theory of Communications," Bell Systems Technical Journal, Vol. 27: 379, 623 (1048), reprinted in C. E. Shannon amd W. Weaver, A Mathematical Theory of Communication, University of Illinois Press, 1949.

[2]   Proakis, J., *Digital Communications,* New York: McGraw Hill, 1983.

[3]   Blahut, R., *Theory and Practice of Error Control Codes,* Reading, MA: Addison-Wesley, 1983.

[4] Clark, G. G., and J. Bibb Cain, *Error-Correction Coding for Digital Communications,* 1981.

[5] *Concise Oxford English Dictionary.*

[6] Berlekamp, E. R., *Algebraic Coding Theory,* New York: McGraw-Hill, 1968.

[7] Birkhoff, G. and S. Maclane, *Abstract Algebra,* Chelsea.

[8] Van Der Waerden, B., *Abstract Algebra,* Ungar.

[9] Peterson, W., and E. Weldon, *Error-CorrectingCodes,* 2nd Edition, MIT Press, 1972.

[10] Lin, Shu, *An Introdution to Error-Correcting Codes,* New Jersey: Prentice-Hall, 1970.

4

Management and the Transporter Interface

4.1 Introduction

We saw in Chapter 2 that the actual transporter in a digital communications system is a channel plus the signal(s) it carries, but that in most circumstances the client interfaces with a virtual transporter—the discrete memoryless channel (DMC)—created by the encapsulation of the channel by a modulator and demodulator. The modulator and demodulator hide from client and destination the implementation details of the actual channel and signal(s). Chapter 3 took this virtualization of the transporter one step further, encapsulating the DMC with an encoder and decoder. By introducing redundant information into the client's data to better withstand the effects of the noise and transient faults occurring in the channel, the result is a more robust virtual transporter, with the same task set but greater reliability/maintainability. In effect, we have an "outer" virtual transporter composed of an encoder and decoder encapsulating an "inner" virtual transporter composed of a modulator, demodulator, and channel.

Whatever the transporter's components, we still need to define its interface. At first blush, this would appear to be a trivial task since all we have discussed is the transporter executing its transport task. What else is there? In a word, management. While coding, by reducing the incidence of faults and/or repairing their effects, affects bandwidth management, this is hidden management: The client of a transporter consisting of just a channel does not see any difference between it and a transporter consisting of an encoder/decoder encapsulating a DMC. The client does not explicitly request that the coder/modulator transform the data; the client "sees" just a transporter to which it sends data (and may receive as well). Now, however, we must consider transporter management that is explicit and visible: The client of the transporter can invoke to request actuation of the transporter (for example, turning the transporter on or changing the service rate); can request to be informed about the state of the transporter; and can request allocation (that is, scheduling) of the transporter. The first two of these tasks involve bandwidth management, while the last is an instance of workload management.

This chapter concludes the virtual transporter hierarchy within the physical layer by discussing management and the virtual transporter defined by the interface between clients, referred to as Data Terminal Equipment (DTE), and transporters, generically known as Data Circuit-Terminating Equipment (DCE). The parallel is with abstract data types and objects. Just as methods define the interactions with the data structure (object), so the tasks defined at the interface (such as Request to Send, Send, and Receive) define the interactions with the transporter and its manager. In this chapter we analyze these interface tasks in terms of the MESA model. These channel interfaces come in many forms and flavors, but in this chapter we focus on two classes. The first are the serial interfaces, such as EIA-232 and V.35, most of which use another interface standard called V.24 that defines the interface functions. The second class consists of the physical layer specifications of the IEEE 802 local-area network standards, some aspects of which we touched on in Chapter 2.

4.2   DTEs, DCEs, and Interface Standards

4.2.1   DTE and DCE

As we just said, in the world of data communications both client and destination of a transporter are referred to as *Data Terminal Equipment* (DTE). In other words, the DTEs are the computers that want to send data to each other. The term *Data Circuit-Terminating Equipment* (DCE) is used to encompass the modulators, encoders, decoders, and demodulators that provide access to the actual channel. In other words, the DCE contains the equipment that, as its name implies, "terminates" the communications circuit (the actual channel) before it reaches the DTE; that is, while the DCE does not include the channel itself, it provides the interface by which the DTE invokes the services of the channel (see next section).

At its simplest, the DCE may be a simple pass-through mechanism (the so-called null modem or modem eliminator); here the channel signals are

the same as the signals that are used between DTE and DCE. Modem elimina-
tors are useful only for very short distance connections between DTEs because
of the signal attenuation difficulties we studied in Chapter 2. Depending on the
desired characteristics of the transporter (DCE–channel–DCE), the DCE may
contain a modulator (with corresponding demodulator in the DCE at the destination), or an encoder/modulator composite (with corresponding demodula-
tor/decoder composite in the DCE at the destination). Obviously, as we saw in
Chapter 3, the latter implementation would have higher reliability and effective
bandwidth (throughput) than the former due to the use of coding, at the price
of greater component cost and complexity. Figure 4.1 illustrates the possible DCEs.

We should note that, at least conceivably, it would be possible to implement a modem eliminator with coding—just an encoder at the client's DCE
and decoder at the destination DCE—but this implementation is seldom if
ever seen in practice.

The DTE–DCE separation reflects a modularization decision. The origins of this separation stem in part from the accidents of history. The telephone
monopolies of years past provided the access equipment to connect users to
their transport networks, and they defined the DCE interface to facilitate this.
Both DTE and DCE were typically customer-premises equipment (CPE) with
the DCE presenting an abstracted channel to the DTE. Manufacturers of
computer equipment (DTEs) simply needed to convert their signals to those
prescribed for communication over the DTE–DCE interface. Given that the
DCEs (modems) were generally supplied by the telephone service provider, this
simplified the lives of computer manufacturers and telephone service providers alike.

4.2.2　Interface Standards

Why standardize interfaces? Because without isolating the implementation
details of the many different types of channels and signals, the challenge of
constructing a data transport system, particularly with modular or off-the-shelf
components, would become horrendous. To remedy this, during the past
40 years the data processing and communications community has defined
a number of interface standards for addressing the needs of different types
of interconnection; as we will see in this chapter, low-speed, short-distance
connections require different types of interfaces than higher speed, longer distance ones.

EIA-232 nicely illustrates the distinction between architecture and implementation in the way it draws its definitions from a family of other standards.
The architectural component is provided by a standard known as V.24, which

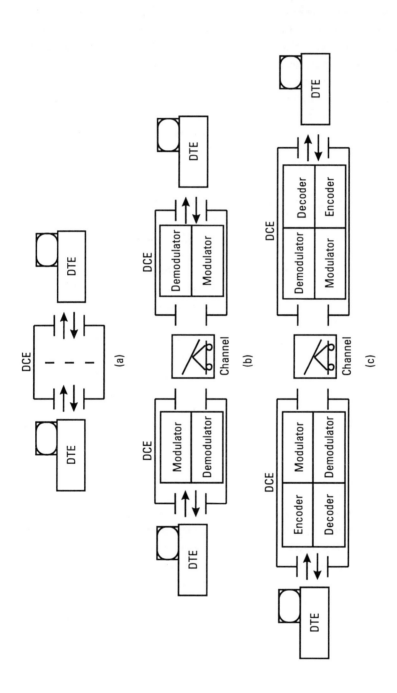

Figure 4.1 DCEs: (a) CDE as null modem/modem eliminator, (b) DCEs with modulators and demodulators, and (c) DCEs with modulators, demodulators, encoders, and decoders.

defines the functions or tasks that the interface enables to be requested. The V.24 standard defines the functional semantics of interface commands and signals used in several other interface standards besides EIA-232, notably V.35. The implementation component of EIA-232 is provided by V.28, which specifies the signal and channel characteristics of the connections between DTE and DCE. If the functional signals defined in V.24 are, by analogy, the methods for accessing our transporter "object," then the signaling standards such as V.28 define the exact syntax of the control blocks and parameters that realize these methods.

We now examine some signaling standards.

4.2.3   Interface Signal Standards

By virtue of its association with EIA-232 interfaces, V.28 is probably the most common interface standard in the world today. It is, however, not the only standard used. V.28, as we will see, is particularly designed for relatively low-speed interfaces where low cost is important. Given the wide variety of DTEs and DCEs and their associated communication requirements, it is not surprising that there are several different standards defining interface signal characteristics which are tailored to address different requirements. Two other interface standards that we look at in this section, V.10 and V.11, are designed to meet the needs of higher speed and/or longer distance signaling.

Notwithstanding these possible variations, all physical layer interfaces between the DTE and DCE do have at least one thing in common: They use baseband (digital) rather than broadband (carrier-modulated) signaling. Recall from Chapter 2 that we said carrier modulation was necessary when signals were to propagate over attenuating channels for relatively long distances; that digital signals may, depending on the signal bandwidth and the channel, propagate adequately for limited distances without any sophisticated mechanisms such as repeaters or modulators. Because both DTEs and DCEs are customer-premises equipment, it is generally assumed that the distances between them are relatively limited, from a few meters to a few kilometers, at most.

An obvious question then is, since by definition a digital computer uses digital signals internally, can these signals be used on the DTE–DCE channel(s)? The answer is no, for two reasons. First, even when computers (DTEs) are colocated with the DCE equipment, though they are close enough to allow digital signals to be used between the two, they are not close enough for the very specialized signals used over the computer's internal busses to propagate successfully between DTE and DCE. The signals used internally within a computer are optimized for channels measuring a few centimeters to a meter at most and transmission of tens of megabytes per second.

Therefore, the DTE must convert its internal digital signals (waveforms) to the digital signals specified by the interface standard being used between it and the DCE; the DCE, in turn, generally converts these waveforms into internal waveforms that its circuitry uses. This is exactly the "language translation" problem we mentioned earlier: Because DTE and DCE convert their respective internal signals into a common *lingua franca*, namely, the interface standard's signals, each is free to use the signals that best fit its needs without worrying about compatibility or the permutations of interconnection.

The second obstacle to using the DTE's internal signals over the DTE–DCE connection is serialization. Data are generally transferred within a computer over parallel channels—for example, 32- or 64-bit-wide data busses. Apart from esoteric interface standards such as the High Performance Parallel Interface (HPPI) defined in ANSI standard X3T9.3/88-023, communications interfaces are serial, meaning 1 bit is transferred at a time. This means that "serialization" of the data used within the DTE is required before it can be transferred to the DCE.

Another concern is the electrical stability of the interface channel(s) connecting the DTE and DCE. A very common complication is the risk of ground potential differences as the distance between DTE and DCE increases. A grounded electrical device has the ground potential of its local ground—where its power circuit is grounded to earth. It is relative to this (local) ground potential that the voltage levels of the signaling waveforms are actuated. Unfortunately, the values of ground potential can vary widely within a relatively short distance so that two electrical devices (such as a DTE and a DCE) may have very different ground potentials. A signal created at +5V relative to local ground may, even without any attenuation or other loss, be measured at the destination as +2V relative to the ground level at the receiver.

It is in part to overcome the effects of such ground potential differences that balanced signaling is used when the distance between DTE and DCE is great and/or the signaling rate is high. Balanced signaling requires two circuits (transporters) to reliably transport a waveform. A mark or space is sent by creating waveforms (signals) on each of the two channels, the difference of which is +5V or −5V, respectively. At the destination these two signals are measured and, by taking their difference, the information being transported is recovered. This reconstruction is another instance of estimation: The plant, namely, the information to be transported, cannot be measured because neither of the transmitted waveforms by itself carries the information. Figure 4.2 illustrates a balanced channel and estimator that reconstructs the original data.

Balanced signaling improves the reliability of the interface channels, but not all circuits require it. If the distance between the DTE and DCE is short, then on these interface circuits a simple, unbalanced signaling can be used.

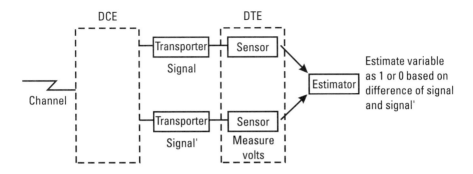

Figure 4.2 Balanced interface with estimator.

Balanced signaling is also advantageous when the channel is to transport high-speed signals. Here we should note that the exact throughput of a channel is inversely proportional to its length: the longer the channel, the greater the attenuation and noise and hence the lower the attainable channel capacity (bandwidth) (see, for example, [1]). Because of its superior noise resistance, balanced signaling is better suited to high-speed signals. When the distance relative to signal bandwidth (speed) exceeds the safety margin of unbalanced signaling, then balanced signaling is used.

The V.35 interface standard, intended for communications above the 20,000-bps limit of EIA-232, illustrates this nicely. Based on the same functions defined in V.24, V.35 is very similar to EIA-232 except for one significant difference: Because V.35 is intended for high speeds and longer distance, balanced circuits are used for carrying the data and timing signals. On the control circuits (see later discussion) the interface uses unbalanced circuits that conform to V.28 specifications. This division works because the control signals are not truly high-speed signals while the data signals are.

Table 4.1 lists the principal serial interface standards and some of their implementation parameters. The limits of rate and distance are, generally speaking, to be used as design targets rather than hard constraints. For example, according to the standard's specification the V.28 (EIA-232) interface should not be used above 20 Kbps but many vendors ignore this; similarly with the distance limit of 15m. The principal balanced interface is the V.11, which can handle signaling rates up to 10 Mbps.

4.2.4 The Logical Structure of the Serial Transporter Interface

Now that we have discussed these implementation characteristics, we want to return to the "architectural" aspects of interfaces, that is, their functional

Table 4.1
Serial Interface Standards

CCITT Standard	EIA Standard	Signaling	Balance	Rate	Distance
V.10/X.26	EIA-423	Voltage	Unbalanced	<100 Kbps	10m
V.11/X.27	EIA-422	Voltage	Balanced	<10 Mbps	Variable
V.28	EIA-232 (electrical)	Voltage	Unbalanced	<20 Kbps	15m
V.31	EIA-410	Current	Not applicable	<75 bps	—

structure. As we said earlier, an interface defines the methods by which services are invoked and, in the case of a transporter, by which they are delivered. That is, a transporter has by definition *two* interfaces: (1) between the transporter and the client and (2) between the transporter and the destination. A destination is not necessarily a client (in the sense that it does not request of the server any transport actuation—see below), but it nonetheless interacts with the transporter. This brings up an important design principle for our transporter interface: universality. Unless we want to define different interfaces for client and destination then the interface must accommodate both the send and receive tasks.

Then there is management. For example, to the extent that the transporter is shared between two or more clients there must be some management server to control (schedule) which of the competing/contending clients will have its RFS executed by the transporter at what time. This is workload management. Other workload management interface tasks may include synchronization of the transmitter and receiver on the individual channels between DTE and DCE.

Bandwidth management, too, must be considered. A request to test the transporter, such as loopback testing, changes the task set of the transporter. Whereas neither the composition of channel and modulator/demodulator into a discrete memoryless channel nor the composition of a discrete memoryless channel and encoder/decoder altered the task set of the basic transporter, at the interface level it may be necessary to consider some bandwidth actuation of kind. For example, a loopback test involves changing the task set of the transporter; if this capability is required, then an actuator must be implemented to execute it.

In the next section we explore management and the transporter interface more completely by taking the functions and signals of the V.24 standard and mapping them on to our model of workload and bandwidth management.

4.3   Managing the Serial DTE and DCE

The interface tasks of the V.24 standard include the transporter's actuations (*Send data* and *Receive Data*) as well as management actuations of the transporter, and monitoring information from instrumentation of the transporter that reports its condition, status, and so on. As one author puts it, "[V.24] is a 'superset' standard. Vendors select the appropriate V.24 circuits for their product, as do the standards groups that publish the V series interfaces" [2]. Note that we discuss only the 100 series of V.24 recommendations; including the 200 series, which specifies certain additional circuits for supporting automatic dialing of telephone numbers, would add little to our discussion.

4.3.1   Managing the Transporter: Bandwidth Control

We are going to divide bandwidth management tasks into control (those tasks that result in changes in the transporter's state and/or parameters) and monitoring (those tasks that measure and/or estimate these variables). The former tasks allow the DTE to request actuations of the transporter's task set and/or service rate(s). For example, this would allow the DTE to specify the speed of the transport actuation. These are bandwidth actuations in that they change the rate at which the composite DCE–channel–DCE server can execute its transport actuation task(s). Table 4.2 lists the V.24 bandwidth actuation tasks and, where they exist, the equivalent identifiers for EIA-232 and EIA-449 (a seldom used successor to EIA-232). We can see in this table that, beyond basic actuations such as turning on and off the transporter and its associated components such as modulators, V.24 provides for selection (actuation) of a limited set of transmitter and receiver parameters. The range of actuations is limited by two factors. The first is that all V.24 signals, including Requests for Service, are binary. These circuits are, in fact, dedicated communications channels with binary waveform sets. The second reason is that, altogether, the total number of V.24 "circuits" cannot be too large, because the physical connectors terminating the cables between DCE and DTE would become unwieldy. Indeed, a principal reason why the EIA-449 standard was never widely adopted was the large connector it used to support more of the V.24 circuits than the earlier EIA-232 standard could with its DB-25 25-pin connector.

4.3.1.1   Actuating the Transporter: Status

The first management actuation to discuss is turning the transporter "on," which we referred to in Chapter 1 as a *status actuation*. Note that not all transporters need to be turned on. The simplest transporters, for example, those consisting of two DTEs connected to a modem eliminator or null modem, do

Table 4.2
V.24 Bandwidth Control Tasks

Circuit Description	V.24	EIA-232/449	Source
Connect data set to line	108/1	CD/TR	DTE
Data terminal ready	108/2	CD/TR	DTE
Data signaling rate selector	111	CH/SR	DTE
Data signaling rate selector	112	CI/SI	DCE
Select frequency groups	124	–/–	DTE
Select transmit frequency	126	–/SF	DTE
Select receive frequency	127	–/–	DTE
New signal	136	–/–	DTE
Backup switching in direct mode	116/1	–/–	DTE
Backup switching in authorized mode	116/2	–/–	DTE
Loopback/maintenance test	140	LL/RL	DTE
Local loopback	141	LL/LL	DTE

not need to be turned on. However, many DCEs have components such as transmitters and receivers that, for reasons of economy, reliability, and so on, should not be left "on." Prior to transmitting, therefore, the DTE must request that the DCE turn these on.

V.24 defines two RFS messages that cause the DCE to trigger this status actuation. The first of these is carried on circuit 108/1 (*connect data set to line*). This RFS "causes the DCE to connect the signal-conversion or similar equipment to the line" [1]. The second message is carried on circuit 108/2 (*Data terminal ready*); this circuit is equivalent to the familiar EIA-232 circuit CD (*Data Terminal Ready*). As with circuit 108/1, this circuit is generally used to transport a RFS to the DCE to actuate the status of the transporter, turning the latter on and off as required.

4.3.1.2 Actuating the Transporter: Rate Selection

V.24 defines two circuits for specifying the rate of the transporter: 111 and 112 (both called *Data signaling rate selector*). These two circuits carry the same message, namely, selecting between two different signal rates, but the originators are different. All of the bandwidth actuation requests listed in are issued by the DTE with the exception of the message carried by circuit 112, which is issued by the DCE. The reason for this asymmetry is simple. The others concern transporter implementation details, such as signal frequencies, that do not affect the DTE, so there is no need for the DCE to request the DTE change its operating parameters. On the other hand, if the rate of the transporter is actuated, for example in response to a request from the local DTE on circuit 111, then the remote DTE must operate at this higher rate otherwise the total transporter (DTE–DCE–DCE–DTE) will fail.

4.3.1.3 Actuating the Transporter: Signal Selection

Recall that in Chapter 2 we stressed the basic transporter was composed of the channel and the signal(s) it carried. It follows, therefore, when considering actuation of the transporter that actuating the signal may be attractive to some users and indeed such actuations are within the V.24 functions. With circuit 124 (*Select frequency groups*) a DTE may request that the bandwidth manager actuate the transmitter to use all or a predefined subset of the available frequency groups. With circuit 126 (*Select transmit frequency*) a DTE may request that the bandwidth manager actuate the transmitter to use one of two frequency bands.

The requested actuations may also affect the receiver within the local DCE. With circuit 127 (*Select receive frequency*) a DTE may request that the bandwidth manager actuate the receiver to use one of two frequency bands. Finally, circuit 136 (*New signal*) carries a request from the DTE to actuate the receiver of the transporter by changing the response times used to decide if a signal has been lost (circuit 109; see section on "Monitoring the Transporter"). This change is useful if, for example, there is an advantage to rapidly connecting and disconnecting from a switched telephone connection.

4.3.1.4 Actuating the Transporter: Switching to a Standby Transporter

We saw in Chapter 1 that highly available systems frequently replicate critical components, either for concurrent execution with the main/primary components or for use as standby reserves that can be switched on in the event the primary components fail. In the case of transporters, the standby components may include alternative transmitters and/or receivers (modulator/demodulators, encoders/decoders, and so on). Such switching constitutes a major actuation of the transporter and it is supported at the DCE interface by V.24. A DTE can request the bandwidth manager switch to standby components using the V.24 circuit 116/1 (*direct mode*) or 116/2 (*authorized mode*). There are no equivalent EIA-232 circuits but there are for EIA-449.

4.3.1.5 Actuating the Transporter: Testing and Loopbacks

The bandwidth manager within the DCE may also support special actuations of the signal path that aid in fault detection and particularly fault isolation. If a

DTE's request for a connection to another DTE fails to elicit any response, then the source (client) DTE may attempt to initiate fault detection. Recall our earlier comment that with the introduction of the DCE, by which we mean in this instance modulators/demodulators or encoders/decoders, we now have four components (the two DTEs and two DCEs) and three distinct stages in the transportation of data: DTE to local DCE, local DCE to remote DCE, and remote DCE to destination DTE. (If the DCE is a null modem then the intermediate DCE to DCE stage is null and we have a two-stage, three-component transporter.)

The challenge of fault identification given this composite transporter is that the server is distributed and the fault may be in any of the three stages. A simple test, however, can aid in determining where the fault is located. If the transporter can be reconfigured (= actuated) so that data the client DTE sends are returned to it by one or other of the DCEs then locating the fault becomes a process of elimination. For obvious reasons, this actuation of the transporter is called *looping back* the data. Another standard called V.54 defines four types of possible loopbacks, including a loopback within the originating DTE (loop 1) to check its circuitry (Figure 4.3).

If the local DTE passes, then the local DCE is actuated to loopback the data (loop 3); if the DTE receives its data from the DCE then it can be inferred that the local DCE as well as the channel between the DTE and local DCE are operating correctly (in nominal condition). Assuming that this test is passed then the local DCE is actuated back to normal or pass-through operation and a remote loopback is effected by actuating the remote DCE (loop 2). Loop 4 can also be used to test the remote DCE and channel. From these remote loopbacks we can determine, using the same reasoning, if the remote DCE and the

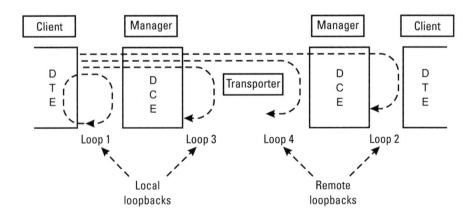

Figure 4.3 Local and remote V.54 loopbacks.

DCE-to-DCE channel are operating correctly. Finally, assuming both local and remote loopbacks are successful, then it can be concluded that the remote DTE is at fault. Figure 4 .3 illustrates both local and remote loopbacks.

Notice that in both types of loopbacks that the task set by the transporter is changed from {local_DTE   remote_DTE, remote_DTE   local_DTE} to {local_DTE   local_DTE}. That is to say, the transport task executed by the transporter is the same whether the loop is before the local DCE (loop 1), after the local DCE but before the channel (loop 3), after the channel but before the remote DCE (loop 4), or after the remote DCE but before the remote DTE (loop 2).

Corresponding to local and remote loopback, V.24 defines two RFSs to the bandwidth managers in the local and remote DCEs. Circuit 140 (*Loopback/maintenance test*) requests the transporter be put into remote loopback, and circuit 141 (*Local loopback*) requests the transporter be put into local loopback. To further specify the type of loopback (1 or 3, 2, or 4 , respectively) the V.54 recommendation specifies using other V.24 circuits; for details consult the standard or [2].

4.3.2 Managing the Transporter: Bandwidth Monitoring

Why provide any information to the client about the state of the transporter? The client in the DTE may contain what is, in effect, its own workload manager. This workload manager decides when and whether to request service from the transporter, basing its decision in whole or in part on the current state of the transporter. This is essentially what happens with connection-oriented communication: The connection establishment phase does not proceed without positive acknowledgment from the transporter (including, for purposes of connection setup, the remote DTE as part of the total transporter). This acknowledgment is, in essence, a form of feedback.

Before a client begins to send data to the DCE, it might be a good idea to check to make sure the transporter (channel, transmitter, receiver, and so on) is on and in nominal condition. Otherwise, the DTE may send data that cannot be forwarded by the DCE due to faults in the channel and/or partner DCE. Provision should be made, therefore, for the DTE to obtain state information about the transporter (including any component servers such as modulators). The key question is how granular should the feedback be? At one extreme, the DCE can merely inform the DTE that the overall state of the transporter is nominal or not. At the other extreme, the DCE can provide information about the signal levels, the noise on the channel, and so on. The constraint here is the same as we encountered with the actuation signals: binary signals and the need to limit the total number of circuits.

Given this limitation, V.24 defines six circuits by which the DCE can send to the DTE information on the state of the transporter. These circuits are listed in Table 4.3 and can be divided into three groups. The first of these groups conveys information about the status of the transporter and its readiness to execute transport tasks. The second group of circuits conveys information about the state of the signal arriving at the DCE from its partner DCE. The last group conveys information about the mode in which the transporter is operating; specifically, is it in standby or test mode?

4.3.2.1 Monitoring the Transporter: Status Instrumentation

Circuit 107 (*Data set ready*), also known as *DCE ready*, is used by the DCE to indicate to the DTE that the DCE has been turned on. This information can be obtained in one of two ways: Either the DCE measures some parameter of the channel and/or its peer DCE, and based on the measurement(s) and/or estimates derived from the measurement(s) decides the actuation has been successfully executed; or the DCE estimates that the composite has been status actuated merely because the DCE has attempted to execute such an actuation. The first is closed loop and the second is open loop.

Beyond this, the use of transitions on 107, or perhaps it would be more accurate to say the use of context (i.e., state), to provide further information increases the information capacity of the circuit. If the circuit's signal goes from ON to OFF during a call then this indicates the occurrence of a fault. Knowing whether the circuit was part of a switched connection or a dedicated connection allows further inference (estimation) as to the type of fault. This is fault isolation. If the two DCEs are communicating using a switched line then the ON → OFF transition during a call indicates a lost call, whereas if the line is nonswitched then the ON → OFF transition during a call indicates a hard failure.

Table 4.3
V.24 Bandwidth Monitoring Tasks

Circuit Description	V.24	EIA-232/449	Source
Ready for sending	106	CB/CS	DCE
Data set ready	107	CC/DM	DCE
Received line signal detector	109	CF/RR	DCE
Data signal quality detector	110	CG/SQ	DCE
Standby indicator	117	–/SB	DCE
Test indicator	142	–/TM	DCE

Circuit 106 (*Ready for sending*), also known as *Clear to send*, indicates if the DCE, by extension including the channel and the partner DCE(s), is either busy and cannot accept data from the DTE or is idle and can accept data. Because this is, in essence, workload scheduling, we defer a more detailed analysis until the next section on workload management. However, note that some simple implementations of DCE interfaces use this circuit as a proxy for circuit 107 and do not include the latter. The DTE can assume, if it receives permission to send via circuit 107, that the DCE is ready.

When both circuits 107 and 106 are used, however, the relationship between the two circuits and their signals is shown in Table 4.4. For the most part it is straightforward: If the DCE is on and ready to operate (107 = 1) but not available (106 = 0), then it can be inferred that the DCE/channel is busy executing a transport actuation task for another client; if the DCE is on and ready to operate (107 = 1) and available (106 = 1), then it can be inferred that the DCE/channel is either not busy executing a transport actuation task for another client or that the transporter supports concurrent multitasking (FDX) execution. The only interesting aspect of the relationship between the *Data set ready* circuit and the *Ready for sending* circuit is if the former indicates the DCE is off but the latter says that the DCE is ready to accept data; in this case, there is clearly a fault in the DCE's instrumentation.

4.3.2.2 Monitoring the Transporter: Signal Instrumentation

V.24 provides two state variables concerning the signal: *Received line signal detector* (circuit 109) and *Data signal quality detector* (circuit 110). *Received line signal detector (RLSD)* carries/transports a measurement/estimate of the reliability of the transporter that moves data from DCE to DCE. For the RLSD (carrier detect) circuit to be meaningful, there must be a sensor in the DCE to measure the signal arriving over the communications channel and a decision mechanism (an estimator) to determine if the signal quality as measured is sufficient. This is a binary decision: Either the received line signal is adequate or it is not. In the simplest case, the criterion is simply whether a carrier was detected or not.

Table 4.4
Data Set Ready Circuit Versus Ready for Sending Circuit

	Data Set Ready = 0	Data Set Ready = 1
Ready for Sending = 0	DCE is off	DCE is busy
Ready for Sending = 1	Fault in DCE instrumentation	DCE is idle

Whereas RLSD informs the DTE if the received signal is within established tolerances, the *Data signal quality detector* circuit informs the DTE if, in the estimation of the DCE, there is a "reasonable probability of error" having occurred in the data just received. If, for example, a decoder in the DCE is unable to decode a received string of symbols (a decoder failure; see Chapter 3), then the DCE can use circuit 110 to signal the DTE that the data it has been sent have probably been corrupted in transit. The signal itself may have been within tolerances but nonetheless its "quality" has been reduced.

4.3.2.3 Monitoring the Transporter: Mode Instrumentation

As we saw in the previous section on actuations, the DCE and by extension the transporter can be in one of five modes of operation, four of which (Off, Ready, Test, and Standby) are generally entered in response to requested actuations originating with the DTE while the fifth, Fault, is entered from either the Ready, Test, or Standby mode as a result of a fault in the transporter. Off is the default state. V.24 defines two instrumentation circuits specifically for signaling whether the DCE is in Test or Standby mode.

Circuit 142 (*Test indicator*) indicates that the DCE (and hence the transporter) is in test mode, meaning that data can neither be sent or received. For example, if a DCE has actuated a loopback in response to a request from one of the DTEs, then this circuit will indicate to all the DTEs that the transporter is unavailable and that any data presented to it should be treated as test data rather than client data. In addition, if a DCE has initiated testing of its components and/or the channel at the initiative of its own bandwidth manager, then circuit 142 is used to signal the consequent unavailability to the DTEs. When the tests are completed, if the condition of the components and channel is nominal (or close enough) then this circuit signals that test mode has been exited and the DCE is ready ($107 = 1$).

Circuit 117 (S*tandby indicator*) indicates that the DCE has switched to backup components. Again, this actuation of the DCE's internal structure may have occurred in response to a request from the DTE (*direct on* circuit 116/1 or *authorized on* circuit 116/2) to switch to alternative circuits or, if the DCE has a bandwidth manager with the autonomy, it may have been initiated by the DCE itself in response to some internal condition it detected, such as an imminent failure of the primary transmitter. In either case, circuit 117 signals to the DTE that the DCE is in backup mode. The DTE may want this information for a variety of reasons related to management. For example it may wish to send an SNMP message, for example, indicating that repair and/or replacement of the primary components is required.

We can summarize these transporter actuations and corresponding feedback information with the state (or mode) transition diagram in Figure 4.4.

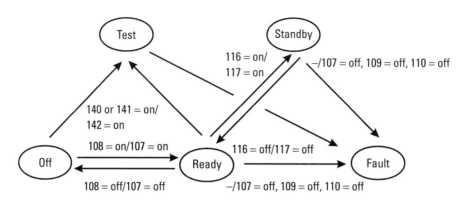

Figure 4.4 Mode transition diagram for a transporter.

The transition arcs indicate the (event/output) pairing that triggers the state transitions and the output messages sent using various V.24 circuits.

4.3.3 Managing the Client: Workload Control

Recall that workload control actuates the arrival of work at a server; and that, in its absence, the client's RFSs (and accompanying plant, if any) will arrive at the server at the rate and in the order in which they left the client (or clients). As such, one of workload management's main concerns is "throttling" (= actuating) the flow of work (RFSs). But if the DTE knows the bandwidth (= service rate) of the transporter for which the DCE is the interface, then why should it ever be necessary to control the traffic from the DTE? In other words, wouldn't the DTE limit itself to what it knew the transporter could handle and thus not overwhelm the transporter?

There are at least three difficulties with this assumption. First, even a simple, single-stage transporter may have a highly variable service rate due to exigencies of the channel and any intermediate components. For example, it is well known that microwave links suffer severe degradation under certain atmospheric and geographic conditions. Likewise, modems and other DCEs may fall back to lower rates when there is too much noise on the channel or when lower speed backup systems are switched on. In the absence of workload control mechanisms, it is impossible to be sure that a client that has been allocated (= scheduled) the channel will not overwhelm it since the channel/transporter's bandwidth may change while the client is using it.

Second, flow control/workload management may be necessary if the destination is unable to receive the transported data as fast as it can be generated by the client and moved by the transporter. With the introduction of DCEs as isolating elements, the transporter has in fact not one but three stages: DTE to DCE, DCE to DCE, and DCE to DTE. However, it may not know what the bandwidth of the destination DTE is. Indeed, this can be highly variable depending, among other things, on other tasks the destination DTE may be executing—such as data arriving at the same destination from other sources (clients). Or, as we will see in Part III, if the DTE-DTE traffic flows over multiple transporters and relays, then congestion in the intermediate components can cause fluctuations in the overall service rate.

Finally, and perhaps most importantly, at any given time some or all of the transporter's capacity may be allocated to other clients, of which the "local" client knows nothing due to the spatially distributed nature of transport actuation and transporter alike. The most obvious example is a local-area network (LAN) that has dozens or even hundreds of stations (clients) that may be seeking its services at the moment. Even something as simple as a bidirectional transporter may have contention problems between the two clients (DTEs) if the channel and/or DCEs can only support transmission in one direction at a time.

It is for these reasons that mechanisms of flow control are used even at the lowest layers of the protocol stack. In the case of LANs, as discussed later, the workload management is mainly located at the Data Link layer; this includes carrier sense and token-passing mechanisms. At the Physical layer of serial transporters, V.24 defines mechanisms by which both the DTE and DCE can start and stop data flowing over the send and receive circuits. We will discuss these mechanisms, including the *Request to send* and *Ready for sending* signals we discussed as part of bandwidth management, and also the mechanisms for synchronizing DTE and DCE.

4.3.3.1 Transporter Task Set and Tasking

We discussed in Chapter 1 that we can characterize a transporter (in fact, any server) by its task set and its tasking. This is important here because the type of workload management a transporter requires will depend on the nature of its task set and tasking. How many clients (DTEs) does it have to contend for its services? How many tasks (transport actuations) can it execute at one time? These parameters determine the complexity of the workload manager's task.

A trivial transporter is one that has a single client. The transporter is necessarily simplex, that is, traffic flows in one direction only. It has only the single client so that means traffic from the reverse direction is impossible. Examples of such simplex transporters include telemetric applications, such as power line or pipeline monitoring systems. However, the overwhelming majority of computer communications involves two (or more) clients competing/contending for the services of a shared transporter.

Beyond the task set, the tasking of the transporter is crucial in determining the applicable workload management. That is, how many transport actuation tasks can the transporter execute at once? The obvious minimum is one, and a server that can execute only one task at a time is referred to as *single tasking*. If the transporter can execute more than one task at a time then we say it is *multitasking;* the distinction we made in Chapter 1 between serial and concurrent multitasking is important and we will return to it in later discussions, but for the moment all we need to note is that a multitasking transporter supports two or more transport tasks in execution.

Reflecting on this, we make a distinction between two types of duplex: half duplex, which can be either single tasking or serial multitasking, and full duplex, which implies support for concurrent multitasking. For example, let's say we have a transporter that has two clients A and B and a task set that includes A $\rightarrow$ B and B $\rightarrow$ A transport actuations. If A can send data to B at the same time as B is sending data to A, then this is a full-duplex transporter. On the other hand, if A sending data to B precludes at that exact moment B sending data to A (or vice versa), then the transporter is a half-duplex transporter. Note that if A and B are allowed to completely send their data to each other without interruption, then this is effectively single tasking, whereas if their respective tasks are "time sliced" and interleaved then this is serial multitasking.

With a full-duplex (FDX) transporter that is bidirectional there is no problem with coordinating (scheduling) the two clients because there is no conflict for common resources. The same is obviously true with simplex transporters. However, a half-duplex (HDX) channel is a shared resource that requires a workload manager to schedule its allocation to competing demands. Likewise if the transporter has more than two clients then even with full-duplex transportation some potential conflict cannot be eliminated; such multi-dropped configurations are common with certain data link protocols that are designed primarily for terminal handling, such as SDLC and Bisync. In these cases the coordination of this sharing is the task of the management entities within the DCEs.

V.24 defines a number of messages and services by which these servers can affect the required workload actuations. Table 4.5 lists these V.24 workload actuations. Note that we have included a number of the V.24 circuits we have already discussed in the context of bandwidth management. The reason for this repetition is, as we will see later, that these circuits affect workload as well as the transporter itself. In addition, there is a set of circuits that concern managing workload specifically for an HDX transporter. They allow a DTE to request the DCE to actuate the allocation of the transportation—that is, "turning around" the transporter.

Table 4.5
V.24 Workload Control Tasks

Circuit Description	V.24	EIA–232/449	Source
Request to send	105	CA/RS	DTE
Ready for sending	106	CB/CS	DCE
Data set ready	107	CC/DM	DCE
Ready for receiving	133	–/–	DTE
Request to receive	129	–/–	DTE
Return to nondata mode	132	–/–	DTE
Transmitter timing	113	DA/TT	DTE
Transmitter timing	114	DB/ST	DCE
Receiver timing	115	DD/RT	DCE
Receiver timing	128	–/–	DTE
Received character timing	131	–/–	DCE

Finally, in this section we discuss the synchronization of the DTE and DCE. Why treat synchronization as a special case of workload management? In part, because there is an inherent relationship of a clock to a scheduler; in fact, every scheduler must have some clock with reference to which it constructs schedules. Given a pair of DTEs and DCEs, each has its own clock.

As the data are transported from, say, DTE A to DCE 1, DCE 1 must know when to sample the signal so that it can know what it is to send to the channel for transportation to DCE 2. In turn, DCE 2 must be synchronized to DCE 1 so that it can know when to sample the channel output; we discussed in Chapter 2 the means of doing this, notably self-clocking codes and clock recovery mechanisms. Finally, the data are transported from DCE 2 to DTE B and DTE B must know when to sample the output.

If these receivers and transmitters are not synchronized, that is, if their respective clock rates are not kept within certain tolerances, errors will occur. V.24 provides mechanisms for either the DTE or the DCE to provide the clock signal to be used by the schedulers in each. For example, a transmitter can send the clock it used to schedule the actuation of the signal to the receiver so that the latter can schedule the measurement of the received signal; or the receiver can provide the clock for a transmitter to schedule its actuations. The effect is the same: the execution of the transport task, or rather its component transmission and reception tasks, is actuated, and this is why we include these circuits under workload management.

4.3.3.2 Workload Actuation: Scheduling the Execution of the Transporter

As we outlined earlier, scheduling the arrival of work at a server depends to a considerable degree on the client. We can make an analogy with automobiles and the rules of the road. If a car's driver ignores traffic rules, there will be no orderly sharing of the pavement, irrespective of how well designed the signals may be. In the same way, fundamentally a DTE can only be "throttled" if it allows itself to be throttled. This isn't to say that a DTE that insists on sending data will be allocated the transporter by the workload manager in the DCE. The latter is still the "traffic cop" and will discard unallocated traffic, but it would obviously be preferable in most circumstances to avoid this.

The key to "well-behaved" DTEs is the recognition that, as with the rules of the road, if one disregards them the result will be chaos. If one DTE monopolizes a transporter then the RFSs of other DTE(s) cannot be executed. There are two possible solutions: (1) The upper layer protocol will self-limit the length of DTE transmission (2) or the DCE can shut down the DTE by using the V.24 circuits for flow control (workload actuation).

What are these controls? Let's start with the two circuits in Table 4.1, which we discussed in the section on bandwidth monitoring, namely, *Data set ready* (circuit 107) and *Ready for sending* (circuit 106). Both of these report on the status and readiness of the transporter, which is indeed a bandwidth monitoring task, but they also are part of the workload actuation process. Just as the *Data terminal ready* (circuit 108/2) is more than a measurement/estimation of the status of the DTE—in the absence of *Connect data set to line* (circuit 108/1) it is also an implicit RFS to the DCE's actuator to turn the transporter on—so do *Data set ready* and *Ready for sending* convey more than just information on the transporter. In particular, *Ready for sending* tells the DTE more than that the transporter is not busy; it is an implicit actuation of the DTE, allowing it to send data to the transporter.

Ready for sending has different meanings depending on the tasking of the transporter under consideration. If the transporter is HDX then *Ready for sending* "measurement" is an open-loop estimate of the transporter/channel (uses timer after RTS). If the transporter is FDX then *clear to send* is a measurement/estimate of the channel based on RLSD. If the transporter is FDX with error detection the *Ready for sending* circuit can convey internal DCE condition (e.g., modem buffers are ready to accept data).

With an HDX transporter or a transporter (HDX or FDX) with more than two clients, the DCE will send *Ready for sending* to a DTE in response to the latter sending a *Request to send* message on circuit 105. *Request to send* is an RFS to be scheduled for allocation of the transporter. The sequence *request to send/ready for sending* is sometimes referred to as *hardware flow control*, as opposed to the use of the ASCII control characters X-ON (x'11') and X-OFF (x'13'), also known as control-Q and control-S, which is a widely used method of software flow control. Unlike hardware flow control, however, software flow control is not character transparent: If the patterns x'11' or x'13' occur in client data then the confusion may cause the DTE to suspend sending unnecessarily.

Note that not all transporters needtoimpose flow control. For FDX point-to-point transporters (bidirectional since there are two clients) there is no reason to actuate the DTEs sending data since each client (DTE) appears to have its own transporter—the concurrent multitasking of the transporter effectively creates two "virtual" transporters. On the other hand, if it is possible that the "bandwidth" of the receiving DTE can be exceeded by the traffic sent from the transmitting DTE there is still need for some throttling. V.24 provides for the receiving DTE to itself actuate the flow using *Ready for receiving* (circuit 133). This backward pressure from the receiving DTE causes its associated DCE to signal the transmitting DCE to throttle back, which will in turn result in the transmitting DTE losing the *Ready for sending* signal. At this point the transmitting DTE should cease sending data until it receives permission to send more.

Reviewing these mechanisms, it is clear that the DTE has three choices:

1. It can just start sending data to the DCE and hope for the best.

2. It can send a *Request to send* (circuit 105) RFS to the DCE and wait for the DCE to respond with a *Ready for sending* (circuit 106) message and only when this is received start sending data to the DCE.

3. It can issue a *Data terminal ready/Connect data set to line* (circuit 108) RFS to the DCE, wait for a *Data set ready* (circuit 107) message from the DCE, then issue a *Request to send* RFS to the DCE and wait for the DCE to respond with a *Ready for sending* message and only when this is received start sending data to the DCE.

Figure 4.5 illustrates this full sequence of "handshaking" between DTE and DCE.

4.3.3.3 Workload Actuation: "Turning Around" Half-Duplex Transporters

Several V.24 workload actuation tasks are meaningful primarily with half-duplex transporters. Circuit 129 (*Request to receive*) allows the DTE to request that the workload manager within the DCE enter the "receive" mode and thus free the channel for a remote DCE to use. This is an explicit request for "turning around" the channel, as opposed to the implicit channel turnaround we discussed earlier. Circuit 132 (*Return to nondata mode*) allows the DTE to request that the workload manager within the DCE enter "nondata" mode, meaning

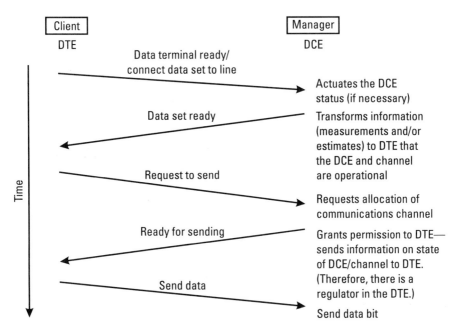

Figure 4.5 DTE–DCE signaling.

that the DCE is effectively off-line; however, the channel is *not* released to another DCE.

Figure 4.6 summarizes the workload actuations we have considered so far.

4.3.3.4 Workload Actuation: Synchronization

The last instance of workload control, albeit at a considerably different level than we have seen up to now, is synchronizing or "clocking" the transmitter and receiver over the DTE–DCE channel(s). This is what we might call *micromanaging* the flow of data between DTE and DCE. The issue is not the starting or stopping of the larger units of data that we have been discussing up to now, but rather when the individual bits (or waveforms carrying those bits) are sent between the two. Such control is necessary because the spatially distributed nature of transporters introduces an inescapable uncertainty between transmitter and receiver. The latter cannot know exactly when the former has created (sent) waveforms over the channel.

And when dealing with digital transporters the discrete (in time) nature of digital signals makes such uncertainty problematic. Scheduling the measurement of the signal by the receiver's sensor is crucial: too early or too late and the probability of error increases as the signal-to-noise level falls, the effects of intersymbol interference rise, and so on. This is exactly the same question we

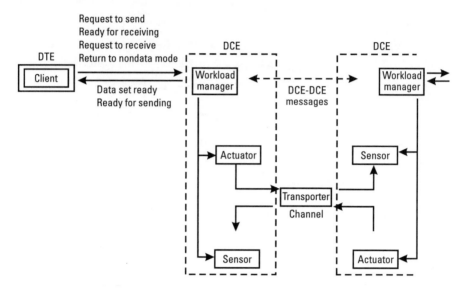

Figure 4.6 Workload management: actuations.

discussed at some length in Chapter 2. If there is a mistiming between receiver and transmitter (whether the receiver is the DCE and the transmitter is the DTE or whether the receiver is the DTE and the transmitter is the DCE), then the transported data may be mismeasured and/or misestimated, resulting in corrupting faults. Synchronization tells the workload scheduler in the recipient (DTE or DCE) when to schedule the measurements of the incoming circuits and/or actuations of the outgoing circuits.

There are three ways to affect this level of workload control. First, DTE and DCE can use asynchronous character framing so the receiver, alerted by start bits, knows when to begin measuring the bits of each character. As we saw in early discussions, asynchronous framing is the least expensive to implement in terms of the circuitry required; it is, however, the most expensive in terms of the overhead. Second, DTE and DCE can use a self-clocking code (such as NRZI, B8ZS, or one of the others discussed in Chapter 2) and an estimator to recover the clock from the code. The receiver reconstructs the transmitter's clock using a phase-locked loop or equivalent estimator. The third approach to synchronizing the transmitter and receiver is to use explicit timing signals from either DTE to DCE or vice versa. This is called *external clocking*, and we discuss in this section the V.24 mechanisms for supporting it.

The situation is complicated by the fact that with a duplex transporter, there are *two* transmitter–channel–receiver composites that may require synchronization. That is to say, for a bidirectional (duplex) transporter there are, in fact, two clocks: the sending clock, which provides the synchronization for the *Transmitted data* transporter that transports data from DTE to DCE (V.24 circuit 103); and the receiving clock, which provides synchronization for the *Received data* transporter that transports data from DCE to DTE (V.24 circuit 104). For each of these two clocks, V.24 provides circuits for either the DTE or the DCE to provide the synchronizing signal to its partner.

The V.24 circuits that provide clocking for the *Transmitted data* transporter (circuit 103) are shown in Figure 4.7. Circuit 113, *Transmitted Signal Element Timing (DTE source),* enables the DTE to provide clocking to the DCE's receiver for the measurement of the data signal carried on circuit 103. Circuit 114, *Transmitted Signal Element Timing (DCE Source),* enables the DCE to provides clocking to the DTE's transmitter. The transmitter uses the clock from the DCE to schedule the actuation of the signal.

The V.24 circuits that provide clocking for the *Received data* transporter (circuit 104) are shown in Figure 4.8. Circuit 128, *Received Signal Element*

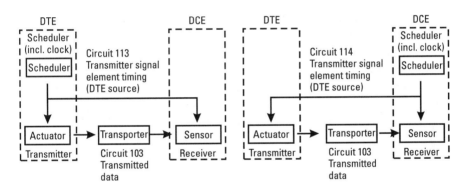

Figure 4.7 Transmitter timing: DTE and DCE sources.

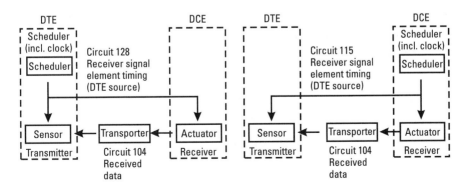

Figure 4.8 Receiver timing: DTE and DCE sources.

Timing (DTE Source), enables the DTE to provide clocking to the DTE's transmitter. The transmitter uses the clock from the DCE to schedule the actuation of the signal. Circuit 115, *received signal element timing (DCE source),* enables the DCE to provide clocking to the DTE's receiver for the measurement of the data signal carried on circuit 104.

As their names imply, these four circuits provide synchronization at the signal element level, which, in the case of binary waveforms as defined by most of the interface standards such as V.28, V.10, and V.11, means the bit level. In addition to these V.24 provides for character level synchronization via circuit 131, *Received Character Timing,* by which the DCE signals the receiver in the DTE when characters are transported; this signal indicates the framing of the character. Table 4.6 summarizes these circuits.

4.3.4 Managing the Client: Workload Monitoring

Unlike the transporter itself, explicit monitoring of the workload is much simpler because the very receipt of data from the DTE is, in some ways, "monitoring." There are, however, two V.24 circuits that do carry information about the workload. The first of these is *Calling indicator* (circuit 125), which informs a DTE that another DTE wishes to establish a connection. This is particularly important in dial-up applications. The second monitoring circuit is *Received data present* (circuit 134), sent from the DCE to a receiving DTE to indicate that the *Receive data* circuit is carrying data from a sending DTE. This information can be used by the DTE to disregard data on that circuit, for example, if a loopback or test is being executed. Table 4.7 summarizes these circuits.

4.4 LAN Interface Standards

We now move on to consider LANs and *their* interfaces. As we'll see in discussing the physical layer portions of the 802.3 and 802.5 standards, the same issues and concepts that we just discussed apropos of serial transporters apply to

Table 4.6

Sychronization Circuits

	DTE Source	DCE Source
Transmit Clock	Circuit 113	Circuit 114
Receive Clock	Circuit 128	Circuit 115 (signal element), Circuit 131 (character)

Table 4.7
Workload Monitoring

Circuit Description	V.24	EIA-232/449	Source
Calling indicator	125	CE/IC	DCE
Received data present	134	–/–	DCE

LANs as well. However, because LANs were designed decades after the serial protocols and reflect advances, for example, in synchronization technology (estimators), there is no need for many of the management circuits we have been discussing and hence the physical layer of LAN interfaces is considerably simpler than their serial counterparts. That is, this simplicity comes because LAN signals are self-clocking; LANs at the physical layer are connectionless; and many workload management tasks are relegated to the data link control protocols.

Notwithstanding their vastly improved rates of transmission and reduced error rates, LANs are still finite in both their bandwidth (service rate) and reliability, and this compels the inclusion of management to accommodate these imperfections. The focus of this section is to map these management tasks on to our model of workload and bandwidth management, exactly as we have done with the V.24 tasks for serial transporters.

4.4.1 IEEE 802.3 Physical Interface

With Ethernet (802.3) some of the terminology used in the world of serial interfaces has changed. The 802.3 standard speaks of end stations as DTEs but, rather than DCEs, the term *medium access unit* (MAU) is used to denote the set of mechanisms that provides the interface to the channel. Outside the standard, however, MAU is rarely used; in practice, the term *transceiver* is used, from the abridgment of *trans*mitter/re*ceiver*. Whatever it is called, the functions of DCE and MAU/transceiver are largely identical: to isolate the DTE from the channel and to provide the interfaces to mechanisms of bandwidth and workload management by which the DTE can request information about the channel and/or actuate changes.

4.4.1.1 MAUs

Positioned between the channel and the DTE, the MAU has three main components: the Attachment Unit Interface (AUI), to which the DTE is attached by means of what is called the AUI cable; the Physical Medium Attachment, which contains functional circuitry (see later discussion); and the Medium Dependent Interface, which provides the interface to the actual channel. Figure 4.9 illustrates the relationship of DTE, channel, and MAU components. Note that the standard makes clear this modularization is functional: An implementation that bundles the MAU with the DTE, say on the adapter card, is perfectly valid; in this case, the AUI interface is hidden and there is no AUI cable.

The 802.3 standard specifies four functions that the MAU must execute, and a fifth that is optional. The first two are the *Transmit* and *Receive* functions, which are self-explanatory. The third is called the *Collision Presence* function, which detects if two or more DTEs are transmitting on the channel simultaneously. The fourth is the *jabber* function, which will prevent a DTE from transmitting for too long by interrupting its transmission. The jabber function is an example of a management task that the MAU will execute automatically, without any RFS from the DTE. The last of these is called the *monitor* function, which actuates the MAU to disable the transmission of data, and is optional to 802.3 AUIs.

Mapping these on to our model of workload and bandwidth management, the collision presence and jabber functions are workload management— collision presence is workload monitoring and jabber is workload control. The monitor function is an instance of bandwidth management. Figure 4.10 shows these components and their mapping to the MAU's workload and bandwidth managers. We also show a *Signal error* message from the bandwidth manager to the DTE; this is not a formal component of the MAU.

4.4.1.2 AUI Channels

The 802.3 equivalent of the DTE–DCE channels we discussed earlier is a set of balanced (two conductors) circuits that form the AUI cable. The reason balanced circuits are used is simple: speed. Just as the V.11 standard uses balanced

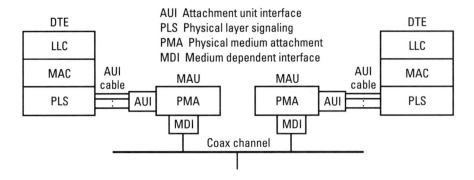

Figure 4.9 IEEE 802.3 DTEs and MAUs.

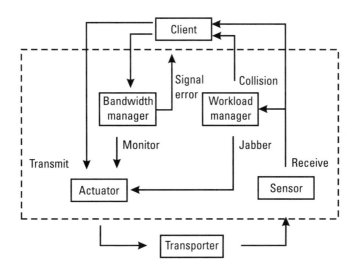

Figure 4.10 MAU components.

circuits for high-speed serial channels, so are AUI circuits used in the 802.3 standard; originally these were running at 10 Mbps but with later versions this has increased first to 100 Mbps and recently into the Gbps range. The AUI cable has four of these balanced pairs: one pair for data into the DTE, abbreviated DI; one pair for data out from the DTE to the MAU, abbreviated DO; a third pair for control (management) signals from the DTE, abbreviated CO; and a last pair for control (management) signals from the MAU, abbreviated CI. There are in addition several power and ground circuits, so that the AUI cable typically uses 15-pin DB-15 connectors.

Two differences stand out between these DTE–MAU interface circuits and the DTE–DCE interface circuits used in serial transporters: more waveforms but fewer circuits. First, whereas the latter employ binary waveforms, with the former, the signaling is ternary for both the data and the control circuits. In addition to the CD0 and CD1 waveforms on the data circuits and the CS0 and CS1 waveforms on the control circuits, respectively, each circuit can carry a third waveform, namely, the idle signal IDL. This third waveform, though, does not compensate for the fact that instead of the several dozen or so management circuits defined in V.24, there are only two management channels (circuits) connecting the AUI to the DTE—one in each direction. Consequently, the number of management actuations is extremely limited as is the granularity of information the MAU can report about the transporter/workload.

We will pass quickly over the two data channels DI and DO because they are relatively straightforward: Each carries data as the Manchester-encoded waveforms CD1 and CD0 (see Chapter 2) and the IDL waveform to indicate no data to be transferred. With the control circuits, however, the limited number of messages (three in, three out) presented an obvious challenge to the designers. The Control In (CI) circuit transports monitoring information from the MAU to the DTE. When the MAU sends an IDL waveform to the DTE on the CI circuit, this carries the *mau_available* message. This message is approximately the equivalent of the *Data set ready* message of the V.24 standard. On the other hand, if the MAU sends a CS1 waveform then this carries a *mau_not_available* message, indicating that the MAU is not ready to connect the DTE to the channel; this message is optional in that not all AUIs need this "instrumentation" to conform to the 802.3 standard.

Part of the reason the *mau_not_available* message is optional is that if there is any difficulty then the MAU will send to the DTE a message called *signal_quality_error* (SQE) by means of the CS0 waveform. To quote from the 802.3 standard:

> The PMA sublayer [within the MAU] may send the *signal_quality_error* message to the PLS sublayer [within the DTE] in response to any of three possible conditions. These conditions are improper signals on the medium, collision on the medium, and reception of the *output_idle* message. [3]

That is to say, the *signal_quality_error* message "lumps" together several very different states which nonetheless are all indicators that the channel is not available at the moment. The first of these, improper signals or signal errors on the medium (that is, the channel), is an instance of bandwidth monitoring: Some component of the transporter (a transmitter or the channel itself) has suffered a fault. The second cause for sending an SQE message to the DTE is to indicate a collision. This, too, can be regarded as a fault but it is properly speaking a workload fault—two or more stations have started transmitting at the same time. This is an inevitable consequence of the distributed scheduling mechanism, carrier sensing, that is the cornerstone of the 802.3 and related protocols. The *output_idle* message, when received from the DTE after the transmission of data by the MAU, indicates the end of transmission (EOT) and triggers the MAU to execute an SQE "self-test," which checks its internal components; an SQE message is sent if there is any anomaly.

The standard elsewhere specifies some additional circumstances under which the *signal_quality_error* message is sent by the MAU to the DTE. If there are no other MAUs on the channel, then this triggers an SQE message. If another MAU starts transmitting after the local MAU has started transmitting, then an SQE message is sent. If the Jabber function of the MAU has had to inhibit transmission then the MAU sends an SQE message [3].

The Control Out (CO) circuit carries bandwidth actuation RFSs to the bandwidth manager in the AUI. Here, too, the options are limited by the single circuit and the ternary waveform set. If the DTE wishes the AUI to function normally it sends over the CO circuit an IDL waveform. In this case, the AUI will send to and receive from the DTE data over the DO and DI circuits, respectively. On the other hand, some AUI implementations provide an optional actuation that the DTE may invoke by sending the AUI a CS0 waveform over the CO circuit. This waveform corresponds to an *isolate* RFS. The AUI will disable its transmitter while leaving the receiver and collision detection mechanisms functioning. This optional state is called *Monitor Mode* and can be used if a DTE suspects that its AUI's transmitter is malfunctioning.

It is important to recognize that there is no equivalent AUI message to V.24's *Ready for sending* (EIA-232, *Clear to send*) by which the MAU might schedule (i.e., allocate to the DTE) the channel. A DTE that wishes to transmit will monitor the channel's activity by the DI circuit, which will carry any data on the shared channel from the MAU to the DTE; if the DI circuit is transporting the IDL waveform, then the DTE regards this by inference as a *Ready for sending*.

Figure 4.11 shows the signals the DTE and MAU exchange over the data and control circuits.

4.4.2 IEEE 802.5 Physical Interface

The 802.5 standard specifies a simpler physical interface than the 802.3 interface. First of all, there are only two interface circuits, one for send and one for

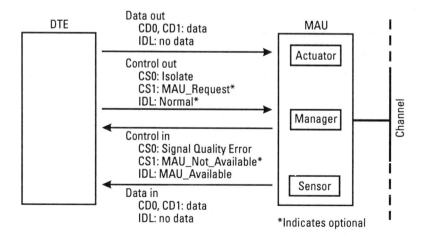

Figure 4.11 AUI channels and signals.

receive; as in the 802.3 standard, these are balanced interfaces. These circuits connect DTEs, *stations* in the language of the standard, not directly to each other but instead by means of a central concentrator that is variously known as a Trunk Coupling Unit (the standard's term) or a Multistation Access Unit (in practice). This results in what is sometimes called a *star-wired ring,* as shown in Figure 4.12. Observe that in this figure each station on the Token-Ring LAN is connected to the next station by a simplex transporter. That is, an 802.5 LAN is a circular concatenation of simplex channels—each has a single client and a single destination, all joined like so many elephants "trunk-to-tail." Because there is no channel shared between two or more transmitters, many of the workload management functions we discussed earlier are not necessary. In addition, there is no instrumentation in the TCU/MAU to report on the state of the channel but this omission is immaterial because every station is, in effect, directly attached to the channel.

So what is the reason for having a TCU/MAU? The basic function of the TCU/MAU is, to quote from the standard, to provide the "mechanism for effecting the insertion or bypass of the station" [4]. This is bandwidth actuation of the LAN: The task set is actuated whenever a station is joined or leaves the ring. We can see in Figure 4.12 the bypass mechanism and that when a station on the token-ring LAN is bypassed then the TCU/MAU connects its upstream

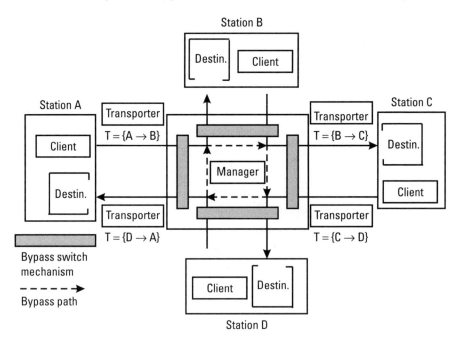

Figure 4.12 Token-Ring stations with trunk coupling unit.

and downstream neighbor stations directly. In part this is designed to accommodate stations that might be powered off, which might otherwise "break" the ring.

The mechanism executes this actuation in response to a request for service from a station. The question is, how does the station send the RFS? Because there are no control circuits, any control messages must pass over the two data circuits. Given that the waveform set contains two nondata symbols (J&K) in addition to the data symbols for 1 and 0 (see Chapter 2) these would seem logical candidates for carrying control messages. However, they are not used for this.

So if there are control messages, how are they carried? They are carried by means of what is called *phantom circuits*. These are created by means of DC voltages, transparent to the various symbols, that a station sends on its send and receive circuits concurrently with the data traffic. In effect, the phantom voltage signals constitute an additional waveform. To quote from the standard again:

> The voltage impressed is used within the TCU to effect the transfer of a switching action to cause the serial insertion of the station in the ring. Cessation of the phantom drive causes a switching action that will bypass the station and cause the station to be put in a looped (*wrapped*) state. This loop may be used by the station for off-line self-testing functions.... The phantom drive circuit is designed such that the station may detect open-wire and certain short-circuit faults in either the receive pair or transmit pair of signal wires. [6]

Thus the TCU/MAU's bandwidth management mechanism also effects a certain amount of fault detection and recovery, in the sense that a malfunctioning station can automatically be removed from the token-ring LAN. We discuss this more when we consider the 802.5 medium access control protocol.

4.5 Summary

In this chapter we completed building our basic transporter by defining its interface to client and destination. We started by defining from first principles what such a generic interface must include to meet the needs of client and transporter alike. The client, or data terminal equipment, interacts with the transporter via the interface presented by the DCE, the data circuit-terminating equipment, via the explicit management visible at the DCE interface. Whatever the details, the DCE–channel–DCE entity is a composite that hides implementation details from the DTE. Isolating implementation details is perhaps the single most important principle of modular system design.

This chapter explored how the V.24 standard defines the functional semantics of interface commands and signals used in several other interface standards, notably EIA-232 and V.35. In addition, we discussed the actual channels between the DTE and DCE, including the waveform sets used on these. In addition, we discussed the V.24 circuits for bandwidth management, including selection of signal rates and frequencies. V.24 also includes numerous circuits that carry information on the transporter's state as well as workload management of a HDX channel.

We concluded this chapter by examining the Physical layer portion of the IEEE 802.3 and 802.5 LAN standards. Among other things, we saw that these are notable for their relative simplicity compared to serial interfaces. Using self-clocking Manchester and Differential Manchester waveforms over balanced high-speed circuits, respectively, the 802.3 interface uses only four circuits with just two for management messages; whereas the 802.5 interface uses just the bare minimum of two circuits, with the only management signaling accomplished by means of a so-called phantom circuit.

References

[1]   McNamara, J., *Technical Aspects of Data Communications,* 3rd ed., Bedford, MA: Digital Press , 1988.

[2]   Black, U., *The V Series Recommendations,* New York: McGraw Hill, 1991.

[3]   IEEE Standard 802.3, New York: Institute of Electrical and Electronics Engineers, 1988, p. 79.

[4]   *IEEE Standard 802.5,* New York: Institute of Electrical and Electronics Engineers, 1988.

Part II
Management in the Data Link Layer

5

Data Link Management: Basics

5.1 Introduction

Having considered the physical layer and its management in the context of the MESA model, we now move up the encapsulation hierarchy to look at management in the data link layer. There is a sharp division of opinions as to how much management should be incorporated into data link protocols (and indeed the higher protocols as well). At its most extreme, the choice is between simplicity/efficiency, on the one hand, and complexity/robustness, on the other. As we will see, with data link protocols the early consensus which favored connection-oriented mechanisms that attempted to create a reliable transporter for upper layer clients has given way substantially to an increasing preference for connectionless mechanisms that offer only "best effort" delivery.

Above and beyond the noise faults, which we considered at some length in Part I, transporters are subject to two additional events that, depending on one's perspective, may or may not be faults: overflow/saturation and "nobody home." Noise faults are amenable to forward error control (FEC), although at the data link layer the primary technique used for recovery is retransmission or backward error control (BEC). Overflow/saturation faults occur when the receiving link station is unable to accept data as fast as the transporter can deliver it from the client, that is, the sending link station. This can be prevented by employing flow control, where the receiver paces the sender. Finally, there are "nobody home" faults, which (as the name implies) occur when the destination is not on-line or is otherwise unable/unwilling to receive data. These can be prevented by requiring a connection be established before the client sends any data: if the connection does not want to recieve data then it will decline or simply ignore a connection request.

This chapter is devoted to discussion of generic data link management issues, including four that define a core management taxonomy: connections versus connectionless operation, framing, flow control, and fault management. We also look at the relationship between transporter topology and data link protocol and the relationship between the data link protocol and its upper layer client(s). Finally, we discuss the overarching protocol management question, namely, the locus of control: Is the protocol based on peer managers or a master/slave dichotomy? The aim is to objectively examine the pros and cons of these options by using the MESA model to shed some light on the management mechanisms required.

5.2   Data Links and Data Link Protocols

5.2.1   What Is a Data Link Protocol?

In Chapters 2–4 we constructed a virtual transporter consisting, first and foremost, of a channel and associated waveform signals, along with the modulator/demodulator, and/or encoder/decoder that, with internal management, comprises the DCEs. Given such a transporter, what more is necessary? In a word, a protocol governing the behavior of the clients (and destinations) with respect to the transporter and with respect to each other. That is to say, workload management. Recall that we said in Chapter 4 that control of a DTE's transmission can be effected by a workload manager within an upper layer protocol (to schedule the transmission of waveforms over the channel).

So what is a protocol? One reference work defines it as follows: "A formal set of conventions governing the format and relative timing of message exchange between two communications terminals" [1]. Another definition is that a protocol is "a formally specified set of conventions governing the format and control of inputs and outputs between two communicating systems" [2].

A protocol is realized by a set of finite state machines, also called protocol machines, located at the DTEs. Part of a protocol specification is the set of input messages and their respective formats; another part of the protocol specification is the set of output messages and their respective formats; and the final part of the protocol specification is the set of the state transitions of a protocol machine that the input messages trigger and which result in the corresponding output messages.

There is a hierarchy of protocol machines within a DTE corresponding to the protocol stack, that is, the layered communication model, implemented at that node. The (input and output) messages exchanged between protocol machines of the same layer are, in the terminology of the OSI model, called *Service Data Units*. Service Data Units are further encapsulated in two types of messages exchanged between protocol machines. In the terminology of the OSI model, the messages carried between the protocol machines at adjacent layers in the same DTE are called Interface Data Units, which consist of Interface Control Information plus Service Data Units. This may seem somewhat confusing but that is what the architects of the OSI model arrived at.

The messages carried between protocol machines at the same layer in the different DTEs are called Protocol Data Units (PDUs). A PDU consists of three parts: header of protocol information, a section we will call the *payload* for carrying the data to be transported (in this case, the Service Data Units), and a trailer of protocol information. Layer 2 PDUs are more commonly known as frames. Figure 5.1 shows the components of the generic PDU.

Note that with some protocols the trailer is omitted, and as we will see when we examine specific protocols the payload is omitted in special messages (PDUs) that are exchanged between state machines. What is delivered to each layer of protocol machine is the corresponding Service Data Unit. Figure 5.2 shows the relationship of Interface Data Units and Protocol Data Units.

A data link protocol is realized by the protocol machines at layer 2. These protocol machines are also known as link stations; hence, in SDLC we speak of the protocol machines as Primary and Secondary link stations and in HDLC we refer to them as Combined link stations. While all link stations contain client and receiver components, what differentiates them (and their respective protocols) are the management servers that implement the protocol mechanisms.

The language of the Open System Interconnect Reference Model in the X.200 standard is clear on the purpose of data link protocols:

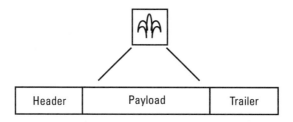

Figure 5.1 Protocol data units: header, payload, and trailer.

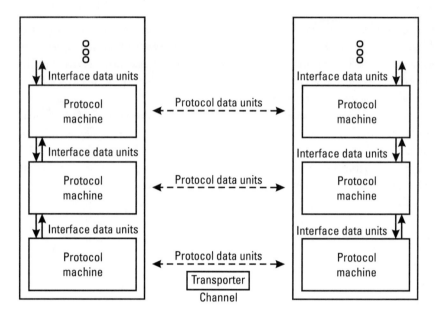

Figure 5.2 Service data units and protocol data units.

The Data Link Layer provides functional and procedural means to establish, maintain and release data-link-connections among network-entities and to transfer data-link-service-data-units. A data-link-connection is built upon one or several physical-connections [3].

The orientation toward connections is unambiguous. Zimmerman, who was a principal author of the X.200 standard, wrote in 1980 that the "purpose of the data link layer is to provide the functional and procedural means to establish, maintain, and release data links between network entities" [4]. Since that time, however, as we discuss later there has been a major change in thinking toward connectionless services.

With the addition of the data link protocol, we are creating yet another virtual transporter on top of the virtual transporter hierarchy we have just constructed within the physical layer. The process is identical to our previous "virtualizations," namely, the embedding of management servers that effectively alter the implementation characteristics of the underlying server, in this case the (already virtual) transporter presented by the physical layer.

There is, though, one significant difference. Up to this moment, we have made a clear demarcation between client and transporter, on the one hand, and between destination and transporter, on the other. The three were assumed to be distinct physical entities, the transporter of course a composite of the DCEs and the channel. Now, however, at each step up the protocol stack, the transporter is composed of the lower level protocol machines within the respective DTEs plus the DCE–channel–DCE composite. In the case of the data link layer, the transporter is composed of the physical layer of each DTE plus the DCE–channel–DCE composite (Figure 5.3).

We can help understand and classify data link protocols with the following "taxonomic" criteria:

- Does the data link protocol require a connection between client and destination to be set up prior to the transportation of data?
- Does the data link protocol use variable or fixed length PDUs?
- Does the data link protocol have any mechanism for throttling the transmission of clients?
- Does the data link protocol have any mechanism such as retransmission for recovery from transient faults?

In addition to these criteria, to characterize data link protocols we will also ask these questions:

- Can the data link protocol support channels with three or more DTEs (so-called multidrop or multipoint topologies) or is the protocol limited to point-to-point topologies?
- Can the data link protocol support multiple upper layer protocols?

We now explore these topics in more detail.

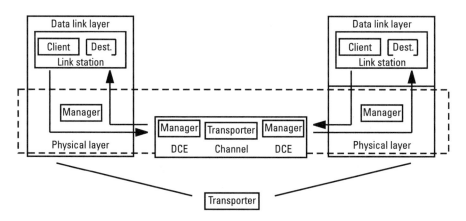

Figure 5.3 Transporter as composite of DTEs and transporter.

5.2.2 Workload Management: Connection Versus Connectionless Operations

As we discussed in Chapter 1, the computer networking community has been divided into two camps over the issue of connections since its earliest days. At the network layer this has taken the form of a long-running debate over the merits of "virtual circuits" versus "datagrams." But even at the data link layer the partisans of connectionless and the partisans of connection-oriented communications have continued their dispute. The former seek simple, efficient protocols unencumbered by much overhead; the latter argue that some added protocol complexity is worth the price if it purchases additional robustness.

Unlikely as it may seem, ultimately this is a philosophical disagreement over what constitutes the transport (locational) actuation of data. Recall that in Chapter 1 we illustrated the issue with the paradox of the tree falling in a forest when no one is present to hear it: Does it make a sound? To the connectionless advocates, data that has been transported has been sent, irrespective of whether any recipient was present to receive it; in other words, the tree makes a sound. On the other hand, to the advocates of connection-oriented communications data is not sent unless the destination is ready and able to receive it, and the best way to ensure this is to secure "permission" from the destination in the form of establishing a connection; to advocates of connection-oriented transport, the tree makes no sound if no one is present to hear it.

What are the differences between connection-oriented and connectionless data links? In connection-oriented data links, prior to sending user data there must be an exchange of management data; this management data initiates/actuates the connection. The management data is exchanged by means of messages with special formats, different from the message formats used to carry the client's data to the destination. If either side of the connection fails to execute its tasks, then the connection establishment fails (Figure 5.4). This allows, first of all, for a destination, be it down, busy, or otherwise unavailable, to decline a connection; the source will not attempt to send data because it now knows from the destination's lack of response that this would be pointless.

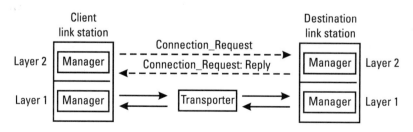

Figure 5.4 Link station connection request and reply.

In this respect, a connection establishment is an exchange, either explicit or implicit, of state information by the two link stations. And because the destination is required to give its permission if the connection establishment is to occur, failure to do so will convey to the source state information about the destination computer much in the way of Sherlock Holmes's dog that didn't bark. Connection establishment, therefore, is "fail-safe" in that a connection will not be brought up with a link station that is incapable of receiving data.

The exchange between source and destination can even be extended to allow a negotiation of connection parameters (Figure 5.5). The station that is seeking to establish a connection may propose certain values for negotiable parameters, such as optional protocol features to be supported by the connection, that is, by the protocol machines. The recipient of the connection request may accept the proposed parameter values, may decline the connection entirely, or may propose its own set of values. Other uses for this exchange of information include security passwords and telephone numbers for dial circuits.

In part, the advocates of connectionless service base their arguments on basic principles of modular design, namely, the decomposition of tasks into their components for the purposes of reuse, seeking what we might call, by analogy, the "greatest common denominator" (see, for example, a text on software decomposition, such as [5]). That is to say, modular design emphasizes the decomposition into subsets based on common functionality. And from the point of view of modular design theory, a connectionless transporter *is* the basic module. After all, a connection-oriented transporter is built on top of a connectionless transporter by the addition of protocol logic in the form of managers; and, clearly, the management exchanges that constitute connection establishment requests, responses, and negotiation (if any) are transported without benefit of connections.

Another argument for a connectionless transporter as the basic module is that not every client (DTE) needs the assurance provided by the connection that the destination is ready and willing to accept data. For example, telemetry

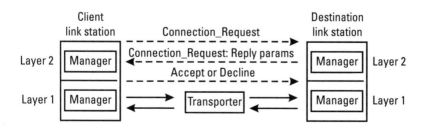

Figure 5.5 Link station negotiation.

data may be sent irrespective of the destination's state; all things considered, just transporting the data is the simplest course of action and the benefits of the implicit feedback that a connection conveys are not worth the overhead of the connection management process (establishment and disestablishment). This is particularly true if the data are sent frequently but not so often that one wishes to dedicate the resources implied by keeping a connection open. A client that is content with the greater uncertainties of connectionless service should not be forced to use a connection-oriented transporter.

All this is not to say that there are no advantages to connection-oriented communication. Foremost of these, perhaps, is that it facilitates the implementation of flow control, fault detection and recovery, and other management tasks. A connection provides a context, a shared frame of reference, by which the management in the source and/or the destination can communicate state information concerning, for example, corrupted data or full buffers. In addition, advocates of connection-oriented protocols argue that, when the volume of data to be transported is sufficiently large and/or the duration of the exchanges is sufficiently great, the overhead due to connection management is small compared to the benefits; with short-lived exchanges, this is not as clear.

In the remaining sections we examine framing, flow control, fault detection and recovery, and other management tasks.

5.2.3 Workload Management: Time Slicing the Transporter

As we have remarked more than once, in the development of data communications many ideas were borrowed from the world of operating systems. Indeed, just as the idea of layering in communications architectures came from the layered approach to the design and implementation of operating systems, so the inspiration behind much of the original research in packet switching was the idea of preemptive multitasking via time slicing the execution of a processor. The aim in both cases was identical: to treat a shared actual server as a serially reusable resource, affording multiple clients the appearance of dedicated virtual servers while at the same time ensuring no client could monopolize the actual server's bandwidth by dividing the client's RFS into multiple pieces, the execution of which was interleaved with those of other clients.

At the data link layer this division (actuation) is called *framing*. We explained in Chapter 1 that time slicing was one form of workload actuation, namely, workload actuation of degree$_2$. An RFS to transport a unit of data (the plant) is replicated into k RFSs and the plant is divided into k plants. When data arrive at the data link layer from upper layer clients (users or applications) it is up to the data link protocol to ensure that the data are transported. Irrespective of whether this involves connections or not, the data link protocol encapsulates the data in frames, another term for data link PDUs. Just as a symbol (and corresponding waveform) is the basic module of a plant for a physical layer transporter, a frame is the basic module of a plant for a data link transporter.

There are at least two issues in framing: the size of the total frame and the size of the frame's header and/or trailer. Technically speaking, the header and/or trailer are overhead, much like the start and stop bits used in asynchronous line encoding. While these (header and/or trailer) must contain fields to carry the information (state) necessary for the sending and receiving protocol state machines to execute, at the same time each extra bit included in these fields constitutes overhead that is consuming transporter bandwidth that could otherwise be transporting client data, so the incentive to keeping down the sizes of header and trailer is obvious. The challenge is finding the optimum balance.

There are two aspects to the size of the total frame: first, the maximum allowable length of a frame; and second, whether the frame length should be variable or not. Allowing variable length frames, up to the maximum dictated by time slicing, accommodates different users and types of data; however, variable length frames require special delimiter flags and/or frame size fields to identify the beginning and end of the frames, all of which adds to the overhead and the total frame length. Most of the data link protocols in use today, both WAN protocols like SDLC, HDLC, and PPP and LAN protocols like Ethernet and Token Ring, allow variable length frames.

With fixed length frames, generally referred to as *cells,* such length and/ or flag fields are unnecessary and hence their associated overhead can be eliminated. In addition, using fixed length frames is advantageous because the latency associated with transporting and processing these is relatively constant; in contrast, with variable length frames, the amount of time required to transport and process the frames at each station is highly variable as it depends on the length of the frame. This variability in latency is called *jitter,* and is particularly problematic for multimedia or constant-bit-rate traffic such as voice and video. Because cells are typically much shorter than frames (53 bytes, compared to 1500 bytes, for example) their latency is low in addition to being constant; and small cell size gives fine control over the granularity of delay since a cell carrying video will not be delayed by a frame carrying a file transfer. The most important cell-based protocol is Asynchronous Transfer Mode/Broadband ISDN, which uses 53-byte cells with payloads of 48 bytes.

Advocates of cell-based protocols argue that another advantage is that with these the frame (cell) handling mechanisms can be much simpler. Taken in conjunction with simplified error handling and flow control mechanisms, all of this facilitates implementation in silicon, allowing very high forwarding rates. For this reason, there is much interest in high-speed data links and networks using cell-based protocols such as ATM. However, recent interest in such alternatives as PPP over SONET/SDH (RFC 1619) indicates that the argument is far from settled. For our purposes, we will regard frame-based and cell-based protocols as merely different implementations or instances of the same thing, namely, workload managers time slicing a transporter.

Finally we note that if the client is an upper layer protocol that is packet oriented, then the data link transporter will receive its plant already in discrete units. There may, however, be a mismatch in the sizes used at the different layers; what is optimal at the data link layer may not be optimal at the network layer, for example. After extracting an upper layer SDU from its corresponding IDU "envelope," the workload manager in the data link layer has three choices:

1. It may encapsulate one SDU in one PDU.

2. It may divide one SDU between two or more PDUs.

3. It may encapsulate two or more SDUs in one PDU.

The second form of encapsulation is called *splitting,* and the last is called *blocking* (Figure 5.6).

5.2.4 Workload Management: Flow Control

The aim of employing flow control at the data link layer is to prevent a destination link station from being overwhelmed by a sending link station that can generate data faster than the destination can accept it; as we will see later in this book, flow control is closely related to congestion control and what is called traffic management. Note that if the recipient lacks capacity to accept a PDU and hence must discard it, this is a fault no less than if the PDU never arrived or arrived corrupted. (Admittedly, such an overflow fault is not in the transporter itself but in the "extended" transporter consisting of the channel/signal pair and the end systems.) Thus flow control can be considered an instance of fault management, specifically fault recovery.

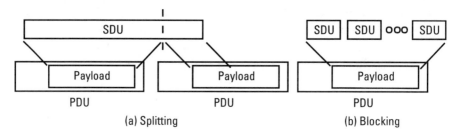

Figure 5.6 (a) Splitting and (b) blocking.

Just as with the question of whether a data link protocol should use connections or not, there are several schools of thought about the question of flow control at the data link layer. Once again it is a debate between protocol simplicity on the one hand versus protocol robustness on the other. One school, not surprisingly the same one that argues for connectionless data link protocols, advocates that in the interests of simple, efficient protocols, no flow control should be employed. If the destination cannot process all of the frames it receives, it should discard the excess; if the frames were important, upper layer protocols will retransmit and may even have workload management mechanisms to control the rate of transmission and avoid more congestion. And, as with connections, the other school argues that flow control brings benefits that outweigh the additional complexity and overhead. By implementing flow control mechanisms, clients can be assured more reliable service and predictable response times and also avoid the inefficient retransmission of frames discarded due to lack of resources at the destination.

Three entities are involved in transport (and flow control) at the data link layer: the sending link station, the transporter (consisting of the physical layer DTEs, the DCEs, and the channel), and the receiving link station. The difficulty in preventing the sending link station from overwhelming the receiving is that the two are spatially disjoint and, without some explicit feedback, the former will be ignorant of those aspects of the latter's state (buffers, processor usage, and so on) that might affect its ability to receive more data. In fact, the feedback need not be so granular: a simple semaphore (yes/no) can suffice to signal the sending link station to stop sending (Figure 5.7).

This is not to say that flow control *must* be feedback based; flow control without feedback is said to be "open loop." A workload manager using open-loop

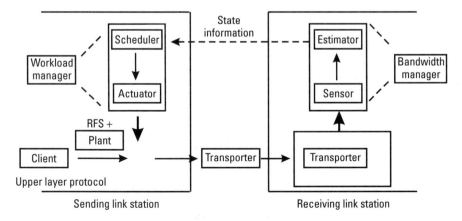

Figure 5.7 Distributed workload management: feedback from destination.

flow control relies on a priori knowledge (measurements and/or estimates) of the nominal bandwidth of the transporter and/or the bandwidth of the receiver, and with this information paces the sender to stay below that rate. As long as the actual bandwidth of the transporter and/or receiver stays at or above the nominal rates assumed a priori, open-loop flow control can work well. Another term for open-loop flow control is timer-based flow control: The workload manager schedules the client's data according to the clock rather than any state information about the destination.

However, as we discussed in Chapter 4, open-loop flow control is vulnerable to uncertainties and variations in the transporter's bandwidth due to noise, faults, and other exogenous factors, as well as uncertainties and variations over the rate at which the destination can receive the transported data. In contrast to open-loop flow control, which is timer based, closed-loop flow control is state based. The workload manager at the client receives periodic feedback about the state of the receiver. This way, as the bandwidth of transporter and/or destination ebb and flow, the workload manager at the client alters the amount of traffic it allows to be sent. Of course, the state feedback that the receiver sends to the client's workload manager is management overhead, consuming transport bandwidth that could otherwise be used to move client data; the trade-off, though, is that the destination will not discard frames that arrive faster than it can accept them, itself an inefficiency if it causes retransmission.

The simplest closed-loop flow control involves the client sending one block of data (frame) to the destination and then waiting until the latter sends permission for more to be sent. This is called "stop-and-wait" flow control. Clearly, this will keep the sending link station from overwhelming the receiving link station; the receiving link station will not return permission to send more until it is ready to receive the frame. After a client has sent a frame of data to the destination, it must then wait (or block) until the destination receives the frame, processes it, and returns via the transporter permission to send more.

Such "stop-and-wait" flow control can be inefficient in its use of the transporter's bandwidth: If the destination is busy or otherwise delayed in sending back permission then the client must wait and, if the transporter's bandwidth is dedicated or otherwise reserved, this is wasted. In addition, a channel may be able to accommodate multiple frames at one time if the channel is long relative to the propagation speed of the signal. For example, consider a satellite channel that is more than 43,000 miles long, with a latency of 0.231 seconds.

If we are transmitting at a rate of 256 Kbps then each bit requires 3.9×10^{-6} seconds and, at the speed of light $c = 186,000$ miles per second, each bit "waveform" occupies 0.726 miles of space. If we assume an average frame length of 125 bytes then that means the 43,000-mile-long channel can accommodate 59 frames at one time. Put another way, we can consider the channel as being composed of 59 "virtual stages" through which frames transit as they pass from transmitter to satellite to destination. Assuming the destination station immediately responded with permission to continue sending, at best a stop-and-wait protocol would transmit one frame every 0.462 seconds since the frame must propagate the 43,000 miles and the permission to transmit must also transit the 43,000 miles, for a total round-trip delay of 0.462 seconds. This represents a utilization of 0.00845 of the channel bandwidth, which is clearly unacceptable.

This example also points out another flaw in "stop-and-wait" flow control. The receiving link station must now be given sufficient priority at the destination system to minimize the waste of bandwidth. This means that other processing may be disrupted in order to service the data link, an unacceptable situation in many or even most instances. The problem is that most if not all receiving systems are executing many tasks in addition to processing incoming frames of data; and as the delay increases until a message can be sent carrying permission for the client to send more data, the transporter efficiency plummets.

The key to solving this problem is that most link stations have memory to accommodate more than one frame; stop-and-wait, it turns out, is unnecessarily protective of memory at the expense of transporter and processor bandwidth. With this in mind it is possible to increase the efficiency of the transporter and reduce the disruption to the receiving system if we modify the flow control to allow multiple frames to be sent before the workload manager must receive permission to send more. This accomplishes two things: (1) It fills the channel and thus increases the utilization of the transporter's bandwidth and (2) it allows the destination link station to receive frames and just store them in buffers without having to drop everything when a frame arrives to send permission for another to be sent. Another advantage is that it reduces the overhead of permission messages by reducing their frequency.

This is an example of pipelining, which exploits concurrency in the transporter and/or destination by allowing multiple outstanding frames. This technique has long been used in the design of high-performance CPUs, where the pipelined architectures allow multiple instructions to be in execution at any given time. One of the key design parameters for a pipeline is its depth: What is the optimal number of stages? The equivalent question for data link flow control is this: What number of outstanding frames is optimal?

The answer is complicated by several factors. First, some link stations (and their systems) will have more memory than others; if the allowable number of outstanding frames is made too large then smaller systems will be overwhelmed and run out of buffers, defeating the flow control. As we will see, the solution to this employed with protocols such as SDLC and HDLC has been to allow two different values for the number of outstanding frames, 7 and 127, to meet the needs of different types of systems and different types of channels. A second complication is that flow control is frequently coupled with fault detection and recovery, particularly in the means of identifying frames for the purpose of fault notification and retransmission. What is optimal for flow control may result in poor fault recovery performance.

We will examine the relationship between flow control and fault management more closely in the next section.

5.2.5 Workload Management: Reliable Versus Best Effort Transportation

5.2.5.1 Simple Versus Robust Data Link Protocols

When we consider fault management at the data link layer, an obvious question is why is it needed if we employ error control coding within the physical layer transporter? The answer is that Shannon's coding theorem only proves that we can build an encoder able to make a channel arbitrarily reliable, not that ideal transporters are realizable. Implementation trade-offs and the law of diminishing returns inevitably result in transporters with finite reliability. Given this fact, the question becomes: What maintenance tasks should a data link protocol execute to increase the reliability and/or availability of the transporter?

As with connections and flow control, the consensus on this has been changing, particularly during the last 10 years. Prior to this, the conventional wisdom about the data link layer was:

> The function of this layer is to convert an unreliable transmission channel into a reliable one for use by the layer above it (i.e., the network layer). The raw data bit stream is organized into frames each containing a checksum for detecting errors. [2]

This is generally effected by retransmission of corrupted frames, hence the central role of workload management. The advantage to reliable data link protocols is that they enable a simplified upper layer protocol to assume that any data passed to it are uncorrupted. Of course, this reliability comes at a price, notably greater protocol complexity. In addition, there is the transport and storage overhead that comes with retransmission or other replication of the data.

But the advocates of protocol simplicity argue that all this has become unnecessary, that fault management is *not* needed at the data link layer, notwithstanding the inevitable faults in the underlying transporter, as transmission facilities have increased in quality (see Table 5.1).

To the advocates of simplicity, the basic service provided by a transporter should be unencumbered with the management complexity and overhead

Table 5.1
Bit Error Rates for Communications Channels

Type of Channel	Bandwidth	BER
Telephone circuit, ca. 1971	300–2400 bps	10^{-4} to 10^{-6}
802.3 Ethernet	10 Mbps	10^{-8}
802.5 Token ring	16 Mbps	10^{-9}
FDDI	100 Mbps	2.5×10^{-10}

needed to respond to faults. Not all clients will care if the occasional fault occurs in the transportation of data, and these should be allowed to avail themselves of a simple transporter.[1] Finally, between these two extremes, a compromise has the data link protocol provide fault detection but not any recovery.

5.2.5.2 Fault Detection

Fault detection, whether as an end in itself or as a means to fault recovery, involves monitoring the transporter. However, as was discussed in earlier chapters, direct instrumentation of distributed parameter servers such as transporters is not really possible. The solution is to use some form of indirect inference mechanisms, that is, estimators. The mechanisms of fault detection used in data link protocols most often used are remarkably similar to those we saw at the physical layer, namely, some form of redundancy data encoding which, when it arrives at the destination, can be used to estimate whether a fault occurred in transit that resulted in corruption of the frame. These can range from simple checksums to outright replication of the entire frame; in this last case, the fault detection works by simply comparing, symbol for symbol, the received frames and detecting inconsistencies.

By far the most important mechanism, however, is the cyclic redundancy check (CRC). This involves the sending link station generating (via an estimator; see Chapter 3) a CRC using the same generator polynomial methods we discussed in Chapter 3. Note that a CRC is also called a frame check sequence

---

1. For example, clients requesting the transportation of multimedia data (audio or video) do not care if a small fraction of samples transported is corrupted by faults in the transporter; the reconstruction processes at the destination are not so fragile that infrequent errors invalidate the audio or video output. Beyond this, such clients might find little use in retransmission of corrupted frames because they would arrive out of sequence and could not be incorporated in any real-time reconstruction of the original data.

(FCS). For example, SDLC uses the polynomial $X^{16} + X^{12} + X^5 + 1$ (the so-called V.41 polynomial [6]) to produce a 16-bit remainder. ATM/BISDN and PPP, on the other hand, use 32-bit polynomials for greater detection capabilities. Irrespective of the polynomial used, the CRC produced is then appended to the frame's trailer and the frame is sent to the physical layer transporter.

Once the frame is delivered to the receiving link station from the physical layer transporter, it is inspected, first, to determine if the frame conforms to the protocol's requirements (length, fields, special delimiting characters, and so on); this will detect any egregious faults that might have occurred in transit or at the sending link station. After this, the receiving link station creates a new FCS/CRC based on the frame as received. By comparing the FCS/CRC that arrived with the frame (from the source link station) with the new FCS/CRC thus created at the destination, yet another estimator in the destination link station can decide if a fault has changed the header and/or payload of the frame. This is summarized in Figure 5.8.

This type of fault detection has parallels outside the world of data communications. In the early days of cereal manufacturing, consumers complained that they had been swindled when the contents did not fill the box to the brim. Manufacturers (the sender) responded by printing the weight of the contents on the box, along with the caution that "contents may have settled during shipment." A consumer (the receiver) thus has a simple fault detection estimator: Weigh the contents and compare. If the weight at the destination equals the weight on the box then there is no fault; otherwise, a fault has occurred either in manufacturing at the sender or in transporting.

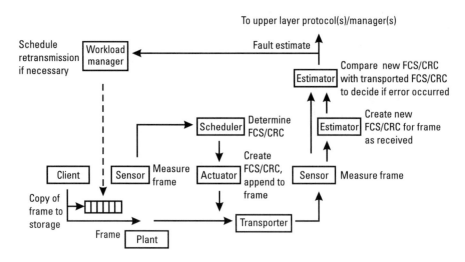

Figure 5.8 Task analysis of fault detection via cyclic redundancy code.

The FCS/CRC estimation is block coding. Recall from Chapter 3 that a key parameter of a block code is the block size, that is, the number of symbols on which the code word is based. When a data link protocol includes fault detection by means of a FCS/CRC, the equivalent parameter is what portion of the protocol's frame is used to calculate the FCS/CRC. Some data link protocols, notably ATM/BISDN, provide coverage only for the header, on the principle that if the header's management data are ensured to be intact then any corruption of the payload can be handled by upper layer protocols if they so desire, whereas if the header suffers an undetected fault this can lead to a cascade of forwarding faults that will impair the performance of the (composite) transporter [7]. Other protocols such as SDLC base their FCS/CRC calculation on the entire frame (minus special delimiters called flags; see later discussion).

5.2.5.3 Fault Recovery: Forward Versus Backward Error Control

Once a fault has been detected, there remains the question of what to do with the information. As we can see in Figure 5.8, a fault detection estimate is produced at the destination on the basis of redundant information sent with the transported data—such as the FCS/CRC. There are three possible destinations for such estimates:

1. They can be sent to an upper layer protocol machine and/or manager.

2. They can be sent to the sending link station to initiate fault recovery via retransmission.

3. They can be used locally to initiate fault recovery if there is sufficient redundant information.

In this section we will concentrate on the last two of these possibilities, which are generically called backwards error control (BEC) and forward error control (FEC), respectively.

We explored forward error control at the physical layer in Chapter 3. There FEC is realized via error control coding, which is basically a sophisticated mechanism for introducing redundancy into the data so that they could withstand faults better. At the data link layer the FEC redundancy is more straightforward: Make n copies of the data to be transported, from which the original data are reconstructed at the destination, most often by simple comparison logic. An advantage of FEC and its concurrent replication is that the impact of any faults can be hidden from destination and sending link stations, assuming that the estimator at the destination is able to decide with sufficient confidence the original frame from the frame(s) that arrived. But because of its overhead (see later discussion) FEC is not often used at the data link layer.

Backwards error control, on the other hand, involves retransmission of one or more copies of the data sent only in the event of a fault. It is for this reason that we say the workload manager in the sending link station makes its decision to introduce redundancy closed loop in the case of BEC and open loop in the case of FEC. That is to say, whereas the workload manager in FEC automatically makes one or more copies of the frame and sends them to the destination link station, in BEC the workload manager is more sophisticated: Given a frame to be sent, it makes a copy and stores it locally; only if it gets feedback from the destination link station indicating that the frame has been corrupted by a transporter fault does it retrieve the copy from storage and send it over the channel.

Obviously, if the fault recovery occurs only in response to feedback from the destination, the nature of the feedback is critical to the operation of the protocol. Here there is an important distinction to be made between positive and negative acknowledgments. Some of the first data link protocols to incorporate feedback for fault detection used feedback only to indicate a fault. If a frame arrived that had been corrupted then a message, called a negative acknowledgment, indicates this was sent to the sender. On the surface this would seem perfectly adequate, but there is a difficulty with using only negative acknowledgments. What if the negative acknowledgment is lost or otherwise distorted beyond recognition? The sender will never learn of the original fault and will proceed on the assumption that all is well.

It is for this reason that the basic BEC mechanism relies instead on positive acknowledgments, which indicate that a frame has safely arrived. If the sending station does not receive such an acknowledgment within a specified interval of time after it sends a frame then it estimates a fault has occurred and initiates recovery procedures consisting, at least initially, of retransmission. This is distributed fault estimation in that there is a redundancy-based estimator in the destination link station that covers (detects) one class of faults while a timer-based estimator in the sending link station covers another class of faults (Figure 5.9).

This is called positive acknowledgment with retransmission (PAR), also known as automatic repeat request (ARQ), and forms the basis for most reliable data link protocols. Two things are worth noting about PAR protocols. First, they introduce a potential instability into the client-transporter dynamics: if an acknowledgment is delayed because the receiving link station (end system) is busy then the sender may wrongly estimate a fault and initiate retransmission. The consequence of this will be to increase the processing load on an already overburdened destination, exacerbating the delays and, in effect, pouring fuel on the fire. Second, PAR is somewhat problematic in a connectionless environment since, in the absence of retransmission limits, a sending link station may

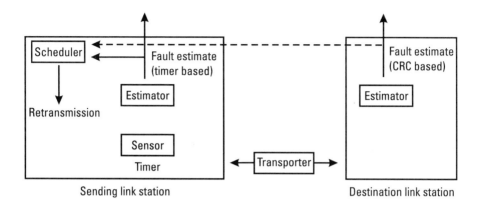

Figure 5.9 Distributed fault estimation.

continue to resend frames to a destination link station that is not even powered up and able to receive frames, on the incorrect inference that an absence of acknowledgments indicates only channel faults.

The relationship between fault recovery and flow control stems from their mutual reliance on feedback from destination to sending link station. If a protocol employs flow control then it already has a mechanism for sending feedback from the destination to the sending link station. The acknowledgment of a frame is also permission to send more data. The obvious next step is to extend this feedback from just simple flow control pacing (permission to send) to include requests for retransmission. The extension is natural: a pacing message requests sending unsent frames while a retransmit message requests sending frames already sent but which arrived corrupted by faults.

As with flow control, the challenge comes once we allow multiple frames outstanding. We must devise a way to identify each of these frames so that the destination can signal the source about which of many frames received have been corrupted by faults. How does the receiving link station indicate in its feedback which frame has been corrupted by a fault and thus requires retransmission? In the case of stop-and-wait flow control, the difficulty does not arise because only one frame is sent at a time. The challenge is to reduce the ambiguity in such feedback by means of some sort of identifier that "tags" each frame so that when the receiving station detects a fault, it can reference the frame by its identifier in subsequent feedback to the sending station.

One means of doing this is with sequence numbers. Although sequence numbers are generated in various ways, the most common is cyclic or modulo numbering. The sequence numbers range from 0 to $2^n - 1$, after which they wrap back to 0. This allows a sequence number to fit in an n-bit field in the frame's header. This is commonly referred to as a *sliding window* (Figure 5.10). Note that there are separate send and receive indicators (Ns and Nr) since the data link protocol is managing bidirectional transporters.

When a receiving link station determines that a frame has been corrupted by a fault during transportation, it is said to *reject* the frame by its feedback, either explicitly with negative acknowledgment or implicitly by sending an indicator message to the sending link station that this frame needs to be retransmitted. Those data link protocols that employ BEC may differ in their granularity of rejection. If a frame is corrupted, does the protocol offer a feedback mechanism such that the individual frame is identified or must a group of frames be rejected? The latter is called *go-back-n,* and the former is called *selective rejection.* Obviously, selective rejection results in fewer frames requiring retransmission. However, the cost is greater buffering at the destination since, depending on the extent of SDU fragmentation, packet reassembly may require storing frames until the missing frame is received.

The overhead that comes with fault management is of two types: transport overhead, that is, the additional transporter bandwidth required by the fault management mechanisms and their communications; and storage overhead, to hold the additional plant and/or state information required. The amount of overhead increases as we move from best effort to fault detection to fault recovery via BEC and FEC (Figure 5.11).

Transport overhead is the most direct. Fault detection by means of a FCS/CRC field introduces some overhead to the transportation of each frame corresponding to the size of the FCS/CRC field. BEC further increases the amount of transport overhead, since the FCS/CRC is still present and now we have the additional overhead of the acknowledgments and of the retransmitted

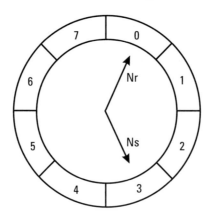

Figure 5.10 The modulo 8 sliding window.

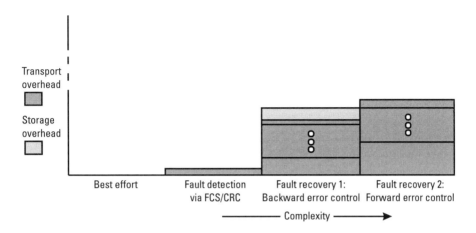

Figure 5.11 Complexity versus overhead of reliability actions.

frames. The exact amount depends on the number of retransmissions, which will depend on the quality of the underlying transporter as well as such factors as the size of the frames. On the other hand, the concurrent replication (sending *n* copies of each frame) of FEC necessarily leads to an *n*-fold increase in the bandwidth of the transporter.

Storage overhead is a consequence of retransmission mechanisms: the sending link station must have buffers (memory) to store the frames that have been sent until they have been acknowledged by the destination link station—or the transporter has been deemed to have suffered a persistent fault that retransmission by definition cannot resolve. Most reliable data link protocols that use retransmission have implementation-dependent limits for the number of times they will retransmit before estimating that a persistent fault has occurred; note that unsuccessful retransmissions are themselves "measurements" from which such estimates are drawn.

Finally, we should note the relationship between fault recovery and framing. We saw in Chapter 3 that error control coding relies on grouping symbols into larger units to gain much of its power. At the data link layer, this also is true. A data link frame constitutes the minimum unit of recovery because it is the unit of retransmission. As such, retransmission brings us back to the question of the maximum size for a frame: As frame size increases, so to does the amount of data that must be retransmitted. In addition, as the frame size grows, so does the probability of a fault that corrupts the frame. These trade-offs can be analyzed mathematically by protocol designers to find the optimal size frame/unit of recovery.

5.2.6 Topology and Addressing

We know by definition that every transporter connects at least one client and one destination; and that, in most instances, there is both a client and a destination at each location. If a transporter connects only two locations then its topology is said to be point to point. The original data networks were built out of bidirectional point-to-point data links. On the other hand, if a transporter connects more than two locations its topology is said to be multipoint (also known as multidropped). With the introduction of Bisync and later SDLC, multidropped data links became common and, since the introduction of LANs, they have become the rule in many campus environments.

Coupled to topology is the question of addresses. With point-to-point data links, there is no need to actually identify the source or destination of each frame: An address is implicit in each RFS (frame). If station A sends data, it can send them to only one location, namely, station B. Likewise, if B sends data then the only destination is A. After all, with only two stations on the data link, if a frame did not originate locally then there is little doubt as to where it must have come from. And because addresses represent additional management overhead, omitting them when possible only increases the bandwidth available for transporting client data. We call such addressing implicit. On the other hand, if the data link is multipoint the addressing must be explicit: when a third station is added, implicit addressing is no longer adequate. If A issues an RFS to the data link and it does not qualify the request with an address then the data link (i.e., the link stations) lacks sufficient information to remove the ambiguity in the request.

Addresses can also be characterized by their span of uniqueness. This refers to the extent to which any other location can have the same address. The larger the span of uniqueness the more stations needing unique addresses and the larger the addresses must be. Obviously, if explicit addresses are used then, with the exception of broadcast and multicast aliases (addresses), these must be unique within the span of the data link. We say such addresses are locally unique. In the early WAN data link protocols, addresses were either implicit or locally unique. There were two reasons for this: first, the bandwidth necessary to transport addresses (in frame headers) is management overhead and WAN bandwidth was both scarce and expensive; and second, globally unique addresses were unnecessary because the scope of a data link protocol is a single data link transporter. Global uniqueness was reserved for layer 3 addresses, where they were necessitated by the nature of multitransporter concatenations.

The advent of LANs, however, brought globally unique addresses which changed this picture in ways that had deep ramifications for the way we concatenate data links; notably, bridging and layer 2 frame switching would not be possible with locally unique addresses. Why did the early designers of LANs opt for globally unique addresses? In part, because they could. The dramatic increase in LAN bandwidth, by three, four, or even greater orders of magnitude over the bandwidths of typical WAN data links, meant that the overhead associated with globally unique addresses was much less important. This is reflected in the 48-bit addresses chosen by the IEEE 802 committee versus, for example, the 8-bit addresses used by SDLC and its daughter protocols.

Taking these factors together, we can see three possibilities:

1. Implicit addresses (example: PPP);
2. Explicit addresses—locally unique (example: SDLC); and
3. Explicit addresses—globally unique (example: 802 LANs).

Addressing raises other issues. First of all, how are addresses determined and what entity assigns them? Second, a link station must decide if it is the destination of the plant. A surprising consequence of explicit addressing on multipoint data links is the requirement it brings for estimation. The question of interest to the link station is this: Is this frame intended for me? Reading (= measuring) the address in the frame is insufficient to answer this. Instead, the question ("Is this frame for me?") can only be answered by comparing the frame address and the station's address and aliases, if any. This means that an estimator is required, since the information (variable) of interest (a binary variable since either the data/information is for the station or it is not—there is no intermediate possibility) can only be estimated.

5.2.7 Workload Management: Scheduling the Data Link

In the discussion of data link protocols up to this point, we have yet to broach the central issue of workload management. How and where is the execution of the workload of the data link scheduled? Because the physical layer transporter is a server with finite bandwidth and finite reliability, we know the requests of competing clients cannot all be executed as they arrive. As such, there must be some workload management for allocating the transporter to clients, and it is the data link protocol that specifies how these mechanisms execute their tasks. All data link protocols, from Bisync to SDLC and from Ethernet to ATM, allocate (i.e., schedule) the physical layer transporter. To do this they contain workload schedulers. The different approaches to such scheduling—centralized or distributed, open loop or closed loop—should not be allowed to obscure this.

The exact workload management required depends in part on the topology of the underlying transporter, its task set, and its tasking. For example, no workload management is required for either simplex (monodirectional) or full-duplex point-to-point data links because in each case there is no conflict for the services of the data link. In contrast, workload management is required for duplex (full or half) multipoint or half-duplex point-to-point data links precisely because two or more clients may wish to use the transporter at the same time; in the latter case this is called "turning around" a half-duplex link. The analogy is with an intersection of two roads. If there is no stop sign or other "rules of the road" allocating the intersection, then an accident is inevitable. Such is also true of shared transporters. The data link protocol enforces the rules of the road for sharing the data link, preventing the otherwise inevitable faults as clients interrupt each other.

Such workload managers can be characterized by two principal parameters: type of control (open versus closed loop) and, when the plant being managed is distributed, locus of control (centralized versus distributed). With centralized management, a single scheduler is responsible for allocating the transporter (data link) to competing clients. With distributed management, on the other hand, each link station contains its own scheduler and these collectively allocate the transporter. We have already seen that such distributed management implementations with both flow control and fault management, particularly where schedulers and/or estimators are located at both source and destination locations (link stations), are common in data link protocols. Extending this to overall scheduling is straightforward, albeit at the price of increased complexity. Indeed, this is an area where the trend has run toward more protocol complexity rather than less.

In addition, these protocol workload manager(s) may be either open loop or closed loop depending on what information about the state of the data link is required by the scheduler(s). Open-loop workload management means there is no measured information on the status (busy or idle) of the transporter, that is, if it is currently executing another station's transport task. Closed-loop workload management implies the opposite: The schedulers receive measurements and/or estimates of the transporter's status. For example, recall that in Chapter 4 we saw that instrumentation is built into Ethernet MAUs precisely to monitor the status of the data link.

Although this theoretically gives four possible types of workload managers, in practice there are really only two: centralized/open loop and distributed/closed loop. With centralized control the workload manager does not need to monitor the transporter's status because it knows, from its past and present scheduling decisions, whether the transporter is busy or idle. Closed-loop monitoring is unnecessary if we assume that all the clients heed the scheduling authority of the workload manager; and if a client transmits without permission this is a fault, meaning that any monitoring is effectively part of fault management.

On the other hand, closed-loop control is integral to most realizations of distributed workload management, notably the LAN protocols such as Ethernet and Token Ring. When two or more autonomous workload managers are scheduling a shared transporter, one of the simplest ways to coordinate their actions is for each to monitor the transporter to see if it is executing another client's task before attempting to allocate the transporter to its own client. The two most common types of distributed control are random-access protocols such as with Ethernet that involve monitoring the transporter directly, and sequential access protocols such as Token Ring or FDDI where the monitoring is of a common semaphore called a *token* that is set by a workload manager when it allocates the transporter to itself. Although such coordination conceivably could be realized via open-loop methods using timers, this type of distributed control is seldom encountered because of potential difficulties in synchronizing stations.

Table 5.2 summarizes the various combinations.

Centralized workload scheduling is also called master/slave because the single scheduler that decides the execution of the data link is the master of the data link, and clients in all the other stations are slaves, dependent on the master for permission to send. Another term for the master is the primary link station (PLS) while the slave(s) are called secondary link station(s). Master/slave workload scheduling functions quite well in terminal-oriented computer networks because of the asymmetry of processing power and storage that typified computing until relatively recently, where the mainframe and/or its communications controllers had much more memory and CPU cycles than was economical to put in the vastly more numerous terminals and/or terminal controllers. Bisync and SDLC are the main data link protocols that use centralized control (Figure 5.12).

The primary advantage to master/slave workload management is the relative simplicity of the data link protocols. The disadvantages, however, should

Table 5.2
Workload Managers: Locus Versus Feedback

	Open-loop	**Closed-loop**
Centralized	Serial (e.g., SDLC)	—
Distributed	—	LAN (e.g., Ethernet)

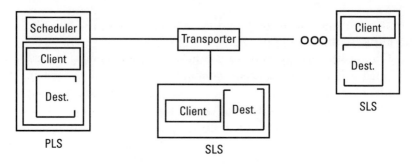

Figure 5.12 SDLC link with workload manager at PLS.

not be discounted. First of all, any centralized mechanism constitutes a single point of failure. If the primary link station fails, then the entire data link is disabled unless there is a fallback mechanism. In addition, master/slave workload management is not conducive to any-to-any data links. If a secondary link station cannot send data to another secondary link station unless given permission by the primary link station, then this introduces additional delay and traffic on the data link. However, the deficiency of master/slave protocols with respect to direct communication between secondary link stations is less important in terminal-oriented network protocols (such as SNA) because of the intrinsically host-oriented nature of the traffic flows.

Distributed workload scheduling is considerably more complicated than centralized scheduling, not least because of its concomitant requirement for monitoring the transporter (feedback) (Figure 5.13).

Distributed workload scheduling is at the heart of so-called peer data link protocols, most notably the LAN protocols such as Ethernet, Token Ring, and FDDI. As we remarked earlier, such peer control falls into two categories:

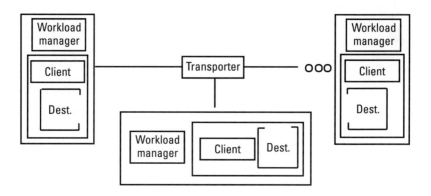

Figure 5.13 LAN with distributed workload managers.

random access and sequential access. These protocols were not possible before the VLSI breakthroughs of the 1970s, because the complexity of their implementations would have made them prohibitively expensive. In addition, as we pointed out earlier, there was really no need for peer protocols supporting any-to-any transport when the computing environment was terminal oriented and all the computing power was located in mainframes or even minicomputers. Only with the explosive growth of desktop computing, first in the Unix workstations and then with PCs and Macs, was the opportunity ripe for LANs. We explore LAN protocols in more depth in Chapter 8.

5.2.8 The Data Link in Multiprotocol Networks

The data link layer is obviously an intermediary between the protocol layers above it and the physical layer below it. The physical layer, from DTE to DCE to channel to DCE to DTE, appears as a transporter to the data link layer. Likewise, the data link layer's encapsulation of the physical layer gives the appearance to the protocol layer above it that it is just a transporter. This is the inheritance property we discussed earlier.

This brings us to the last topic in this chapter, namely, the question of supporting more than one upper layer protocol. Data links were originally part of monoprotocol networks, that is, networks that are homogeneous in both the end systems and intermediate systems. When there is just one type of upper layer protocol then all the data passed to and from the data link protocol can be treated the same. This, in turn, simplifies the data link protocol. Multiplexing of multiple upper layer clients is completely invisible to the data link protocol, as is the demultiplexing at the receiving end. The receiving link station merely passes the contents of the information frames to upper layer protocol, where the demultiplexing is handled by looking into the packet headers. With the progenitor of modern data link protocols, SDLC, this was not a problem because its use was proprietary to SNA and hence used only in monoprotocol networks.

On the other hand, when two or more upper layer protocols are sharing a data link, the invisibility of the upper link protocols cannot be preserved for one simple reason: the data link protocol machines cannot parse the upper layer SDUs carried in the payload of data link frames to determine to which protocol they belong. To a data link protocol machine, an IP packet and an SNA packet are indistinguishable; they are merely random bits that cannot be resolved into packet headers, address fields, and so on. And, unlike the monoprotocol case, upper layer demultiplexing cannot be handled without some additional information since it is impossible to tell a packet's protocol type by inspecting the bits.

This became an issue with the emergence of multiprotocol end systems and intermediate systems (routers). Once these came on the scene in the mid-1980s the problems posed required a solution, which was provided by the Xerox Palo Alto Research Center's networking group. This was the use of a protocol type field, an additional field in the data link PDU header that identifies the type of upper layer protocol SDU being carried in the payload of the frame. A data link protocol that supports a protocol type field in effect creates virtual data links, each dedicated to a given upper layer protocol. Figure 5.14 shows a PPP link that is transporting upper layer protocol data units for three different protocols. In effect, each of these upper layer protocol clients thinks it has a dedicated data link when in fact all are multiplexed over a single PPP data link.

5.3 Summary

In this chapter we have developed a management taxonomy for the classification and understanding of data link protocols, although we again note that

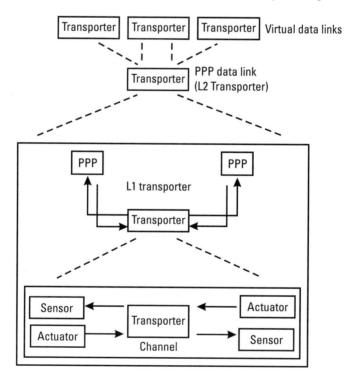

Figure 5.14 Virtual data links.

much of the taxonomy is applicable to protocols at any layer. We also briefly explored the hierarchical structure of layered communications architectures and the role played by data link protocols in such hierarchies.

The taxonomy, as we saw, involves four basic criteria: (1) whether the data link protocol requires connection establishment prior to sending data; (2) the nature of the time slicing that the protocol effects, that is, variable versus fixed length PDUs; (3) whether flow control is implemented, enabling a destination to throttle (workload actuate) a source; and (4) whether fault management is effected and if so to what level (fault detection or fault recovery). In addition, to fully characterize a protocol we saw that it is necessary to discuss the topologies (of the physical layer transporter) that it can manage as well as its ability to support multiple upper layer protocols. We also discussed the implementation of the management mechanisms involved: centralized versus distributed and open loop versus closed loop and the trade-offs entailed in the various combinations.

References

[1] *IEEE Standard Dictionary of Electrical and Electronics Terms,* 4th ed., New York: Institute of Electrical and Electronics Engineers, 1988, p. 746.

[2] *Van Nostrand Reinhold Dictionary of Information Technology,* 3rd ed., New York: Van Nostrand Reinhold, 1989.

[3] *ISO Standard X.200,* Section 7.6.2 International Standards Organization,

[4] Zimmerman, H., "OSI Reference Model—The ISO Model of Architecture for Opens Systems Interconnection," *IEEE Trans. Commun.,* Vol. COM-28, No. 4, Apr. 1980.

[5] Myers, G., *Reliable Software Through Composite Design,* New York: Van Nostrand Reinhold, 1975.

[6] *Cypser, R. J., Communications Architecture for Distributed Systems,* Reading, MA: Addison-Wesley, 1978, p. 388.

[7] de Prycker, M., *Asynchronous Transfer Mode,* 3rd ed.,Chester, U. K.: Ellis Horwood, 1996, p. 61.

6

Data Link Management: SDLC and HDLC

6.1 Introduction

Now that we have examined in some detail the generic characteristics of data link protocols, it is time to examine some actual protocols used in real-world networks. Given the diversity of data link protocols, we will divide our examination into several parts. The first part, which occupies the balance of this chapter, focuses on the so-called "serial" protocols and specifically on the family of bit-oriented protocols that includes IBM's Synchronous Data Link control (SDLC) and the High-level Data Link Control (HDLC).

The protocols represent different design trade-offs. This is not surprising given the different eras in which they were developed: SDLC is a product of the late 1960s/early 1970s (pre-VLSI), whereas HDLC was developed in the mid- to late 1970s as memory and CPU power decreased dramatically in cost. In addition, SDLC was designed to manage a multipoint link of terminal controllers and other devices with limited resources (processing power and storage). As a consequence, SDLC opted for centralization of most decision-making and management authority. SDLC is also a product of its times in that it is designed to run over very noisy channels; as such, much of SDLC's management is concerned with fault detection and recovery.

Next we examine HDLC, which as we indicated was designed when the advances in integrated circuits had greatly reduced the costs of distributed control. HDLC is a superset standard, including SDLC as a protocol subset. We focus on the HDLC subset known as Asynchronous Balanced Mode (ABM). HDLC/ABM is a peer protocol connecting two link stations over a point-to-point channel. The two link stations are symmetric; neither is master nor slave of the other. In part, HDLC/ABM's importance stems from its incorporation into a variety of other standards, notably the X.25 family of protocols, where it is known as LAP-B and serves as the data link protocol; narrowband ISDN, where a variant called LAP-D is used over the D channel; and even modems, where a variant called LAP-M is used by smart modems to negotiate such features as compression and encryption.

With both protocols we will seek to identify the management tasks involved. We will explore the protocols and map their management tasks onto our client/server and workload management/bandwidth management model.

6.2 SDLC

Synchronous Data Link Control (SDLC) was introduced in 1974 as the serial data link protocol for IBM's Systems Network Architecture (SNA), one of the first comprehensive layered communications architectures developed. We note that SDLC owes much of its design to an earlier IBM protocol called Bisync, a so-called *byte-oriented* protocol. Byte-oriented protocols are intimately coupled to character sets; in contrast, bit-oriented protocols are transparent to character sets, and hence are better suited to handle the transmission of digitized media, computer programs and data, and so on.

In terms of the data link taxonomy we established in the last chapter, SDLC is connection oriented, it uses variable length frames, it employs flow control, and it provides both fault detection and fault recovery. SDLC is a master/slave protocol. Each SDLC data link has a primary link station (PLS) and one or more secondary link stations (SLSs). The PLS has overall control of the data link, meaning the physical layer transporter and the SLSs. SDLC supports both point-to-point and multipoint data link topologies; however, SDLC in multipoint topologies does not allow direct communication between secondary link stations.

Part of the reason that SDLC was built on a rigid master/slave model is because of its intended use as a terminal-handling protocol. In addition, given the expense of processor bandwidth and memory, at the time (early 1970s), there was considerable cost advantage to having the protocol's management "intelligence" centralized in one link station rather than replicated at each link station. And the designers of SDLC sought a data link protocol that was robust, albeit at the price of considerable complexity. In the words of one of SNA's early architects, with SDLC the

> ... objective is to give the effect, so far as the other layers of the communication system are concerned, of a highly reliable link, even though

retransmissions may be needed. This requires that the DLC level itself be able to detect errors and to correct those errors without outside interference [i.e., management]. [1]

6.2.1 SDLC Task Set and Tasking

6.2.1.1 Transporter Topology and SDLC Task Sets: Nominal Versus Effective

To define the task set of an SDLC data link, we first start with the task set of the transporter (that is, the channel) being managed by the data link protocol. The task set of a bidirectional point-to-point link is $\{A_1 \rightarrow A_2, A_2 \rightarrow A_1\}$ where A_1 and A_2 are the two link stations. With a multipoint link that provides any-to-any transportation between k stations A_j, $j = 1, 2, \ldots, k$, there are $k(k-1)$ tasks in its task set: $\{A_i \rightarrow A_j, \rightarrow A_j \rightarrow A_i, i \text{ and } j \text{ different}\}$. SDLC, however, is not an any-to-any protocol. It is logically a hub and spoke topology, with only the PLS able to communicate with every other station. With SDLC, therefore, the task set is $\{PLS \rightarrow A_1, PLS \rightarrow A_2, \ldots, PLS \rightarrow A_k; A_1 \rightarrow PLS, A_2 \rightarrow PLS, \ldots A_k \rightarrow PLS\}$ where $A_1, A_2, \ldots, A_k$ are the secondary link stations; this is because the secondary link stations cannot have direct communication (see Figure 6.1).

Because SDLC is connection oriented, we need to distinguish between what we will call the *nominal* task set of the SDLC data link versus what we will call its *effective* task set. The nominal task set of an SDLC data link consists of the transport actuations between the PLS and the SLS(s) attached to the communications channel. The effective task set, on the other hand, consists of the transport actuations between the PLS and the SLS(s) attached to the communications channel that have connections to the PLS. At any given moment, these are the only tasks the SDLC data link can execute.

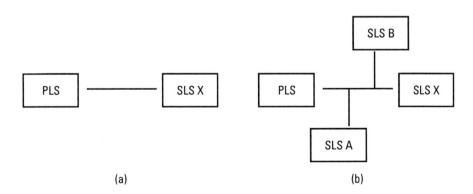

(a)   (b)

Figure 6.1 (a) Point-to-point versus (b) multipoint SDLC data links.

Clearly, the effective task set is a subset of the nominal task set. The nominal task set is determined by the topology of the data link, in particular, the addresses of the SLS(s), whereas the effective task set is determined by the topology of the data link and the state of the SLS(s). Whereas the nominal task set is relatively static, the effective task set changes as secondary link stations are activated and deactivated; or suffer faults and are "repaired" by management intervention. These are the principal bandwidth management tasks of the PLS. In the next section we discuss the mechanisms by which the PLS activates and deactivates link stations; fault recovery is discussed in a later section.

6.2.1.2 SDLC Tasking: Two-Way Simultaneous Versus Two-Way Alternating

In addition to supporting point-to-point and multipoint topologies, SDLC data links support two types of multitasking: sequential, or half-duplex, meaning one transport task at a time, and concurrent, or full-duplex, meaning (in this instance) two transport tasks at a time. In SDLC these are referred as Two-Way Alternating (TWA) and Two-Way Simultaneous (TWS), respectively. TWS SDLC (full-duplex) means that concurrent transport is possible in both directions (Figure 6.2). In this case, because of SDLC's task set one of the tasks necessarily is from the PLS to an SLS, and the other is from an SLS to the PLS.

With a master/slave protocol like SDLC, what does this mean? Consider a point-to-point SDLC link. Because with SDLC the secondary station cannot send without permission of the primary station, this limits one advantage of full-duplex communications, namely, unconstrained bidirectionalism. With SDLC's master/slave control, the principal advantage of full-duplex communication is that the PLS can send at any time, including sending control information to the secondary station concerning communication in the reverse direction (i.e., from secondary to primary). The PLS does not have to share the transporter with any SLS, which it must when the data link is TWA. This in itself is not an inconsiderable advantage.

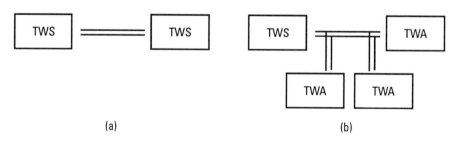

(a)   (b)

Figure 6.2 Two-way simultaneous stations on (a) point-to-point and (b) multipoint data links.

Support for a TWS operation requires several things. First, the PLS must be multitasking, that is, able to send and receive simultaneously. On a point-to-point link, the same holds true for the SLS. If the PLS is able to send and receive at the same time, then its partner SLS must be able to as well or the capability will be wasted. On the other hand, on a multipoint data link the PLS may be capable of TWS operation while the SLSs are capable of only TWA operation; the PLS simply sends to one station while it receives from another. (This does not violate the restriction that only one poll is possible at a time since the PLS never polls itself.)

The second requirement for exploiting TWS SDLC is that the channel it uses should be FDX. The tasking of the data link is not independent of the tasking of the underlying physical transporter. Recall from earlier chapters that, using standard line driver technology, a two-wire facility cannot support concurrent multitasking; the two lines can carry HDX signals from A to B or from B to A but not both signals simultaneously. To achieve concurrent multitasking, that is, FDX, the solution generally involves a four-wire facility, where one pair of wires carries the A $\rightarrow$ B signals while the other pair carries the B $\rightarrow$ A signals. An alternative is to use frequency-division multiplexing to split the signal "space" into two or more bands, each of which can carry signals independently of the other.

Four scenarios correspond to the four possible combinations of SDLC tasking (TWA and TWS) and channel tasking (HDX and FDX) (see also Table 6.1):

1. Single-tasking communications channel/Single-tasking link stations: TWA link stations connected by an HDX channel. Because the link stations cannot handle concurrent sending and receiving, nothing is lost by using the HDX channel.

Table 6.1
SDLC Versus Channel Tasking

Communications Channel	SDLC Link Station	
	TWA	TWS
HDX	Scenario 1	Scenario 3
FDX	Scenario 2	Scenario 4

2. Single-tasking communications channel/Multitasking link stations: Theoretically, nothing prevents implementing a TWS link station running over an HDX communications channel. However, an HDX channel vitiates the advantage of a TWS link station, namely, that it can concurrently (simultaneously) send and receive data. Therefore, this is seldom used.

3. Multitasking communications channel/Single-tasking link stations: If all the stations on an FDX communications channel are TWA, then at any given time the channel will be executing only one task. This wastes the channel's capability to execute two tasks concurrently, namely, from the PLS to an SLS and from an SLS to the PLS.

4. Multitasking communications channel/Multitasking link stations: TWS stations on a FDX channel permit full exploitation of their respective capabilities for concurrency. If a channel is point to point, then both the PLS and SLS must be TWS. On the other hand, if the channel is multipoint, then the PLS must be TWS but the SLSs may be TWA.

6.2.2 SDLC Frames

SDLC link stations exchange (via the underlying transporter provided by the physical layer) frames (messages) ranging from RFSs to actuate the state of a link station to an RFS to send state information to data intended for upper layer protocol clients. For reasons best known to the protocol designers, in SDLC the frames (messages) are referred to as "elements of procedure."

The distinction made in SDLC between primary and secondary link stations carries over to the frames sent between the two types of link stations. When a frame is sent from the PLS it is called a *command*. Likewise, when a frame is sent from an SLS it is called a *response* (Figure 6.3). What this means is that certain frames that either the PLS or SLS can send may be either commands or responses depending on the originating station.

SDLC frames all follow the generic structure shown in Figure 6.4. Every frame is delimited in front and back by an 8-bit field called the *flag*; the flag consists of a reserved pattern, which is b001111110 (= 0x7E). This pattern is

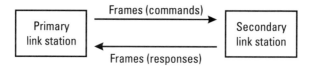

Figure 6.3 Commands and responses.

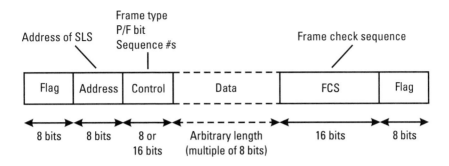

Figure 6.4 SDLC frame format.

never permitted in client data and an SDLC station will "bit stuff" to prevent this. When six consecutive 1's are detected in client data then a 0 is inserted. At the receiving link station the corresponding inverse operation is performed to restore the client's data to its original form. The importance of bit stuffing is that it ensures SDLC is data or character code transparent; client data can be in ASCII, EBCDIC (a character set used by IBM's mainframes), or even in no character set at all like programs or digitized signals (audio, video, and so on) because the SDLC sending link station will ensure via bit stuffing that it not contain the only truly reserved bit pattern, namely, the flag.

The flag in SDLC is critical for several reasons. First, unlike some protocols that include a field specifying the PDU's length, SDLC relies on the flags to allow the receiver to parse beginning and end, thereby enabling the receiver to reconstruct the frame's length. A second reason for the use of this particular flag is related to the line encoding. In conjunction with NRZI line encoding, the leading flag ensures that a receiver will have sufficient time to recover (estimate) the clock before important data are received. (Recall that NRZI inverts the waveform on a binary zero; too many consecutive 1's precludes a transition, and the receiver's clock reconstruction mechanism can lose synchronization.) Note that SDLC (and SNA) did not initially recognize a clear demarcation between the data link and the physical layers, and the initial SDLC specification coupled the protocol with the NRZI line encoding we discussed in Chapter 2.

6.2.2.1 SDLC Addresses

Following the flag, every frame has an 8-bit field for the address of the secondary link station. This is a very important point about SDLC frames: They contain a field for only one address, not the two addresses (source and destination) that one would expect. This works because SDLC permits transport only between the PLS and an SLS, never two SLSs. The PLS address is always implicitly present, whereas the SLS address that is explicitly present is, depending on the direction of traffic, either the client (source) or destination of the frame.

SDLC addresses are typically statically assigned when individual SDLC devices are configured by a network operator or technician. Note that two SDLC addresses are reserved: the all-zeros address (0x00), which is used for test purposes, and the all-ones address (0xFF), which is used for SDLC broadcasts to those SDLC devices that support this capability.

6.2.2.2 The Control Field

Following the address field is the control field, which has several subfields. The control field indicates which type of SDLC frame it is. The subfields include a field that identifies the type of frame, a 1-bit field called the Poll/Final bit and, depending on the type of frame, zero, one, or two sequence number fields for use in flow control and fault recovery.

There are three types of SDLC frames (Table 6.2). The largest in number are called *unnumbered* (also known in earlier SDLC documents as *nonsequenced*) frames; SDLC defines 13 unnumbered frames. These are management frames exchanged by link stations primarily for purposes of managing the connections. The second type are *Supervisory* frames, three frames which are exchanged for purposes of flow control and fault recovery. The last type of frame is called an *Information* frame (I frame); this is the only frame that can carry normal user data on an SDLC connection, and constitutes the great majority of frames exchanged over a typical SDLC data link.

Note that this frame identification uses a form of Huffman encoding (see Chapter 3): I frames, which are the most common, use a single-bit identifier while supervisory and unnumbered frames use 2 initial bits augmented by 2 and 5 bits, respectively. This means that there are four possible Supervisory frames, of which SDLC uses three; and that there are 32 possible unnumbered frames, of which SDLC uses 13. We discuss these frames and their role in managing an SDLC data link in the next section (Table 6.2).

The Poll/Final (P/F) bit in the control field of every frame is so called because it means two separate things depending on whether the frame is sent from a PLS, in which case it indicates a poll or no poll, or from an SLS, in which case it indicates a final frame. That is:

1. When the frame originates at the PLS the P/F bit is used to indicate either that the PLS is retaining the transporter for itself to send further frames or that the PLS is finished with the transporter and that it has been allocated to the destination SLS.

Table 6.2
Bits in Control Field of SDLC Frames (Modulo 7 Sequence Numbers)

	Bit 0	Bit 1	Bit 2	Bit 3	Bit 4	Bit 5	Bit 6	Bit 7
Control Field for Unnumbered Frames: No Sequence Numbers	1	1	U	U	P/F	U	U	U
Control Field for Supervisory Frames: Received Sequence Numbers	1	0	S	S	P/F	N(R)	N(R)	N(R)
Control Field for Information Frames: Sent and Received Sequence Numbers	0	N(S)	N(S)	N(S)	P/F	N(R)	N(R)	N(R)

2. When the frame originates with an SLS then the P/F bit (often shortened to the Final bit) is used to indicate that the frame in question is the last to be sent by the SLS (or not).

We will discuss the P/F bit more when we explore workload management.

The last subfields of the control field are dedicated to sequence numbers. Two uses for such sequence numbers are to identify the frame being sent for purposes of acknowledging its receipt or requesting its retransmission and to acknowledge the receipt of SDLC frames that have arrived without fault. These are designated, respectively, the Ns and Nr sequence numbers. In addition to their role in fault management, sequence numbers play a part in flow control. The fact that the quantity of sequence numbers is finite imposes a worst-case flow control on a station sending frames. Once it has exhausted the sequence numbers the station must stop sending until its partner link station sends acknowledgment for some or all of the outstanding frames.

Unnumbered frames lack either Nr or Ns numbers, in part because there is no room in the control field after setting aside the bits needed for identifying the type of frame. These frames are referred to as Unnumbered or nonsequenced precisely because they lack sequence number identifiers. Supervisory frames, too, lack sequence number identifiers; on the other hand, they do carry Nr sequence numbers so they can acknowledge frames received at the same time as they carry their particular management messages pertaining to flow control. Finally, the I frame carries both Ns and Nr numbers. The Ns provides an unambiguous identification of the frame being sent and the Nr number acknowledges frames that have been received.

The number of bits in the sequence number is obviously a crucial parameter in determining how well the protocol manages links. If too few bits are used then a link station will be forced to frequently halt sending, approaching the worst case of stop-and-wait behavior. The original SDLC specification used 3-bit sequence numbers, which meant that there could be seven outstanding frames. However, as we discussed earlier, when SDLC began to be deployed over high latency channels, most notably channels using geosynchronous satellites to relay signals, the result was very poor utilization of total channel bandwidth (capacity). Most of the time, the channel was idle. The transmitting station had blocked after sending seven frames since it was waiting for acknowledgments that would not arrive until the receiving station had received the frames and sent its own acknowledgments.

On the other hand, if too many bits are used then the management overhead per frame becomes unacceptable, as does the buffer (storage) requirements for link stations holding frames awaiting acknowledgment. As a compromise, SDLC allows two different sequence number schemes: modulo 8, which uses 3-bit sequence numbers; and modulo 128, which uses 7-bit sequence numbers. The control field of an I frame, which includes two sequence numbers as well as the P/F bit and the type bit, will either be 8 (2 × 3 bits + 2 bits) or 16 bits (2 × 7 bits + 2 bits) in length. This is why, as we can see in Figure 6.4, the length of the control field may be either 8 or 16 bits depending on the modulo number used.

Including the Nr sequence number and the P/F bit in the control field are instances of "piggybacking," whereby a single frame effectively serves two or more purposes. In earlier protocols such as Bisync, polls and acknowledgments required separate frames, increasing the management overhead and reducing the bandwidth available to carry client data. By including acknowledgments and polls in the SDLC control field, the protocol's designers reduced the need for separate frames just to poll and/or acknowledge, thereby increasing the data link's efficiency.

6.2.2.3 Payload and Trailer Fields

Following the control field is an optional information field for carrying client data, in the case of I frames, or management data, in the case of certain unnumbered management frames. Some unnumbered frames, however, lack any information field, as do all the supervisory frames. When data fields are present SDLC requires that they be a multiple of 8 bits in length. The maximum frame size determines how large the data field can be, and this is typically a configuration parameter set when defining the link stations.

The next field is the Frame Check Sequence (FCS), which is present in all frames. The FCS is a 16-bit Cyclic Redundancy Polynomial similar to that discussed in Chapter 3. The purpose of the FCS field is to provide a second level of fault detection beyond that offered by the Physical layer. We discuss the FCS and its use in fault detection later.

The frame is terminated with a flag identical to the flag that was used at the beginning. Note that another use for flags is to keep the line up; even when the PLS has no data to send, it will often send flags to one or more SLSs to keep receiver estimators synchronized and link stations up. This becomes an issue when SDLC frames are encapsulated in so-called payload PDUs for tunneling (see Chapter 12).

Table 6.3 lists all of the SDLC frames. We discuss these and their roles in the next three sections.

Table 6.3
SDLC Frames

Frame Name	Abbreviation	Frame Type
Request initialization mode	RIM	Unnumbered
Set initialization mode	SIM	Unnumbered
Set normal response mode	SNRM	Unnumbered
Request disconnect	RD	Unnumbered
Disconnect	DISC	Unnumbered
Frame reject	FRMR	Unnumbered
Unnumbered acknowledgment	UA	Unnumbered
Beacon	BCN	Unnumbered
Configure	CFGR	Unnumbered
Disconnect bode	DM	Unnumbered
Exchange station identification	XID	Unnumbered
Unnumbered poll	UP	Unnumbered
Unnumbered information	UI	Unnumbered
Test	TEST	Unnumbered
Receive ready	RR	Supervisory
Receive not ready	RNR	Supervisory
Reject	REJ	Supervisory
Information	I	Information

6.2.3   SLS Modes of Operation

Link stations are finite state machines (protocol machines). As such, SDLC link stations are characterized by their states, the transitions between those states, and outputs that occur in response to (are triggered by) various input messages. In the case of secondary link stations, which can have only one connection (to the primary link station) the station's state is identical to the state of the connection. (The PLS is stateless in the sense that it is always "up.")

There are four states, three of which are called *modes,* and the messages that trigger state transitions include the frames exchanged by link stations or external events such as activation of a device or link station by an outside manager. The three SDLC modes are called the Normal Disconnect Mode, the Initialization Mode, and the Normal Response Mode. We discuss these in detail next.

Normal Disconnect Mode

A link station in Normal Disconnected Mode (NDM) is powered on but unable to send or receive client data. Note that NDM does not imply that the station is completely disconnected from the data link. According to IBM's specification, "A secondary station in NDM cannot receive or transmit information or supervisory frames." This does not preclude unnumbered (i.e., management) frames. In fact, an SLS in NDM will stay in NDM until the PLS sends it one of several unnumbered frames that request the actuation of the link station's mode to either the Initialization Mode or Normal Response Mode. (More on these in a moment.)

Initialization Mode

Initialization Mode (IM) is a transitory state for the link station. The link station will not stay in IM but rather will move to Normal Response Mode, assuming no faults occur during initialization that preclude this. Not all link stations are required to transit the IM but instead move directly from NDM to NRM.

Normal Response Mode

An activated SDLC link station, able to send and receive client data, is said to be in Normal Response Mode (NRM). When an SLS is in NRM this means that a connection is up between PLS and SLS. The importance of NRM is that only an SLS in NRM will be polled by the PLS, and this is the only way it will be allocated (scheduled) the transporter. A station not in NRM cannot send or receive client data. In effect, if an SLS is not in NRM then the corresponding transport tasks are not in the task set of the data link.

Finally, an SLS can be in a special state called Frame Reject, which, strictly speaking, is not an SDLC mode but rather an exception condition that is nonrecoverable within the connection. The exception requires either the PLS to undertake management intervention (for example, resetting the SLS with a mode management frame) or an outside manager (programmed or human). We discuss the Frame Reject state when we consider fault recovery later.

6.2.4 Connection (Mode) Management: Bandwidth Actuation

As we indicated earlier, one of the principal tasks of the PLS as bandwidth manager is managing connections with the SLSs attached to the channel. Part of the process by which a PLS activates an SLS involves sending one or more frames requesting the SLS to actuate its mode. The PLS uses unnumbered (management) frames included in SDLC for just this purpose. In fact, the majority of SDLC frame types effect mode actuations of one sort or another. A measure (or, to be consistent with the MESA terminology, an estimate) of the complexity of a data link protocol can be obtained from the number of management frames relative to the number of frames that actually carry client data. In SDLC, 16 frames are, in some respect, management frames and only 2 frames (the *Information* frame and a related *Unnumbered Information* frame) carry client data.

Figure 6.5 shows this exchange between the bandwidth scheduler within the PLS and bandwidth actuators in both the PLS and SLS with which a connection is to be established (or taken down). Note that management of

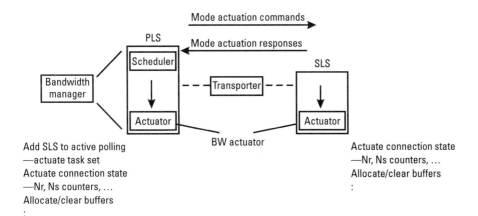

Figure 6.5 SDLC mode actuations.

a connection is inherently distributed. Although scheduling is centralized, the actuation occurs in both locations. That is because on each side of the connection, actuating a connection entails buffer allocation for holding frames, setting or clearing Nr and Ns counters, and so on.

This is part of the challenge of managing transporters. If both sides do not cooperate then failure is inevitable. The primary link station issues commands to the secondary link station(s) that are requests for service concerning mode actuation. At each secondary link station, actuators change the mode of the station from "disconnected" to "initializing" to "normal" and back among them.

The mode of an SLS is just a proxy for what we are really interested in, namely, the state of the connection between the PLS and that SLS; in other words, the effective task set of the data link. Given this, mode (connection) actuation alters the effective task set of the transporter (the SDLC data link). This is bandwidth actuation of kind, and the scheduler in the PLS that initiates the actuation is part of a bandwidth manager. To reiterate, an SDLC data link's static configuration only defines the nominal data link; the effective data link will vary according to which stations have active connections versus those that are just powered on and those that are turned off.

6.2.4.1 Configuring the PLS

But before the PLS can schedule these actuations, it must know which SLS stations are attached to the channel—that is, it must know the topology of the data link and from this the nominal task set. After all, prior to mode/connection actuation there must be some way for the PLS to learn the addresses of these devices, otherwise it cannot send them the requisite mode actuation frames since it lacks their respective addresses. There are two ways for the PLS to learn this information: by manual configuration or automatic "discovery" of the SLSs.

The former approach relies on network personnel defining for the PLS device a static configuration of SDLC SLSs on data link; when an SDLC SLS device is added or removed, the PLS configuration must be changed. Because, as we remarked earlier, a network operator or technician typically must configure the SDLC addresses for the individual SDLC SLS devices, requiring manual definition at the PLS is not an extraordinary onus. Automatic discovery entails the SDLC PLS discovering the link's topology via exhaustive polling of all SLS addresses. There are 254 possible addresses, and the PLS can periodically cycle through the range of addresses to see if any station responds. Clearly, this will discover any SLSs but at a considerable cost of bandwidth.

6.2.4.2 SDLC Mode Actuation Frames

Let's look at the various SDLC frames involved in the PLS's actuation of the data link's connections/SLS modes. Table 6.4 lists the SDLC command and response frames that can be exchanged by the PLS and SLS(s).

Just as there are three (nominal) modes, so are there three types of actuations: initialization, normal response, and disconnect. Corresponding to these, the PLS can send to an SLS a *Set Initialization Mode* (SIM) frame, a *Set Normal Response Mode* (SNRM) frame, or a *Disconnect* (DISC) frame, respectively. We now discuss these in turn.

Set Initialization Mode

Some secondary link station devices need to execute preliminary tasks such as loading software modules before they can proceed to full readiness. By sending the SIM command frame the PLS schedules the execution of these tasks. To quote from the specification: "This command initiates system-specified procedures for the purpose of initializing link-level functions.... The primary and secondary station Nr and Ns counts are reset to 0" [2].

Set Normal Response Mode

A PLS sends the SNRM command to an SLS that is ready to begin sending and receiving data. Again, to quote from the specification: "This command places the secondary station in normal response mode (NRM) for information transfer. ... The primary and secondary station Nr and Ns counts are reset to 0. No unsolicited transmissions are allowed from a secondary station that is in NRM. The secondary station remains in NRM until it receives a DISC or SIM command" [2].

Table 6.4
SDLC Frames: Bandwidth/Mode Actuations

Request initialization mode	Response
Set initialization mode	Command
Set normal response mode	Command
Request disconnect	Response
Disconnect	Command

Disconnect

When the PLS has finished sending and receiving data from a link station, it may wish to tear down the connection and remove the SLS from the polling schedule. This is accomplished by sending a DISC command to the SLS: "This command terminates other modes and places the receiving (secondary) station in disconnected mode.... A secondary station in disconnected mode cannot receive or transmit information or supervisory frames" [2].

Table 6.4 also lists two SDLC response frames that can be sent by an SLS to the PLS to request actuation. Secondary link stations cannot just actuate themselves. An SLS lacks the authority to schedule the actuation of its own mode (and by implication, its connection to the PLS). Instead, the manager in the SLS must request that the manager in the PLS schedule the actuation and send the corresponding mode actuation command frame. An SLS that needs to be initialized will send a *Request Initialization Mode* response to the PLS to elicit an SIM command. An SLS that wants to tear down a connection will send a *Request Disconnect* response. Note that there is no response frame requesting that the PLS send an SNRM; an SLS that wants to be actuated into NRM will instead send an RIM response and the PLS will follow the SIM with a SNRM. In all cases the PLS is the master of the data link.

The reason for this is that the primary link station must alter its polling to reflect any mode change in secondary link stations. This is logical because the PLS's workload manager, which is responsible for scheduling access to the shared transporter by means of polls (permission to send) sent to the SLS(s), cannot do its job effectively if it doesn't know the topology of active link stations, that is, the effective task set. The alternative would be to proceed open loop with respect to topology and poll every possible SLS address, which is clearly inefficient and will result in many unproductive polls and hence wasted bandwidth (wasted both in terms of the polls carried as well as the opportunity cost of unnecessary delays inflicted on those stations present and that have data to transport).

SDLC requires that the mode actuation commands be acknowledged by the SLS recipient. However, the acknowledgment mechanism used here is different than that encountered earlier, where we were acknowledging numbered (sequenced) I frames. There the solution involved the receiving station returning the sequence number of a successfully transported I frame. Clearly this will not work with the various unnumbered (nonsequenced) mode actuation commands. Instead, to acknowledge these, SDLC provides a frame called *unnumbered acknowledgment* (UA).

Figure 6.6 is a state/transition diagram describing the behavior of an SDLC secondary link station. The convention for reading such diagrams is that

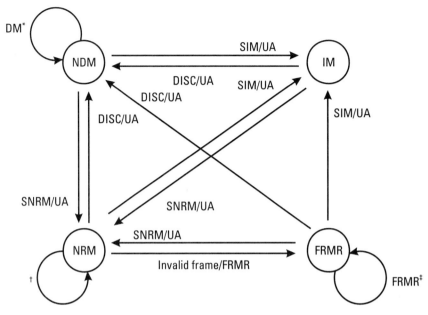

Notes:
*An SLS in the DM state will remain there unless the PLS sends either
a SIM or SNRM; any other frame will elicit a DM response
†An SLS in the NRM state will remain there unless the PLS sends either
a SIM or DISC frame; or a fault causes the SLS to enter the FRMR state
‡In SLS in the FRMR state will remain there unless reset by the PLS
any non-mode actuating frame received by the SLS will cause
the SLS to send a FRMR response

Figure 6.6 SDLC state/mode transitions.

a state transition will be denoted with an arc and a label of the form "A/B"
where A is the input message that triggered the transition and B is the output
(if any) that the finite state machine produces. Here, for example, we see that
an SLS in the NDM state will, in response to a SIM command, move to the
IM state and issue a UA response; in response to an SNRM command, move to
the NRM state and issue a UA response; and, in response to any other SDLC
frame, stay in the NDM state and issue a *Disconnect Mode* (DM) response to
tell the PLS that the SLS is not in a state to respond to frames other than SIM
or SNRM.

Likewise, an SLS is in the NRM state will, in response to a SIM command, move to the IM state and issue a UA response; in response to a DISC
command, move to the NDM state and issue a UA response; and, in response

to any other SDLC frame, stay in the NRM state and follow the protocol for responding to I frames and so on. An SLS in the IM state will, in response to an SNRM command, move to the NRM state and issue a UA response; in response to a DISC command, move to the NDM state and issue a UA response; and, in response to any other SDLC frame, stay in the IM state.

We defer discussion of the FRMR failure state and its associated state/transition behavior until the section on SDLC fault recovery.

6.2.4.3 Connection Actuation: Setup Sequences

A connection can be actuated in at least three ways: (1) The PLS can initiate a connection by sending an SNRM; (2) the PLS can poll (either with an RR or an I frame with the poll bit set) an SLS that is in NDM and, after learning the latter's mode status via the DM sent by the SLS, send an SNRM; or (3) the SLS can initiate a connection by sending a RIM in response to a poll from the PLS. As we remarked earlier, no frame exists that allows the SLS to request a direct actuation to NRM (an RNRM frame); the SLS must request actuation to IM and rely on the PLS to follow through with an SNRM. Note that in all of these scenarios the SLS must issue a UA for the task to be considered complete. Figure 6.7 illustrates the three scenarios.

Again, in terms of state transitions not every secondary link station must pass through the Initialization Mode before going into the Normal Response Mode. However, if an SLS sends a RIM response the PLS cannot send an SNRM command immediately but must first send an SIM command and then an SNRM once the SLS sends a UA indicating completion of the initialization.

As with bringing up a connection and thus actuating an SLS to NRM, tearing down a connection and (thus actuating an SLS to NDM) can be initiated by either PLS or SLS but the scheduling authority rests with the PLS (Figure 6.8).

After a PLS issues a DISC the "expected response is UA" [2]. This means that any data (I frames) that the SLS has yet to send must be either discarded or buffered until the connection is reestablished. On the other hand, it is implementation dependent whether an SLS that issues an RD may have to stay in NRM until the PLS has sent all the data it wishes.

6.2.5 Bandwidth (Server) Monitoring

Having covered mode/state actuation, it is time to look at state instrumentation. The state information provided by such instrumentation enables the PLS to manage the data link more effectively than in its absence (i.e., open loop). We have already seen several instances where an SLS will send its state to the PLS if it receives an RFS/command that is inappropriate to its present mode.

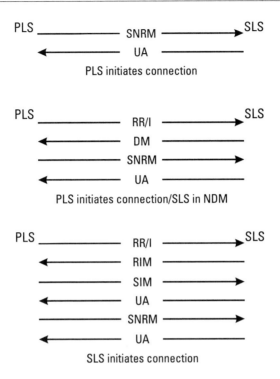

Figure 6.7 Connection actuation: setup.

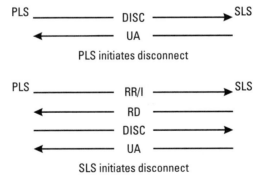

Figure 6.8 Connection actuation: teardown.

For example, we saw earlier that if an SLS is in Disconnect mode and it receives any command other than a SIM or an SNRM then the SLS will send a DM response to the PLS. In addition, the detection of channel faults by means of the Frame Check Sequence and/or transmission timer is also an instance of bandwidth monitoring (Figure 6.9); we discuss this in more detail later.

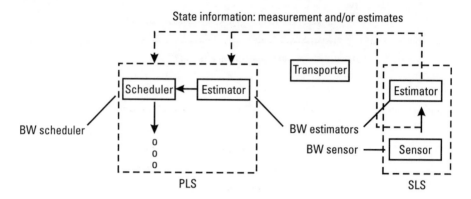

Figure 6.9 Bandwidth monitoring.

There are a total of four response frames defined in SDLC that can be categorized as bandwidth (server) monitoring, whereby an SLS will send to the PLS information describing its current state. These four frames are listed in Table 6.5 and discussed next.

Disconnect Mode (DM) is sent by a secondary link station when it is in Normal Disconnect Mode and it receives a frame other than an SNRM or a RIM. *Exchange Station Identification* (XID) is used by the link stations to exchange configuration parameters and for negotiation of these. *Test* (TEST) carries a data field the contents of which is a special pattern known a priori by all stations. *Frame Reject* (FRMR) indicates a nonrecoverable fault has occurred and that the connection between link stations is terminated.

Because we have already discussed the *Disconnect Mode* response sufficiently, so we move on to the *Exchange Station Identification* command. This may be sent by a PLS for one of several purposes. If a node has only fixed parameters then an XID can only exchange authentication information. Otherwise the exchange of XIDs can be used to select among possible services by specifying different parameter values and configurations.

Table 6.5
SDLC Frames: Bandwidth Monitoring

Frame reject	Response
Exchange station identification	Command/response
Test	Command/response
Disconnect mode	Response

The *Test* frame is used by the PLS to check the channel and the SLS. The PLS will send a *Test* command frame to an SLS to elicit a *Test* response. The *test* command may optionally contain an information field; if it does, then the SLS is supposed to include the contents in its *Test* response. By comparing the sent and received frames the PLS can get an estimate of the reliability of the information transfer.

We discuss *Frame Reject* in more detail in the section on fault management. Suffice it to say for now that if an SLS is in the FRMR state and it receives any command other than a SIM, an SNRM, or a DISC then it will send an FRMR response to the PLS. The FRMR response will contain an information field that includes the cause of the FRMR condition.

6.2.6 Workload Management

Workload management in SDLC has two components: allocation of the transporter to one (TWA) or two (TWS) link stations; and flow control, enabling a destination link station to throttle the arrival of I frames to prevent its being overwhelmed. Although the PLS participates in both types of management, an SLS participates only in flow control; it has no role in allocation of the transporter.

6.2.6.1 Allocating the Transporter

After one or more connections has been set up by the PLS and SLSs exchanging the various mode-actuating frames, the next task of the PLS is the allocation (i.e., scheduling) of the transporter. As with the connection management phase, the PLS is responsible for this. However, unlike connection management this is a workload management issue: Which link station of which active connection is going to be the client of the transporter? The PLS schedules the transporter based on two parameters of the data link: (1) the effective task set of the data link (Which SLSs are in NRM?) and (2) the tasking of the data link (Is it TWA or TWS?).

We have discussed the effective, as opposed to nominal, task set of the data link at some length. The effective task set highlights the fact that the workload scheduler in the PLS is closed loop in its allocation decisions. It will only send permission to use the transporter to an SLS that it knows to be in NRM, and this information is based on feedback from the SLS(s), notably the UA returned after an SNRM, that indicates it is in NRM and ready to send and receive data.

How is this permission conveyed? It is conveyed with the Poll/Final bit in the control field of a frame the PLS sends to an SLS. As has been mentioned, earlier data link protocols, notably Bisync, required a separate frame to carry a poll from the master link station. With SDLC, a poll can be piggybacked on any command frame. By setting the P/F bit = 1 the PLS schedules/allocates the transporter to an SLS.

On a TWA data link, for example, steady-state operation (no mode actuations) may consist of just I frames being exchanged between PLS and SLSs, with the last frame in each exchange "toggling" the P/F bit to alternate use of the data link. Only if the PLS has no data to send would it use a frame other than an I frame, notably the *Receive Ready* (RR) supervisory frame (more on this frame in the next section), to poll an SLS. Likewise, if an SLS receives a poll (any command frame with the P/F bit set) and it has no data to send it will respond by sending an RR with the P/F bit set indicating it is returning use of the data link to the PLS for reallocation.

The Poll and Final bits are both indications that a client is finished with the transporter. In other words, the P/F bit is a semaphore. When a frame originates at the PLS the P/F bit is a semaphore indicating that the PLS is either using the data link or releasing it to the SLS addressed in the frame. Likewise, when the frame originates at an SLS the P/F bit is a semaphore that indicates whether the SLS is still using the channel (transporter). Consider an SLS that has k frames to send. The PLS sends an RR or an I frame with the P/F bit set. The SLS sends the first $k - 1$ frames with the P/F bit not set (P/F = 0). The kth frame, on the other hand, is sent with the P/F bit set (P/F = 1); this indicates that the SLS is finished with the channel (transporter), and the PLS can schedule the transporter to itself, to another SLS, or that SLS again.

6.2.6.2 Controlling the Flow Between Link Stations

The second instance of workload management in SDLC is flow control. As we discussed in our earlier survey of generic protocol issues, flow control is designed to keep a receiving link station from being overwhelmed by a sending link station when the former lacks the capacity or resources to accept frames as fast as the latter can send them. But with flow control, unlike the other management mechanisms we have considered such as mode actuation and transporter allocation, the management is symmetric: the SLS has the same control over its receipt of additional frames from the PLS as the PLS has over its receipt of additional frames from any SLS. For this reason, within the context of flow control, the distinction between primary and secondary in link stations is unimportant, and for the balance of this section we simply discuss flow control in terms of source and destination link stations.

Flow control in SDLC is effected in two basic ways (Figure 6.10):

1. The *Receive Ready* (RR) and *Receive Not Ready* (RNR) supervisory frames; and

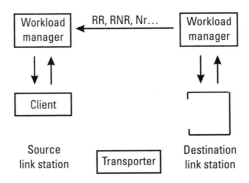

Figure 6.10 Link stations: source and destination.

2. The acknowledgment of frames received via Nr fields in "counter-flow" frames.

The first mechanism involves the use of two supervisory frames: *Receive Ready* and *Receive Not Ready*. The RR and RNR enable and disable, respectively, a link station from scheduling any RFSs to the transporter. When a link station sends an RNR to another link station (secondary or primary), the recipient suspends itself from sending any more to the sender of the RNR; the recipient's workload manager will not schedule any requests for service. The suspended workload scheduler may resume scheduling when the link station that sent the RNR sends an RR.

We have seen the *Receive Ready* frame used by the PLS to carry a poll when it has no data to send in an I frame; and that the SLS likewise uses the *Receive Ready* frame to respond to a poll when it has no data but must respond to a poll. However, the primary purpose of supervisory frames, and RRs and RNRs in particular, is as traffic signals. An RNR is a red light, signaling a sending link station that it should not send any more I frames; and an RR is a green light, signaling a sending links station that it is free to resume sending I frames.

Supervisory frames are unlike unnumbered frames in that, although both are management frames, supervisory frames include Nr sequence numbers in their control fields. This means that supervisory frames can be used to acknowledge frames received. On the other hand, supervisory frames lack any information field; in contrast, some unnumbered frames such as XID can carry state information. And unlike either unnumbered or information frames, supervisory frames are unacknowledged. Finally, note that supervisory frames may be either commands or responses—they may be sent by either a PLS or an SLS. Table 6.6 lists SLDC's three supervisory frames. The third supervisory frame,

Table 6.6
SDLC Frames: Workload Actuations

Receive ready	Command/response
Receive not ready	Command/response
Reject	Command/response

Reject (REJ), also affects the scheduling of the workload manager but because its use is confined to fault notification we postpone further discussion of it until the next section.

Complementing the role of the RR and RNR frames, the second type of flow control involves the use of acknowledgments to inhibit the sending of frames. By withholding acknowledgments, a receiving link station will eventually force its sending link station partner to stop sending. Each workload scheduler keeps track of available sequence numbers. When a frame is acknowledged its sequence number becomes available for reuse. On the other hand, if no frames are acknowledged then the available numbers will be gradually depleted and the link station will be forced to stop sending frames that require sequence numbers—that is, I frames.

There is a potential problem with flow control effected by withholding acknowledgments. Because each link station in a connection maintains a copy of the unacknowledged I frames it has sent, it may run into memory problems. But SDLC has a provision to enable a link station to clear its buffers: An RR sent by a link station that has been "flow controlled" will force its partner link station to respond with acknowledgment of the frames it has received. For a PLS that has tens or even hundreds of connections open, such freeing of buffers is of great importance. A link station that wants to keep inhibiting incoming frames can return this Nr acknowledgment via an RNR and thus maintain its "red light" to its partner sending more frames. Although the sending link station will still be blocked, it would be able to clear its buffers of frames awaiting acknowledgment.

Recall an earlier discussion in which we said a protocol that allows multiple outstanding frames is, in effect, pipelining the transporter it is managing. The *nominal* depth of the pipeline is determined by the number of sequence numbers—7 and 127 being the most common because these can be accommodated with 8- and 16-bit control fields, respectively. The size of the pool of available sequence numbers measures the occupancy of the transporter pipeline. At the extreme, when the pool is empty (i.e., the sliding window is closed) the pipeline is full. When an RNR is received its effect is to actuate the *effective* depth of the pipeline to its current level of occupancy. In other words, when an RNR is received the pipeline is automatically deemed full and will remain that way until an RR restores its nominal depth.

To summarize, the decision logic used by the workload manager in scheduling frames is as follows:

1. Has the destination link station sent an RNR? If yes, then stop; no I frames may be sent until an RR is received (if the destination station is an SLS the PLS may send an RR to elicit new state information and a possible RR). If no, then go to question 2.

2. Is there an available sequence number? If no, then stop; no I frames may be sent until an acknowledgment or rejection is received (if the destination station is an SLS the PLS may send an RR to elicit new state information and a possible acknowledgment/rejection). If yes, then use the next sequence number(s) to send the queued frame(s) up to the number of sequence numbers available (allowed by the sliding window).

Figure 6.11 shows a flowchart for this process.

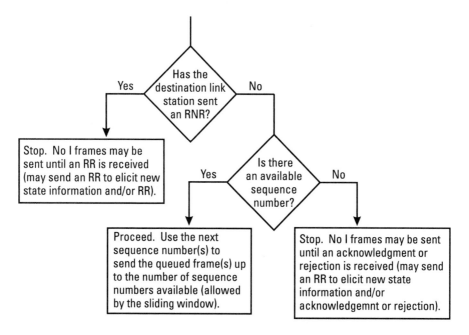

Figure 6.11 Flowchart of flow control decision logic.

We should touch on the relationship between the workload scheduler in the PLS that controls the allocation of the transporter (what we will call the outer scheduler) and the workload scheduler that controls the sending of individual frames (the inner or flow control scheduler). The two schedulers may be coupled or decoupled. If an SLS has sent feedback (either by an RNR or by withholding acknowledgments) to inhibit the PLS from sending data then the outer scheduler should not allocate the channel to itself since the PLS cannot send data to this SLS, although the PLS may send nonsequenced frames or it may send frames to another SLS. If the outer scheduler is aware of the state information used by the inner scheduler (i.e., is coupled) then the outer scheduler would not allocate the transporter the PLS. On the other hand, if the two schedulers are decoupled then the outer scheduler will not know of this state information and will make its scheduling decisions oblivious to the state information available to the inner scheduler.

6.2.7 Fault Detection and Recovery

As we discussed earlier, foremost among SDLC's design objectives is robustness. The protocol creates a virtual transporter (composed of a pair of SDLC link stations managing a transporter itself composed of the channel/signal plus physical layer managers) that appears to upper layer clients as a highly reliable server. To do this, SDLC uses several mechanisms for fault detection and recovery. We discuss these in this section. Most notably, as a positive acknowledgment with retransmission (PAR) protocol, an SDLC link station will resend a frame that has arrived at the destination link station having been corrupted by a channel fault.

By definition, retransmission is the responsibility of the workload manager in the sending link station. However, as with flow control, this is a cooperative process. The workload manager in the sending link station generally relies on the bandwidth monitoring server(s) in the receiving link station to detect faults and communicate this information (Figure 6.12). This is classic backward error control. The exception is timer-based retransmission—a so called "dead-man's switch," which is an instance of forward error control. We look at this more in the next section.

6.2.7.1 Types of Faults

When a stream of bits is passed to the receiving link station by its corresponding physical layer, the link station's bandwidth monitor must decide which faults, if any, occurred to prevent the arrival of a frame of uncorrupted data. The faults can be broadly divided into three categories. First, the link station estimates if a fault has occurred on the basis of the frame's conformance to

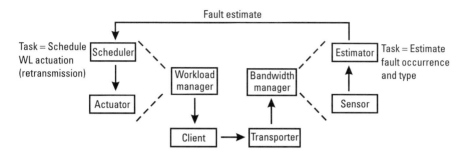

Figure 6.12 Retransmission as backward error control.

SDLC specifications: leading and trailing flags; address, control, and FCS fields; and so on. If, when the bit stream is scanned, no leading flag is found then there is no frame; likewise, if no trailing flag is found (possibly serving as the leading flag of the next frame) then there is no frame. We will call these types of faults, which do not involve frames, or at least not *complete* frames, recoverable protocol faults.

Next are those channel faults that result in an FCS discrepancy. These are the simplest to detect and to recover from. This type of fault, namely, when the frame that arrives over the channel has been corrupted by a fault in one or more of the physical layer components, is detected by comparing the pretransit FCS with the FCS calculated from the arrived frame. The bulk of SDLC faults are FCS faults, and the recovery involves retransmission.

The remaining faults require more complicated means of diagnosis (estimation) and recovery. These faults do not originate in the channel but rather in the sending link station and its misconfiguration vis-à-vis the receiving link station. The difficulty with detecting these faults is precisely that the FCS does not catch them. The FCS, as calculated at the sending link station, matches the FCS calculated on arrival, but the frame is still invalid. Several fault scenarios need to be considered. Either the frame does not conform to SDLC formats or the frame's arrival violates SDLC protocols. An example of the first type of fault is seen when a frame type indicates that it should be of some length (for example, an SNRM) but, as delimited by the flags, it is longer. Obviously, it is somehow malformed. The same is true if a frame arrives that is out of sequence (for example, the correct frame is a UA but an RR is sent instead). When an SDLC frame arrives with correct FCS fields but which is nonetheless invalid this is a nonrecoverable protocol fault.

Figure 6.13 shows the complete decision logic used by the bandwidth monitor to decide which faults, if any, have occurred. We now examine in more detail these faults and their respective management requirements.

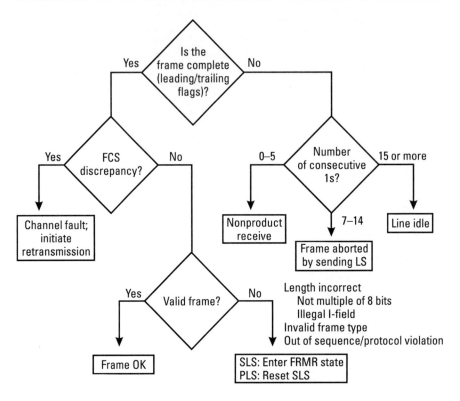

Figure 6.13 Flowchart of fault estimation decision logic.

6.2.7.2 Recoverable Protocol Faults

The first type of fault that the bandwidth estimator checks for is an incomplete frame. It does so by checking for the presence of a leading and a trailing flag. Depending on which (if either) flag is missing there are three possible fault scenarios: (1) The link station, having been allocated the channel, never even starts to transmit; or (2) it never stops; or (3) it stops before a complete frame has been sent. Included, therefore, within this category are those faults where a link station's hardware fails (always off or always on) as well as those where more than six consecutive ones are present in the bit stream.

Logically, if no flag is found in the arriving bit stream (meaning six consecutive ones are never detected) then there are two possibilities: More than six consecutive ones are found or fewer than six consecutive ones are found. As it happens, the precise number is important and means different things. If a stream of bits arrives and contains no block of more than 5 ones this is called a *nonproductive receive condition*. If a stream arrives that contains 7 to 14 ones then this is called a *frame abort condition.* If a stream arrives with 15 or more consecutive ones then this is called a *line idle condition.*

A nonproductive receive condition is detected at the destination link station when an SLS transmitter is "stuck" or is otherwise sending garbage. That is,

> When bits are being received that do not result in frames, a nonproductive receive condition exists. This condition could be caused by secondary station malfunctions that cause continuous transmission. The primary station must provide a time-out period when nonproductive receive occurs. The usual time period is in the range of 3 to 30 seconds. If the nonproductive receive condition continues after the time-out, the problem is normally not recoverable at the data link control level and must be handled by some method above the data link control level. [2]

The PLS will not poll the SLS during this time-out period; if at the end of this hiatus the SLS is still in nonproductive receive then the fault is deemed nonrecoverable and will require the intervention of exogenous management servers (such as a technician to replace the offending circuits).

Recall that a flag is precisely six consecutive ones (b001111110 = 0x7E) and that SDLC has a prohibition on more than five consecutive ones appearing elsewhere in a frame. In certain circumstances, however, SDLC allows the deliberate violation of this prohibition. When a sending link station decides to prematurely terminate a frame before it is complete (i.e., has the F/A/C[data]FCS/F fields), it may do so by sending a minimum of 7 consecutive zeros (but fewer than 15), in violation of the zero insertion. The receiving link station (PLS or SLS) contains abort detection logic that will allow a link station to detect if its partner link station has prematurely terminated transmission.

However, if the link station sends 15 or more consecutive ones then the data link is said to be idled. The idle timer will detect that an SLS has "gone away," that is, suffered a fatal fault or simply been turned off. With NRZI coding, a channel that carries no data will never have a waveform transition and hence its output signal will be interpreted at the destination as a string of ones. The danger, of course, is that without a waveform transition, the destination clock recovery mechanism will not be able to maintain synchronization.

6.2.7.3 FCS Faults

As we discussed in Chapter 3, error control coding (ECC) used at the physical layer is designed to correct the most common channel faults and to detect those faults that cannot be corrected. We also saw in Chapter 4 that many

DCE–DTE interface standards include a circuit/signal that allows the DCE to indicate to the DTE's physical layer that, for example, a decoder failure has occurred and that the data being sent over the receive channel is probably corrupted. The DTE's physical layer then may or may not use an internal mechanism to signal the data link layer of the occurrence of this fault. Such mechanisms are outside the scope of SDLC's protocol definition.

Instead, SDLC relies on the detection provided by the Frame Check Sequence/Cyclic Redundancy Check result appended to the trailer of every SDLC frame. Most channel faults that are not corrected by the physical layer's ECC mechanism result in a frame with an FCS that is different than that of the frame sent by the client; and, in fact, the FCS can provide an additional level of protection by detecting erroneous corrections performed by the ECC decoder. We discussed in the last chapter how producing such transporter fault estimates is accomplished by means of comparing the two FCSs; and that this bandwidth (server) estimation is the responsibility of the bandwidth monitoring mechanisms at the destination link station.

When the estimate is made that a newly arrived frame has been corrupted, the receiving/destination link station can send this information to the sending link station in one of two ways: explicitly, via a *Reject* (REJ) frame; or implicitly, by withholding an acknowledgment for the frame and relying on the timer-based fault estimator to decide a fault had occurred. *Reject* is the third supervisory frame used by SDLC and should not be confused with the *Frame Reject* frame used to signal nonrecoverable faults (see below).

Unfortunately, *Reject* provides a somewhat coarse recovery. If the middle (or first) frame of a sequence of frames is faulty then an REJ of that frame necessitates the retransmission of all subsequent frames, whether or not these arrived corrupted. Clearly this is very inefficient, but remedying this had to await HDLC, which introduced a fourth supervisory frame called *SREJ*, for *Selective REJect*.

Implicit fault notification relies on another component of an SDLC link station: a timer that is set when a frame is transmitted. The timer that triggers retransmissions in the absence of ack/rej is yet another fault detection mechanism. Note that such timers do not schedule the retransmission—that task is the province of the retransmission timer.

With both explicit and implicit fault notification, the key is the sliding window. Consider, for example, the SDLC workload scheduler shown in Figure 6.14. The workload scheduler keeps track of which frames are outstanding and which frame numbers are available for frames to be sent. If the recipient link station sends an acknowledgment with Nr = k (for k = 2, 3, 4, 5, or 6) then that will cause the workload scheduler to adjust the window and the sequence numbers up to $k-1$ modulo 8/128 will be available for reuse. If Nr = 6 then all

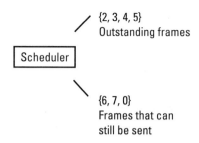

Figure 6.14 SDLC workload scheduler.

outstanding frames have been acknowledged and the recipient link station is indicating it is expecting the next frame to have Ns = 6.

6.2.7.4 Nonrecoverable Protocol Faults

Certain other protocol events were also perceived by the designers of SDLC to be so severe or otherwise indicative of such fundamental problems that they were deemed nonrecoverable. That is, these faults are considered nonrecoverable because retransmission will not effect recovery. In these instances the SDLC state machine enters the FRMR state and stays there until it is reset by the PLS with a mode actuation command. This is the fourth, nonmode state of an SDLC state machine. As with the other three states, this applies only to secondary link stations.

An SLS in the FRMR state will, at the next opportunity (i.e., when polled) send to the PLS a fault notification in the form of an FRMR frame; and until the link station has been reset every frame sent to the link station will be discarded and an FRMR frame sent back to the PLS. As with the other fault notifications, this is an instance of bandwidth management, specifically bandwidth monitoring.

With nonrecoverable faults the bandwidth monitor goes beyond simple fault detection and attempts to provide fault isolation. It does this by including in the information field of the FRMR frame state information that may aid the PLS and/or exogenous manager in diagnosing (isolating) and repairing the source of the fault. (Recall that we said earlier certain unnumbered frames carry information fields.) Figure 6.15 shows the contents of the information field. First is the Command field of the rejected frame. By comparing this with the Command field of the frame as sent (assuming a copy of the frame was kept) this provides a first level fault diagnosis mechanism: If the Command field sent is different than the Command field returned in the FRMR frame, then the most likely cause is a channel fault that escaped detection by the FCS (rare but not impossible).

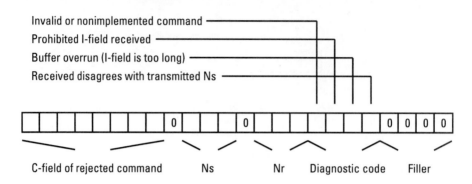

Figure 6.15 Information field of FRMR field (after Figure 3.2 in [2]).

Following this, the information field contains the SLS's current state with respect to its sent (Ns) and received (Nr) sequence numbers. This allows the PLS to compare with its Ns and Nr state and detect any major discrepancy. For example, if the SLS is expecting frame 5 (Nr = 5) and the PLS has sent through frame 0 (the PLS's Ns = 1) then four frames are missing. An SLS that is repeatedly out of step with the PLS is most likely indicative of a systemic fault in the transporter (channel) that needs repair.

Last, the SLS includes a diagnostic code in the FRMR indicating four possible causes of the received frame being rejected. Two of these diagnostic bits convey what is essentially state information: Either the SLS has suffered a buffer overrun, indicating that flow control has failed, or the two link stations have different Ns values. More interesting, however, are the last two bits; these indicate outright protocol violations, specifically an invalid frame type (unimplemented command) and a prohibited I field.

Recall that, although the 5 bits used to identify unnumbered frames allowed 32 possible unnumbered frames, basic SDLC (modulo 7) defined only 13; and of the four possible supervisory frames, SDLC only defined three. If a frame from the PLS arrives that contains a command field indicating its frame type is one of these unused combinations then the SLS sets the corresponding diagnostic bit, sends the FRMR frame, and enters the FRMR state. Likewise, as we indicated earlier, only the I frame and certain unnumbered frames (including, obviously, the FRMR frame) are allowed to include an information field. If the receiving SLS determines, by parsing the frame's flags and thus inferring (estimating) its length that a frame is longer than allowed then it concludes (estimates) that the frame includes an illegal I field. As with unsupported frame types, the SLS sets the corresponding diagnostic bit, sends the FRMR frame, and enters the FRMR state.

6.2.7.5 SDLC Recovery Parameters

The number of times retransmission will be attempted is one of several implementation-dependent parameters that characterize an SDLC link's behavior; some implementations are fixed while others allow considerable latitude in setting these parameters. An example of a more configurable implementation is provided by IBM's AS/400, its flagship minicomputer with more than 1,000,000 systems sold; the AS/400 allows users to specify many fault detection parameters, including those listed in Table 6.7.

6.2.8 SDLC Loop Topologies

For the sake of completeness, we should mention a seldom deployed topology that may be used with SDLC called a *loop* (Figure 6.16). An SDLC loop is constructed using one-way communications channels that connect each station to the next much in the fashion of Token-Ring LANs (see the next chapter). In an SDLC loop only one station transmits at a time; and the SLS transmit sequence is determined by their order in the loop relative to the PLS/loop controller.

SDLC has three frames that are used only with loop topologies. These frames are used in fault isolation and loop configuration. They are listed in Table 6.8.

Table 6.7
SDLC Recovery Parameters

Nonproductive receive timer	How long to wait before deciding a nonproductive receive condition exists (how many bits)
Idle timer	How long to wait before deciding a line is idle
Connect poll timer	How long to poll a link station
Poll cycle pause	How long to wait between poll cycles
Frame retry count	How many times the link station will resend
Fair polling timer	How to apportion bandwidth
DSR drop timer	Physical layer timers (see Chapter 4)
CTS timer	Physical layer timers (see Chapter 4)
Abort Detect	When to decide an SLS has aborted
Recovery limits—count limit, time interval	How many times to attempt link recovery and how often

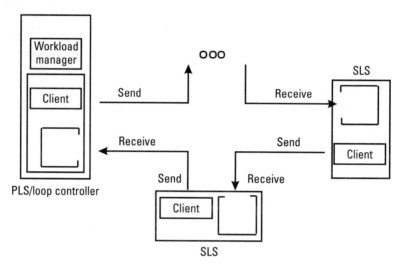

Figure 6.16 SDLC loop topology.

Table 6.8
SDLC Loop Frames

Unnumbered poll	Command/response
Beacon	Command/response
Configure	Command/response

6.2.9 Relationship of SDLC to Other Layers

To summarize what we have covered so far, Figure 6.17 shows the various SDLC frames and the components of the link station. At the interface to upper layer protocols, we see that the SDLC link station receives data (SDUs) to be encapsulated in SDLC frames; and that the contents of I and/or UI frames were passed upward. We also see that the various SDLC frames are passed to workload and bandwidth managers by a command processor that parses frames out of bits as they are passed upward from the physical layer DTE. Finally, there is an interlayer management component consisting of semaphores and such that signal between the two layers.

As we mentioned in Chapter 5, SDLC does not include any protocol type data between layer 2 and layer 3. Because the protocol is limited to SNA networks, this omission has not proven serious.

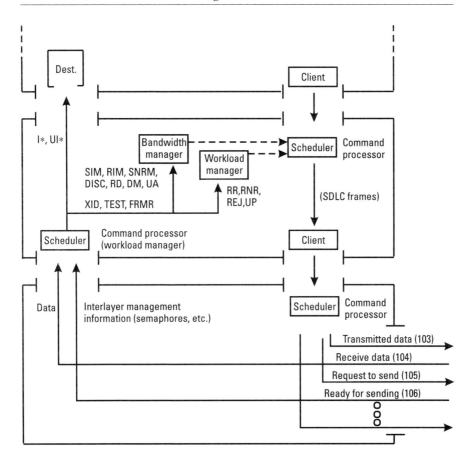

Figure 6.17 SDLC frames: destinations.

6.3 HDLC

6.3.1 SDLC and HDLC

After its introduction SDLC was criticized by many in the data communications community, principally over the PLS/SLS dichotomy and the associated master/slave control. While such mechanisms were accepted as inescapable for multipoint data links (until the advent of the first LANs—see Chapter 7), it was argued that with point-to-point links not only was peer control possible but also much more consonant with the high-availability/distributed management orientation of the newly emerging packet networks and internets. After all, a master controller implies a single point of failure, and this was contrary to the direction in which networking was evolving.

In part as a result of the criticism, the High-Level Data Link Control (HDLC) standard (actually, family of standards) came to include not just SDLC's NRM but two additional modes of operation, the Asynchronous Response Mode (ARM) and the Asynchronous Balanced Mode (ABM), each of which increased the autonomy of the secondary link station. In fact, with the ABM the distinction between primary and secondary link stations is erased completely and we speak instead of the combined link station. The ARM lies between the NRM with its strict master/slave management and the ABM with its complete peer symmetry. Thus with HDLC we have a spectrum of autonomy.

But nothing we have discussed about SDLC's frame structure, frames, or state machines needs to be discarded when we come to HDLC. For our purposes, we will consider HDLC to be a superset of SDLC that builds on SDLC's basic structure to include more "peer" connections.

6.3.2   The Locus of Control in HDLC: NRM, ARM, and ABM

So, if to its critics the principal deficiency of SDLC was its inability to support fully balanced, independent data transfer between equals, how did HDLC's designers remedy this? They created a symmetric state, the Asynchronous Balanced Mode, which eliminated the differentiation of link stations into primary and secondary; indeed, both ABM stations have addresses. Whereas with the NRM, the locus of control for most of the decision making resides in the primary link stations and the only control that is symmetric between primary and secondary link station is flow control and fault detection/recovery, with the ABM the link stations are completely symmetric in their abilities. And a link station in the ARM is halfway between the two: Fewer management tasks (scheduling and estimating) are centralized but the primary/secondary dichotomy nonetheless remains.

This increased autonomy, however, comes at a price, namely, support for multipoint topologies. The reason stems from the term *asynchronous* in the mode names, which is clearly in contrast to *normal*. *Asynchronous* in this instance has nothing to do with the physical layer coding or broadband ISDN/Asynchronous Transfer Mode but rather with the autonomy of each link station to send data without being polled or scheduled. (This is a further overloading of an already overloaded term.) Unlike a link station in the NRM state, a link station in the ARM or in ABM states can, once a connection is extant, send data without waiting for a poll. And, even if the underlying transporter/channel is FDX, with three or more link stations there is the inevitability of a collision between link stations that decide to start sending data at the same time. As we will see in Chapter 7, it was only with the advent of closed-loop management in the LAN protocols that this difficulty was to be overcome.

If that is the distinction between asynchronous and normal modes, what is the difference between response and balance modes? The answer again is in terms of the locus of control. A link station in a response mode, normal or asynchronous, lacks the schedulers and actuators to actuate its connection to another link station; in fact, a link station in NRM or ARM is necessarily a secondary link station, dependent on a primary link station for management of the connection. It is only with the one balanced mode, namely, the ABM, that either link station (recall we are limited to point-to-point links) has the control mechanisms to change its connection to the other link station. By devolving control to what had been the secondary link station until the two became equal in all their capabilities, the ABM designers created the combined link station.

6.3.3 HDLC Modes and Classes of Procedure

Technically speaking, the various modes (NRM, ARM, and ABM) pertain to the states of link stations—secondary link stations in the case of NRM or ARM and combined link stations in the case of ABM. (Recall we remarked earlier that the PLS in SDLC is stateless, and this is also true with the PLS in an ARM data link.) The actual protocols corresponding to the three modes are called *classes of procedure* (cf. elements of procedure, also known as frames). These classes of procedures are called the Unbalanced Normal Class (UNC), the Unbalanced Asynchronous Class (UAC), and the Balanced Asynchronous Class (BAC). These correspond, respectively, to the Normal Response Mode, the Asynchronous Response Mode, and the Asynchronous Balanced Mode. Obviously, the response modes (NRM and ARM) are part of unbalanced (UNC and UAC) classes of procedure.

As we indicated earlier, HDLC is not a single standard but rather a family of standards specified in a set of five documents. The first of these standards specifies HDLC's frame format; this is the familiar Flag-Address-Control Structure. The remaining standards define the various classes of procedure.

In the rest of this section we concentrate on the Unbalanced Asynchronous and Balanced Asynchronous classes since HDLC's Unbalanced Normal Class is identical to the SDLC, which we have just discussed at some length. However, we should note that, notwithstanding the differences in their respective modes, the three protocols have much in common. As we have already remarked, all three use the same frame structure and indeed most of the same frames (more on the frames later). In terms of our protocol taxonomy, all three protocols are connection oriented; all three support variable length frames; all employ flow control to prevent receiving link stations from being overwhelmed by sending link stations; and all attempt to detect and recover from faults. And the UNC protocol (SDLC) is not completely master/slave: with regard to flow control and fault management it relies on peer mechanisms, just as the UAC and BAC protocols.

So where do the asynchronous protocols differ from the normal protocols? The first and perhaps most conspicuous area is in the scheduling, that is, allocation, of the channel. With the unbalanced normal class of procedures (i.e., SDLC) the PLS controls that link a station or stations are allowed to use the channel at any given time; if two stations are using the channel at once then obviously one is the PLS itself. With the asynchronous classes of procedure (unbalanced and balanced), on the other hand, no link station is a master. In the case of the unbalanced asynchronous class of procedure this means the SLS can schedule itself just as readily as the PLS; in the case of the balanced asynchronous class of procedure, there is no distinction between the two link stations sharing the channel. This is what *asynchronous* in this context means: the link stations operate independently.

A consequence of such asynchronous control is the loss of multipoint topology support. As we have already noted, only SDLC/UNC supports effective multipoint operations. UAC does allow multipoint topologies but it is multipoint in name only. Just a single secondary link station can be in ARM at any given time, the remainder being in what is called Asynchronous Disconnected Mode (the asynchronous analog of Normal Disconnected Mode):

> Because of the asynchronous nature of secondary station transmissions when ARM is utilized in a multipoint environment, only one secondary station can be activated (on-line) at a time. Other secondary stations on the multipoint link must be kept in a quiescent disconnected mode (off-line) so as not to interfere with any transmission in progress. [3]

In other words, while the nominal task set under UAC management of a multipoint transporter with k secondary link stations includes multiple pairs of tasks of the form PLS $\rightarrow$ SLS$_i$/SLS$_i$ $\rightarrow$ PLS ($2k$ tasks in all), the effective task set can contain only one such pair at any given time. Support under the UAC protocol for multipoint topologies is, in essence, a fiction.

This brings us to another area of major difference among the three protocols: connection management. The UAC/multipoint fiction can be maintained precisely because, under the UAC protocol, the PLS still retains control over mode actuations, and hence the PLS can ensure that one and only one SLS is in ARM; no SLS can self-actuate a connection. In effect, when the PLS wants to send data to (or receive data from) an SLS it must actuate any existing connection down and bring up the connection to the target SLS; once the connection is up the SLS can send data without its use of the transporter being scheduled by the PLS—that is, no polling is required. This, by the way, also illustrates why with BAC not even the fiction of multipoint communication can be maintained: Each (combined) link station can actuate modes as well as schedule its own use of the transporter, and there is no central manager to keep multiple stations from actuating connections and starting to send data.

Table 6.9 summarizes the differences among the three protocols in the principal areas of link management. Note that with regard to fault detection/recovery and flow control, all three classes of procedure use peer or symmetric management.

6.3.4 HDLC Elements of Procedure

Given the fact that the design of HDLC was driven by a variety of factors, a monolithic protocol was unlikely to satisfy the many competing claims for its favors, and this led to the decision to include three classes of procedure. To increase its flexibility further, the committee that designed HDLC opted to organize the various frames in terms of a base set of frames, to be recognized by all stations that implement a given class of procedure, and a set of optional frames that may or may not be implemented, depending on the requirements of the data link and its management. The resulting "base and towers" was highly modular and could accommodate many different types of data links.

Part of this modularity is that all three classes of procedure use not just the same frame structure but, for the most part, the same basic frames. There were exceptions, however, most notably the mode setting command frames. Obviously, it is problematic to use the SNRM command with either UAC or BAC protocols. To remedy this, HDLC introduces two new mode setting frames called *Set Asynchronous Response Mode* (SARM) and *Set Asynchronous Balanced Mode* (SABM), which are analogs of the SNRM command frame.

The base sets of frames are largely the same for all three classes of procedure and consist of commands and responses, although in the case of the BAC either combined link station (CLS) could issue either a command or a response.

Table 6.9
Management Tasks: Locus of Control

Type of Management	UNC	UAC	BAC
Mode actuation	Master/Slave	Master/Slave	Peer
Transporter allocation	Master/Slave	Peer	Peer
Fault detection/recovery	Peer	Peer	Peer
Flow control	Peer	Peer	Peer

Another complication that comes with BAC is that, unlike UNC or UAC where the PLS has no address, each of the combined link stations does have an address. And whereas with the unbalanced protocols the PLS sends command frames with the address of the destination SLS and an SLS sends response frames with its own address in the address field, with BAC a combined link station sends a command frame with the destination CLS's address and a response frame with its own address.

So what is in the base set? For all three protocols it consists of command and response I frames, as well as command and response workload management frames—RR and RNR—and the respective mode actuation commands corresponding to each class of procedure—SNRM for UNC, SARM for UAC, and SABM for BAC, as well as the DISC command frame and the UA, DM, and FRMR response frames. These base sets were deemed sufficient to provide functional data link management, albeit with certain omissions.

Implementers who are not happy with the base functionality can augment this with additional frames defined in various option sets. Complementing the 3 base sets for the three classes of operation, there are no fewer than 14 option sets or towers, yielding thousands of possible protocols within the overall HDLC framework. Some of these option sets add individual frames while others modify some aspect of all frames such as the addresses or FCS fields.

For example, option set 1 adds the XID command and response frames. Option set 2 adds the supervisory REJ command and response frames, while option set 3 provides the more granular supervisory SREJ command and response frames. Option sets 4 and 6 define the unnumbered information and unnumbered poll frames, which we touched on briefly with regard to SDLC loops. As we will see in the next section, the UI frame has taken on increased importance in conjunction with Internet WAN protocols such as PPP. Option set 5 provides the RIM response and SIM command frames. Option sets 11, 12, and 13 round out HDLC with the RD and TEST frames we discussed earlier in SDLC, as well as a new frame called RSET that is used to reset the Ns counter at the sending link station and the Nr at the receiving link station; this provides a more granular reset capability than that of the SIM or other mode actuation commands.

Three option sets affect all frames. Option set 10 extends the sequence numbers from 3 bits (modulo 8) to 7 bits (modulo 128). In addition, option 10 also introduces new extended mode actuation command frames: SNRME, SARME, and SABME, which replace their corresponding nonextended cousins. Option set 14 replaces the 16-bit FCS polynomial with a 32-bit polynomial, providing greater fault detection for clients that require greater reliability. And option set 7 modifies the address field, which as we have seen is ordinarily 8 bits in length. With option set 7, however, the address field is arbitrarily extensible. Multiple octets may be employed to increase the number of available addresses, facilitating character-based addressing for multidrop data links.

Last, two of the option sets do not add frames but in fact remove them from the base set. Option set 8 calls for the removal of response I frames, and option set 9 removes command I frames. These option sets are used with the BAC protocols to provide an additional means of detecting and isolating faults. If a CLS that is operating under BAC with option set 9 (8) receives an I frame that does not carry the sending (receiving) station's address, that is, it is a command (response) I frame, then the receiving station knows that there is a fault in the data link, either due to a loopback or in the configuration of the partner CLS. Note that with either of the unbalanced protocols the elimination of command I frames would silence the PLS while eliminating response I frames would silence any SLS. And it is difficult to see using both options 8 and 9 with any protocol since this would eliminate sending any data.

A common abbreviation is to indicate the class and the option sets supported—for example, UNC,1,2,5,13 is standard SDLC. It includes XID, REJ, RIM, SIM, and RD, with 3-bit sequence numbers, 8-bit addresses, and 16-bit FCSs. BAC,2,8 is one of the most common implementations of the peer HDLC protocol, which adds REJ frames but excludes response I frames; it is known as *Link Access Procedure Balanced* (more on this later).

Figure 6.18 depicts the base and towers for the various HDLC base and option sets.

6.3.5 HDLC Multiprotocol Support

With all the various option sets it may be surprising that something is still missing from HDLC, but none of these options remedies a key omission, namely, that the frame format specified in ISO 3309 does not include a protocol type field. A protocol type field indicates the nature of the client upper layer protocol being carried in the payload of the HDLC I frames. When multiplexing multiple upper layer clients that all use the same protocol, their respective packets can be disambiguated using information in the upper layer protocol headers (for example, the TCP port numbers). However, when the protocols are different this crucial information is missing: The payload contents that the data link passes upwards may be an SNA packet or an IP packet, and without a protocol type field in the frame itself there is no way to remove ambiguity about the packets.

This omission was not significant before the emergence of multiprotocol networks; and there was no need for another field in the frame header, with the overhead that this represents, if there was no need to indicate the type of protocol being carried. Just as SDLC's frame has no provision for a protocol type

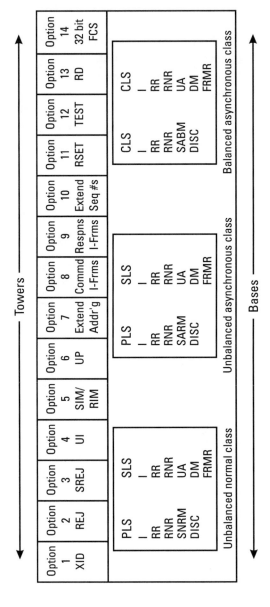

Figure 6.18 HDLC classes: bases and towers.

field because they carry only SNA packets, neither does HDLC, and for the same reason—private networks that used HDLC were monoprotocol. Even the deployment of public data networks based on X.25 did not change this, because one protocol still was the exclusive client of the data links, namely, X.25. And, notwithstanding the flexibility that comes from its base and towers modularity, HDLC is nonetheless a product of that era.

However, as multiprotocol routers started to be deployed in the mid-1980s, it became increasingly common for vendors to implement proprietary extensions of HDLC in which a protocol type field was inserted in the frame, most often but not always immediately following the control field. The most common practice was to adopt the 2 octet protocol type field that Xerox had defined for use with Ethernet, the first truly multiprotocol data link (see Chapter 8) (Figure 6.19).

This had the added benefit of being able to use the so-called DIX protocol type numbers (for Digital Equipment/Intel/Xerox, the consortium that popularized Ethernet) which had been already defined for Ethernet frames to carry common protocols including IP, IPX, and XNS. (We discuss network layer protocols in Part III.) The culmination of these modifications was an entirely new protocol, the Point-to-Point Protocol (PPP) which we will consider in the next chapter.

6.4 Summary

This chapter has applied the data link management taxonomy developed in Chapter 5 to the family of serial protocols that includes SDLC, HDLC, and derivatives such as LAPD, LAPF, and LAPM. In so doing we have seen that many tasks executed by these data link protocols are in fact management tasks.

And we saw that perhaps the most defining characteristic of SDLC management was the central role played by the primary link station. A PLS manages

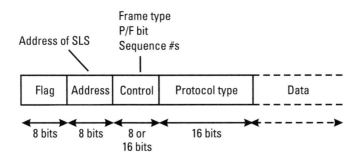

Figure 6.19 HDLC header with protocol type field.

the data link and its connections, and managing connections means managing the state of the SLSs. The PLS is also responsible for workload management, that is, scheduling the execution of the SDLC data link, which is effected by polling centralized/open-loop control. Polls can be issued by the PLS with an RR or they can be piggybacked onto another frame, typically an I frame carrying client data from the PLS to an SLS. Acknowledgments can also be piggybacked in the control fields of supervisory (RR, RNR, REJ) and I frames.

With HDLC, actually a superset protocol that includes SDLC as one mode of operation—the Normal Response Mode—we saw that control was more symmetric. By defining two additional modes of operation, namely, the Asynchronous Response Mode and especially the Asynchronous Balanced Mode, the designers of HDLC moved to introduce greater autonomy to the secondary link station. We saw that with HDLC/ABM we no longer speak of primary and secondary link stations but rather combined link stations. While ARM and ABM move the link management toward control that is distributed, it is still open loop. The price of this autonomy was the loss of multipoint topology support; only NRM (i.e., SDLC) supports multipoint data links. Distributed control of multipoint topologies is possible only with closed-loop control, which only comes with the LAN protocols (see Chapter 8).

References

[1]   Cypser, R. J., *Communications Architecture for Distributed Systems*, Reading, MA: Addison-Wesley, 1978, p. 373.

[2]   *IBM SDLC General Information*, GA27-3093-2.

[3]   Carlson, David E., Bit-Oriented Data Link Control, IEEE Transactions on Communications, April 1980, p. 112.

7

Data Link Management: The Point-to-Point Protocol

7.1 Introduction

In this chapter we complete our survey of serial data link management with the Point-to-Point Protocol (PPP). Besides being the WAN protocol for the Internet, PPP also represents some of the latest thinking in data link protocol design and implementation. Highly modularized, PPP consists of a base protocol and a set of ancillary protocols providing a framework for extensibility that has allowed for the definition of new protocols and features not included in HDLC, let alone SDLC, including authentication, link quality monitoring, encryption, and compression.

Before discussing these, however, we first review the requirements that drove the development of what was then called the Internet Standard Point-to-Point Protocol, starting with support for multiple upper layer protocols. As well as discussing what PPP was intended to accomplish, we discuss why earlier protocols, mainly proprietary variants of HDLC, were rejected. We also discuss the management philosophy pursued by the PPP architects. As we will see, this has a decidedly "Internet" orientation, notwithstanding the fact that it, like SDLC and HDLC, is connection oriented.

But where those protocols realized management with dozens of specialized frames, PPP itself is a "light" protocol that has a single frame type. Building on its modularized management approach, most of the services provided by those SDLC and HDLC management frames are instead realized by PPP's

Link Control Protocol (LCP). As we will see, at its most basic the LCP provides configuration management services for PPP link stations, much like an elaborate form of XID exchanges. However, by design a PPP's LCP is much more powerful and able to accommodate new options and features than any XID exchange.

Next we spend time discussing authentication and link quality monitoring. These are major additions, innovations absent from earlier protocols. We then look at the PPP network control protocols. These enable the negotiation of configuration parameters for various network protocols, another feature absent from HDLC or SDLC. Each network protocol has its own associated NCP, and this chapter examines the NCPs for IP, IPX, and layer 2 bridging. Likewise PPP defines control protocols for link stations to negotiate compression and encryption, and we discuss these as well.

We conclude by examining the PPP Multilink Protocol (MP) for bonding multiple data links into a higher bandwidth composite. PPP MP, along with the Bandwidth Allocation Protocol (BAP) and the Bandwidth Allocation Control Protocol (BACP), offer the greatest bandwidth flexibility among serial protocols.

7.2   Requirements for the Internet Standard Point-to-Point Protocol

As we mentioned in our discussion of HDLC and multiprotocol support, when multiprotocol networking took off in the 1980s, various internetworking vendors began shipping proprietary versions of HDLC to interconnect their routers; these vendors were driven initially by the need to support protocol type fields but soon various other modifications were incorporated from connectionless service to security and so on. Not surprisingly, one casualty of such proprietary proliferation was interoperability: It was impossible to attach a router from vendor A to a router from vendor B over a WAN link and hope for successful exchange of data.

In response to the growing complications this posed for interoperation, the Internet Engineering Task Force (IETF) started working on what was then called Internet Standard Point-to-Point Protocol (ISPPP), subsequently shortened to the Point-to-Point Protocol. The design rationale for PPP was laid out in a working paper subsequently published as RFC-1547, "Requirements for an Internet Standard Point-to-Point Protocol" [1]. The committee that worked on these requirements laid out some basic design predicates, the most important of which were the following:

1. No multipoint topology support;
2. Simplicity and efficiency in the protocol;
3. Highly reliable channels—no error recovery necessary but error detection required;
4. Peer management—no master/slave mechanisms;
5. No flow control—a receiving link station will discard packets it cannot handle;
6. Multiple upper layer protocols;
7. Extensibility; and
8. Negotiation between link stations of configuration parameters.

In addition, the decision was made after drawing up the working papers to exclude simplex and half-duplex data links. An immediate consequence of these requirements and objectives was that the ISPPP would not execute any workload management. Without fault recovery there was no need to retransmit frames; without flow control there was no need to actuate the rate at which a link station sent data; and without simplex or half-duplex transporters, or multipoint topologies, there was no need to schedule mechanisms such as polling to allocate the channel to contending clients. That is why everything that we will consider about PPP involves bandwidth (i.e., server) management.

7.2.1 Why Existing Protocols Were Rejected

RFC-1547 also documents why the existing serial protocols failed to meet these requirements. The prevailing Internet serial protocol, called the Serial Line IP (SLIP) protocol [2] and which was distributed in the BSD 4.3 Unix, was rejected because it lacked support for multiprotocol operation, was not extensible, and had no error detection. The HDLC data link standards (ISO 6256 and so on) were also rejected because they lacked support for multiprotocol operation, as well as for containing too much in terms of error correction, flow control, and so on.

RFC-1547 also catalogs the proprietary serial protocols from Cisco, Wellfleet, Proteon, and others, and why these were rejected. These were typically proprietary versions of HDLC, connectionless, and used only UI frames to carry data between routers. However, each differed from the others and all were missing one or more aspects of the desired solution, such as error detection and extensibility. Not to be discounted, either, was the political difficulty of endorsing one vendor's solution over another.

7.3   PPP Base and Towers

So, after laying out the requirements and rejecting existing serial protocols, what did the IETF committee design? To meet the objectives of simplicity and efficiency, on the one hand, and extensibility, on the other, the committee opted for a stripped down base protocol and numerous option sets (the familiar "base and towers" approach) realized as additional protocols. And like HDLC, PPP is actually defined by a family of standards, which, as Internet specifications are known, are called Requests for Comment (RFCs). The core PPP protocol was originally defined in 1989 with RFC-1134 and was most recently revised in 1994 as RFC-1661 [3].

As we said, this modular approach is similar to HDLC and its option sets. But whereas HDLC option sets generally added frames or modified frame parameters such as the size of the sequence numbers or addresses, with PPP the option sets define additional mechanisms, called *control protocols*, rather than additional frames. Unlike SDLC or HDLC, PPP does not use a multitude of frame types. Instead, the only PPP frame is the UI (unnumbered information) augmented with a protocol type field that may be 8 or 16 bits in length, although typically is the latter (Figure 7.1). RFC-1661 requires that all protocol numbers be odd and further restrictions that we discuss later.

To quote from the PPP standard (RFC-1661),

PPP is comprised of three main components:

1. A method of encapsulating multi-protocol datagrams;
2. A Link Control Protocol (LCP) for establishing, configuring, and testing the data-link connection;
3. A family of Network Control Protocols (NCPs) for establishing and configuring different network-layer protocols. [4]

The first of these components, then, is the UI frame with a protocol field. The second and third, namely, the link and network control protocols, handle the management tasks served by the frames left out of PPP. The former define the data link protocol parameters that are optional and subject to negotiation between the two link stations. Likewise, the latter are used to negotiate protocol-specific options corresponding to the particular upper layer protocol

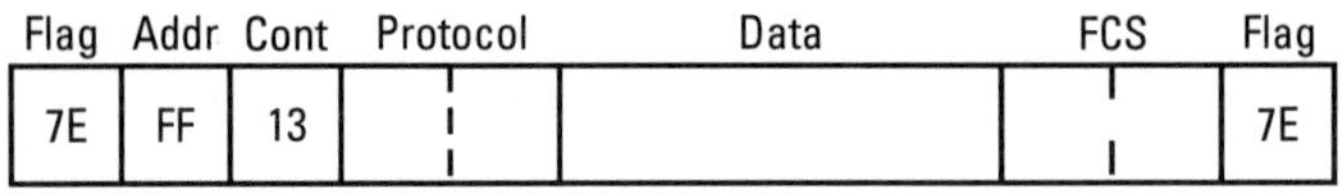

Flag	Addr	Cont	Protocol	Data	FCS	Flag
7E	FF	13				7E

Figure 7.1 PPP frame format.

in question (for example, IP Control Protocol, AppleTalk Control Protocol). The control protocols are intended to make PPP as self-configuring as possible, reducing or even eliminating the need for management intervention.

Although PPP incorporates many features from HDLC, there are considerable differences. First, while the PPP frame structure mimics the ISO 3309 standard, the inclusion of the protocol type field makes the frames fundamentally different. Secondly, PPP uses a completely different approach to management. Whereas HDLC relies on a proliferation of management frames such as SNRM(E)/SARM(E)/SABM(E), DISC, DM, and XID, PPP delegates this responsibility for bringing the link up and down, negotiating parameters, and so on, to the various control protocols.

These control protocols, LCP and NCPs, are to PPP merely other payload protocols to be encapsulated in the UI frames (Figure 7.2). Because the protocol field is 2 octets, more than 65,000 protocol types can be specified so there is little danger of running out of valid numbers. In addition to the link control protocol and network control protocols, these obviously include the upper layer protocols (ULPs).

7.3.1 PPP and the Data Link Taxonomy

Where does PPP fit in our data link taxonomy? PPP uses variable length frames. It does not employ flow control. A link station will send frames unless an upper layer protocol inhibits sending more data. In terms of fault management, PPP provides only fault detection but not fault recovery; as with flow control, any management action is left to an upper layer protocol. In fact, as we mentioned earlier, PPP contains no workload management component at all. And with regard to bandwidth management, meaning configuration management as well as link actuation up and down, PPP is a peer protocol in every

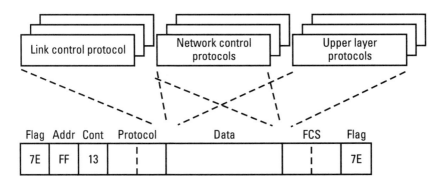

Figure 7.2 PPP payload protocols.

aspect. Table 7.1 lists the taxonomy criteria and compares PPP to SDLC and HDLC (BAC).

The classification of PPP with respect to connection orientation may seem somewhat problematic given that the exchange of PPP frames is not inhibited until a connection a la SDLC or HDLC is established, since PPP uses UI frames and only UI frames. On the other hand, a PPP link is not "up" until the LCP successfully executes its negotiation. And even then a PPP link station will not send the PDUs for a given protocol unless and until the corresponding control protocol has been successfully executed—meaning, the successful negotiation of configuration parameters. Finally, the PPP standard refers numerous times to "connections." Conclusion: PPP is connection oriented.

7.4 PPP and Multiprotocol Networking

To fully understand PPP we need to step back and look at multiprotocol networking for a moment. For purposes of illustration, consider a PPP link joining two systems (end or intermediate) that are each running three upper layer protocols—say, IP, IPX, and AppleTalk (AT). The PPP data link will be a component of three different protocol stacks (Figure 7.3). Each of these PPP links is a virtual transporter that is executed by the actual PPP link.

In particular, because of the actuating role played by PPP control protocols, we must nuance (again) our definition of a transporter's task set. The tasks in the task set must be qualified with respect to the ULPs that are carried. Thus the nominal task set of our example PPP data link would be:

$$\left\{ S_1^{\text{IP}} \to S_2^{\text{IP}}, S_2^{\text{IP}} \to S_1^{\text{IP}}, S_1^{\text{IPX}} \to S_2^{\text{IPX}}, S_2^{\text{IPX}} \to S_1^{\text{IPX}}, S_1^{\text{AT}} \to S_2^{\text{AT}}, S_2^{\text{AT}} \to S_1^{\text{AT}} \right\}$$

Table 7.1
Serial Data Links: Summary

Taxonomic Criterion	SDLC	HDLC	PPP
CO or CL	CO	CO	CO
Reliable or best effort	R	R	BE w/ FD
Flow control	Yes	Yes	No
Fixed or variable size frames	Variable	Variable	Variable
Master/slave or peer	M/S	Either	Peer

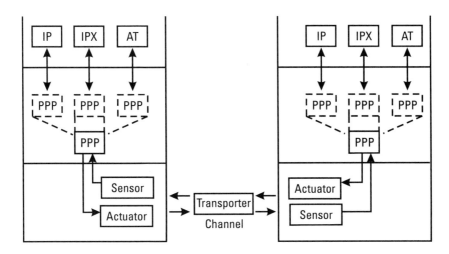

Figure 7.3 PPP link stations with ULPs.

where the link stations are S_1 and S_2. The effective task set, however, depends on the control protocols that have been executed. Prior to execution of the LCP, the effective task set is empty. Even after LCP execution the effective task set is empty with respect to the ULPs. It is only once an NCP executes that the corresponding tasks are in the effective task set.

Because successful execution of the link and network control protocols alters the effective task set of the transporter being managed by the data link protocol, it is clear that these are bandwidth management mechanisms. The control protocols as well as the ancillary protocols concerned with such areas as authorization, compression, encryption, and so on, are all focused on configuration management. (Recall from Chapter 1 that we saw configuration management as an instance of bandwidth management.)

7.5 Management in PPP

7.5.1 PPP Management Philosophy

The overall management philosophy for PPP was clearly laid out in RFC-1547 (the PPP requirements document):

> The internetwork layer (IP) is a fairly simple, almost stateless protocol providing an unreliable datagram service. The data link layer need provide no more capability than the IP protocol; no error correction, sequencing, or flow control is necessary. [5]

This perspective is reiterated with regard to flow control:

> Flow control (such as XON/XOFF) is not required. Any implementation of the ISPPP is expected to be capable of receiving packets at the full rate possible for the particular data link and physical layers used in the implementation. If higher layers cannot receive packets at the full rate possible, it is up to those layers to discard packets or invoke flow control procedures. As discussed above, end-to-end flow control is the responsibility of the transport layer. Including flow control within a point-to-point protocol often causes violation of the simplicity requirement. [5]

Such is also true of error correction:

> It is the consensus of the Internet community that error correction should always be implemented in the end-to-end transport, but that link error detection in the form of a checksum, Cyclic Redundancy Check (CRC) or other frame check mechanism is useful to prevent wasted bandwidth from the propagation of corrupted packets. Link level error correction is not required. [5]

As we said earlier, because PPP is limited to FDX channels and there is no error correction (typically entailing retransmission) or flow control, there is no occasion for any workload management. Everything that we will consider about PPP involves bandwidth (i.e., server) management, in particular the negotiation of connection parameters via the LCP and the monitoring of the link's condition via the Link Quality Monitoring (LQM) protocol (more on this later).

7.5.2 Configuration Negotiation

In addition to simply actuating the link's effective task set and enabling the transportation of various ULPs, most of the control protocols allow the PPP link stations to negotiate certain options. This is similar to HDLC negotiations using XID frames, except it takes place using the packets of the respective control protocols. The two PPP link stations on a data link each have certain properties/parameters that can only be exercised if both sides agree. As we will see when we examine some of the more important control protocols, the options vary considerably depending on the protocol being controlled.

In all cases, however, because PPP is a peer protocol, no fiat (master/slave) control can be exercised by either link station. Instead, every protocol and option must be negotiated between the two PPP link stations by means of the various control protocols. If agreement cannot be reached, then the protocol and/or option cannot be used. For this reason, almost every PPP control proto-col, link and network, follows the negotiation pattern shown in Figure 7.4. The four "commands" shown are all configuration control commands—bandwidth management in our scheme of things. The exceptions to this offer–counteroffer sequence are authentication protocols (discussed later).

The negotiation between PPP link stations to start the PPP connection occurs in three distinct phases, although the second phase is optional and may be omitted if the services provided (authentication and/or link quality assessment) are not required. The three phases are (1) link control protocol, (2) authentica-tion and link quality management, and (3) network control protocol(s).

The LCP negotiation must successfully complete before the link stations can negotiate authentication and LQM; and if authentication and/or link qual-ity management protocols are employed, these negotiations must complete suc-cessfully before the negotiations of the network control protocol(s), mandatory for each upper layer protocol the link is to transport, can begin. Note that in all instances the configuration control commands are not frames. Rather they are options exchanged in protocol-specific packets carried as payload in the infor-mation fields of the UI frames PPP uses for encapsulation.

7.6   The Link Control Protocol

As we indicated earlier, many of the management tasks that HDLC accom-plished with specialized control and unnumbered frames have been abstracted in PPP into optional link control protocols. All link level control protocols such as the LCP have protocol numbers that are in the range from 0xC*** to 0xF***. LCP's protocol number is 0xC021.

LCP packets have an open-ended structure. The first three fields are fixed. These are an 8-bit *Code* field, an 8-bit *Identifier* field, and a 16-bit *Length* field. The *Code* field contains the packet type code. The *Identifier* field contains a sequence number used by the two link stations to correlate requests and responses. The purpose of the *Length* field is to allow the PPP link station to

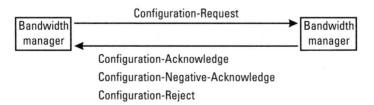

Figure 7.4 PPP link station option negotiation.

parse the packet. The last field consists of zero or more options, each of which is specified with a uniform format consisting of an 8-bit *Type* field, an 8-bit *Length* field, and a variable length data field consisting of zero or more octets. Figure 7.5 shows this encapsulation hierarchy. The kind of LCP packet is determined by the code in the packet header. Table 7.2 lists the set of LCP codes.

The PPP standard organizes the principal codes (and the corresponding LCP packets) into three classes [4]:

1. Link configuration: Configure-Request, Configure-Acknowledge, Configure-Negative-Acknowledge, and Configure-Reject;

2. Link termination: Terminate-Request and Terminate-Acknowledge; and

3. Link maintenance: Code-Reject, Protocol-Reject, Echo-Request, Echo-Reply, and Discard-Request.

The Link Configuration packets are those we saw in the negotiation exchange illustrated in Figure 7.4. These packets are, of course, exchanged in initializing the data link but may also be sent at any time during the link's operation to change the connection's parameters "on the fly." Let's look at these in more detail:

- Configure-Request (Code 0x01): Informs a PPP link station that its counterpart is willing to bring up a PPP connection with the

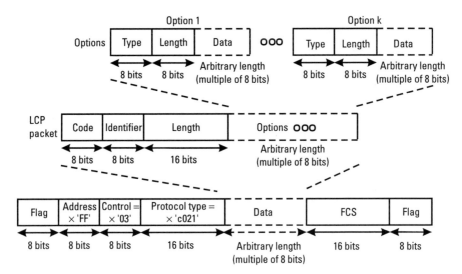

Figure 7.5 LCP packet structure.

Table 7.2

Link Control Protocol Codes

Configure-Request	01
Configure-Acknowledge	02
Configure-Negative-Acknowledge	03
Configure-Reject	04
Terminate-Request	05
Terminate-Acknowledge	06
Code-Reject	07
Protocol-Reject	08
Echo-Request	09
Echo-Reply	0A
Discard-Request	0B
Identification	0C
Time-Remaining	0D
Reset-Request	0E
Reset-Reply	0F

accompanying options in effect (we discuss the options in a moment). Each PPP link station sends a Configure-Request packet to its counterpart; in other words, there are *two* option negotiations that occur with PPP data links, one for each direction of traffic. The options agreed on for one direction do not have to match those for the opposite direction.

- Configure-Acknowledge (Code 0x02): Accepts all of the proposed options and their suggested values that were sent in a Configure-Request packet.

- Configure-Negative-Acknowledge (Code 0x03): Used by a PPP link station to indicate that some values of the options are unacceptable and to propose alternative values for these. If the alternatives proposed are acceptable to the receiving link station then it will send a new Configure-Request packet with the options modified as per the Configure-Negative-Acknowledge packet.

- Configure-Reject (Code 0x04): Used by a PPP link station to reject one or more options entirely, if the link station does not support that option irrespective of its parameter values.

Before discussing the remaining LCP packets, we should talk about the LCP various options that are the subject of these negotiations. As we indicated earlier, each control protocol is used to manage the parameters of the protocol—in the case of the LCP, these concern the link connection. The negotiation between PPP link stations is over which options will be used with the connection between them. When the Configure-Request packet is sent, it contains zero or more options that the PPP link station is willing to accept for the connection being actuated. Table 7.3 lists the most common negotiation options for the LCP.

Table 7.3
LCP Options

Vendor extensions	00
Maximum receive unit	01
Asynch control character map	02
Authentication protocol	03
Quality protocol	04
Magic number	05
Reserved (not used)	06
Protocol field compression	07
A and C field compression	08
FCS alternative	09
Self-describing pad (SDP)	0A
Numbered mode	0B
Multilink procedure	0C
Callback	0D
Connect time	0E
Compound frames	0F
Nominal data encapsulation	10
Multilink–MRRU	11
Multilink–short sequence number	12
Multilink–endpoint discriminator	13
Proprietary	14
DCE identifier	15
Multilink plus procedure	16
Link discriminator	17

According to a book on implementing PPP, "Options 01, 02, 03, 05, 07, and 08 are nearly universal. Options 11, 12, and 13 are common in MP [multilink PPP] implementations. The others are rarely used" [6]. The Maximum Receive Unit (option 01) is used by PPP link stations to negotiate the largest frame they will accept. The Magic Number (option 05) is a four octet number that should be chosen as randomly as possible, and which is used by each link station to disambiguate its LCP packets from those originating at its partner; the magic number is used to help detect accidental loopback conditions, for example.

The Protocol Field Compression (option 07) and A and C Field Compression (option 08) options are designed to reduce the overhead of the LCP packets by reducing the size of the respective fields. The former reduces the size of the protocol type field from two to one octet. The latter is based on the fact that in most cases the Address and Control fields are constant values (0xFF and 0x03, respectively). As for the MP and authentication options, we defer discussion of these until later.

A successful negotiation of LCP parameters effects not just configuration management but also actuates the status of the PPP connection. Once agreement is reached on the LCP options for each direction of the link, the PPP state machines in each link station transition from the initial state to the open state. Put another way, the Link Configuration packets, in addition to configuration management (the actuation of PPP parameters), also effect status actuation. Of course, this is only half of the status actuation process. The obvious question is this: How is a PPP connection torn down? Because PPP is a peer protocol, neither link station can unilaterally decide to tear down the connection. Instead, the teardown must be negotiated, and the LCP has two Link Termination packets which are exchanged to accomplish this:

- Terminate-Request (Code 0x05): When a link station wishes to terminate its connection, it sends to the other link station a Terminate-Request packet;
- Terminate-Acknowledge (Code 0x06): A link station that has received a Terminate-Request packet and agrees to the actuation will respond with a Terminate-Acknowledge packet. At this point the connection is down.

As with the other protocols we have discussed, various fault conditions can arise, and a data link protocol should have a mechanism for detecting and isolating. PPP includes several Link Maintenance packets that are used to isolate faults and to convey various fault estimates. These are as follows:

- Code-Reject (Code 0x07): When a link station receives an LCP packet with an unknown LCP code, it is possible to infer (estimate) both that a fault has occurred and the cause: The two link stations must have different versions of PPP. The receiving link station sends back a Code-Reject packet to inform the originating link station that a version problem exists. The original LCP packet is included in the Code-Reject packet in the data field.

- Protocol-Reject (Code 0x08): As with Code-Reject, when a link station receives a control protocol packet for a protocol that is unsupported it is possible to infer (estimate) both that a fault has occurred and the cause: The two link stations support different upper layer protocols. The receiving link station sends back a Protocol-Reject packet to inform the originating link station that a problem exists.

Figure 7.6 shows the fault estimation process within a receiving link station's bandwidth manager.

We note that the Code-Reject and Protocol-Reject packets serve the same purpose as HDLC's FRMR frame—to indicate that an unrecoverable configuration error has occurred and some intervention (= actuation) by a network administrator is required. With both Code-Rejection and Protocol-Rejection, quoting from the PPP RFC, "it is unlikely that the situation can be rectified automatically" [4].

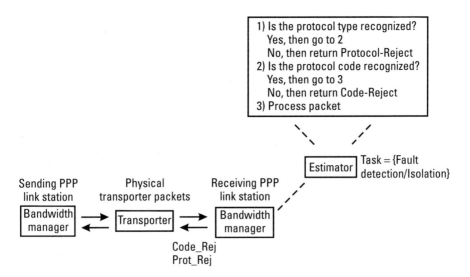

Figure 7.6 Fault estimation: Protocol-Reject and Code-Reject.

Next we come to loopback packets. We saw in Chapter 4 that loopbacks were a special type of bandwidth actuation of use in fault detection, wherein the task set is transformed from $\{S_1 \rightarrow S_2, S_2 \rightarrow S_1\}$ to $\{S_1 \rightarrow S_1, S_2 \rightarrow S_2\}$. To quote from RFC-1661, "LCP includes Echo-Request and Echo-Reply Codes in order to provide a Data Link Layer loopback mechanism for use in exercising both directions of the link. This is useful as an aid in debugging, link quality determination, performance testing, and for numerous other functions" [4].

A PPP link station that wishes to test the link can do so by sending an Echo-Request (Code 0x09) packet, containing a field for a magic number and an arbitrary data pattern. The receiving link station must reply by sending an Echo-Reply (Code 0x0A) packet containing the same data pattern. By comparing the two packets the sending link station can estimate the link quality; and by measuring the response time an estimate of link performance can be made. Either link station can initiate a loopback once the link is Open (up). Note that, unlike the loopback actuation at the physical layer, the loopback that results from the exchange of Echo-Request and Echo-Reply packets is just a temporary actuation, lasting only for the duration of the exchange. Some PPP implementations will periodically send Echo-Request packets on an idle link to check that it is still up.

Finally, the Discard-Request (Code 0x0B) packet is provided in the LCP to enable a link station to send a frame of data to its remote peer strictly for the purpose of testing its own transport mechanisms; on receipt of a Discard-Request packet, a link station simply discards the packet.

7.7 Authentication and Link Quality Monitoring Protocols

Although it may seem that the LCP covers every possible requirement, in fact many management tasks have been deliberately omitted. Instead, it has been left to other link protocols to provide these for PPP link stations that want them. These protocols fall into several distinct categories. In this section we look at two of these, namely, the protocols that provide some sort of authentication service, ensuring that only valid remote locations can connect, and a protocol that helps monitor link quality, providing measurements for use in management and planning.

7.7.1 Authentication Protocols

With PPP being deployed as the preferred Internet serial protocol for users seeking remote access to Internet Service Providers (ISPs) as well as to corporate intranets, there has been an overriding security interest in ensuring that only authorized users are allowed access. Many manufacturers of remote access concentrators have implemented security mechanisms that rely on proprietary authentication protocols; an example is the Shiva Password Authentication Protocol (SPAP), published as an information RFC to help others (browser vendors, emulator vendors, etc.) who might need to connect their products to Shiva systems.

As with PPP itself, the IETF has sought to discourage the proliferation of proprietary protocols by defining its own authentication protocols. The original RFC covering authentication was RFC-1333, which included two protocols: the Password Authentication Protocol (PAP, protocol code 0xC023) and the Challenge Handshake Authentication Protocol (CHAP, protocol code 0xC223). CHAP has been revised with RFC-1994, which omits PAP as too ineffective, but a new protocol, called the Extensible Authentication Protocol (EAP), is being finished that may replace them both. The particular protocol used on a link is determined during the LCP negotiation stage, where option 03 specifies the protocol type code of the authentication protocol a PPP link station wants to be used.

All authentication relies on establishing the legitimacy of an unknown party by means of some token, password, or other information that only authorized parties would know. PPP authentication protocols are executed after the LCP negotiation concludes successfully but before any NCP protocol negotiation. This limits the information an unauthorized user could retrieve from (or insert into) a secured system. Also, unlike the LCP and most other PPP control protocols, there is typically no "negotiation" involved in authentication: either the unknown PPP link station is authorized to make a connection or it is not. The authentication decision is a binary estimation.

Each authentication protocol defines a set of packets that are exchanged between two PPP link stations for the purpose of establishing this legitimacy. For example, PAP defines three packet types: Authenticate-Request (Code 01), which is sent by the link station seeking authentication and which contains the station's password; and Authenticate-Ack (Code 02) and Authenticate-Nak (Code 03), which are sent are by the authenticating link station to either confirm or deny permission to open a connection. Figure 7.7 shows the PAP authentication process.

CHAP, on the other hand, defines four packet types: Challenge (Code 01), Response (Code 02), Success (Code 03), and Failure (Code 04). With CHAP, the authenticating link station creates a random key (the *challenge*), which it sends to the authenticatee in a Challenge packet. Next, both link stations perform an identical look-up operation using a hash algorithm; the choice of which algorithm to use may be the subject of an LCP negotiation, but CHAP currently requires use of the MD5 algorithm [7]. The authenticatee

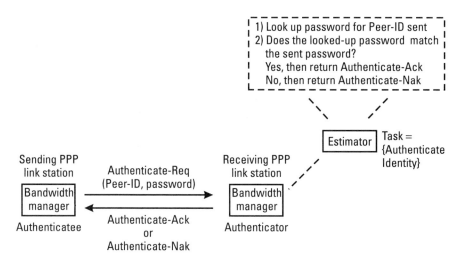

Figure 7.7 PAP authentication process.

returns this in a CHAP Response packet, and the authenticating estimator compares the two values. If there is a match then the authenticating link station responds with a CHAP Success packet; otherwise it sends a CHAP Failure packet, and tears down the connection. Figure 7.8 shows the CHAP authentication process.

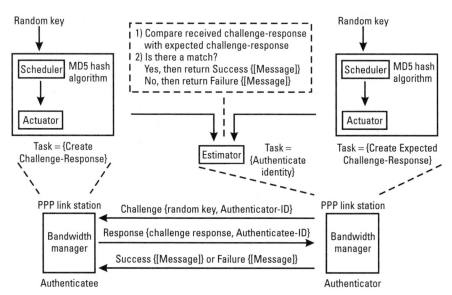

Figure 7.8 CHAP authentication process.

Note that an important distinction between PAP and CHAP is which side, authenticator or authenticatee, initiates the authentication process. With CHAP, the authenticating link station controls (schedules) the authenticating process by sending the challenge packet; any attacker will only get to make as many attempts as the authenticating link station allows. With PAP, on the other hand, the authenticatee takes the initiative when it sends the Authenticate-Request packet. This allows repeated attempts by a potential attacker. Other distinctions between the two protocols include the fact that PAP can only be run at the initiation of the PPP connection, whereas CHAP can be run throughout the connection's existence, enabling repeated challenges to reduce potential damage from an unauthorized user. For a more detailed discussion of PPP security, see [6].

7.7.2   Link Quality Monitoring

Recall we saw that the LCP Echo-Request and Echo-Reply packets can be used to test the condition and performance of a PPP link both during PPP initialization and throughout the existence of the connection. This information is useful, for example, to routers that may wish to select links that have greater reliability, as well as to network managers tracking link problems and initiating repairs. Beyond the use of Echo-Request and Echo-Reply packets, however, PPP makes provision for using a so-called "quality protocol" to further assess link reliability and performance. To quote from RFC-1661, "On some links it may be desirable to determine when, and how often, the link is dropping data. This process is called link quality monitoring" [4].

As with the use of an authentication protocol, the use of a quality protocol is the subject of LCP negotiations. Link quality monitoring is configured during the LCP negotiation using option 0x04, Quality Protocol. This option allows the link stations to specify which quality protocol they are willing to use and the reporting period [4]. Currently, the only quality protocol defined is the Link Quality Monitoring protocol (protocol code 0xC025), which is defined in RFC-1989 [8].

The LQM protocol defines only a single type of packet. This is called a Link Quality Report (LQR), and carries a number of measurements that characterize the link's performance. These include SNMP MIB-II interface counters that record the number of octets and packets transmitted and received by that link station as well as statistics on the number of discarded packets, errors, and so on. In addition, each LQR includes three LQM-defined counters that conform to MIB formats and which record the number of LQRs sent and received and the total number of good octets. By comparing subsequent LQRs, a link manager can infer if the link quality is decreasing, increasing, or staying constant.

7.8 Network Control Protocols

Link Control Protocol

Assuming the LCP has successfully opened the connection between PPP link stations and that the authentication and LQM protocols, if in use, have likewise been successfully configured, the next (and final) stage in actuating the PPP link's task set is the execution of the network control protocols (NCPs) corresponding to the upper layer protocols to be transported over the link. The NCPs are, to quote from the PPP standard (RFC-1661), responsible for "establishing and configuring different network layer protocols" [4].

This begs the question, however, of why a data link protocol should be concerned with the configuration management of upper layer protocols. After all, no other data link protocol we have considered has included such a mechanism, let alone one with provision for negotiating parameters. The answer can be traced in part back to a requirement in RFC-1547 (PPP Requirements) for Network Layer Address Negotiation:

> The ISPPP must allow network layer (such as IP) addresses to be negotiated…. Many network layer protocols and implementations are required to know the addresses at both ends of a point-to-point link before packets may be routed. These addresses may be statically configured, but it may sometimes be necessary or convenient for these addresses [to] be dynamically ascertained at connection establishment. [5]

Given this requirement, the designers of PPP could have opted for a monolithic protocol that encompassed the network protocols requiring such negotiation capabilities, but this would have violated the modular architecture they had embraced, for example, with the LCP. By including configuration management for upper layer protocols in separate network control protocols, they had a solution that was extensible. As new protocols (for example, IPv6) were defined, new NCPs were defined for their management. We should note that PPP allows for protocols that typically have low volume traffic running over PPP links without first executing an associated NCP; these protocols have distinct protocol type codes that inform the PPP protocol machine that no NCP is necessary.

Specifically, in PPP's numbering scheme, those protocols that have no associated NCP have protocol codes in the range from 0x4*** to 0x7*** whereas the protocol codes range for the "normal" upper layer protocols themselves range from 0x0*** to 0x3*** and the protocol codes for the network control protocols themselves range from 0x8*** to 0xB***. As should be obvious, an NCP type code is related to its protocol's type code by simply adding 0x8000. Table 7.4 lists some of the most common network protocols, their

Table 7.4
Protocol Numbers: Network and Control Protocols

Protocol	Protocol Type Code	NCP Type Code	NCP RFC(s)
IP	0021	8021	1332 (1172)
OSI	0023	8023	1377
XNS	0025	8025	1764
DECNET IV	0027	8027	1762
AppleTalk	0029	8029	1378
IPX	002B	802B	1552
Bridging	0031	8031	1638 (1220)
Banyan Vines	0035	8035	1763
NetBIOS	003F	803F	2097
SNA over LLC2	004B	804B	2043
SNA/APPN HPR	004D	804D	2043
IP version 6	0057	8057	2023

associated control protocols, and the NCP's RFCs (obsolete RFCs are noted in parentheses).

The majority of network control protocols are very simple, their RFCs constituting little more than a few pages. Because their basic purpose is to negotiate upper layer protocol[1] parameters, the NCPs resemble the LCP and its negotiating mechanisms. By and large, the NCPs use the same packet types as the LCP, notably the link configuration (Configure-Request, Configure-Acknowledge, Configure-Nak, and Configure-Reject) packets; link termination (Terminate-Request and Terminate-Acknowledge) packets; and, of the link maintenance packets, the Code-Reject packet. In the balance of this section, we briefly survey several of the more important NCPs and the parameters they negotiate.

7.8.1 IP Control Protocol

The first NCP we look at is the IP Control Protocol (IPCP), defined in RFC-1332 [9]. It may seem surprising that, notwithstanding the importance of

---

1. We refer to "upper layer protocols" rather than "network layer protocols" because some of the NCPs are for configuring nonlayer 3 protocols such as NetBIOS.

IP as a protocol, relatively few options are negotiable with the IPCP (see Table 7.5). On the other hand, as was mentioned earlier, because IP was designed to be a very simple protocol providing an unreliable service, it is not too unexpected that its options are limited. Indeed, it can be argued that one of the principal reasons for IP's success has been its relative simplicity compared to such protocols as SNA or OSI.

Note that, of the IP options listed in Table 7.5 only the first three are from RFC-1332 itself. The DNS and NBNS options are defined in RFC-1877, "Name Server Addresses with IPCP." The last option is defined in a draft Internet standard for mobile IP [10]. This incrementalism again highlights the extensibility of the PPP architecture.

Looking at the RFC-1332 options, the first and third concern the exchange of IP addresses between PPP link stations, and the second is used to select which if any IP compression protocol will be used. Option 01, IP-Addresses, involves sending both source and destination IP addresses (for a total of 8 octets; see Chapter 10); because of negotiation convergence problems, this option has been, in the parlance of RFCs, "deprecated" and replaced with option 03, IP-Address. With this replacement option an IPCP Configure-Request packet will carry a 4-octet *local* IP address that the sending system wishes to use on the PPP link. The receiving link station can accede to this suggested IP address with a Configure-Ack or it can reject it, say, because of an address conflict, by replying with a Configure-Nak.

The reason for IP compression (not to be confused with PPP compression, which we discuss in the next section) is that in many circumstances thesize of the IP and TCP headers can be reduced from 40 octets to 3 or fewer octets; the option does not compress the data contents of the TCP segments, and is only applicable to IP packets carrying TCP segments. With IPCP option 2, the

Table 7.5
IPCP Options

IP-Addresses	01
IP-Compression-Protocol	02
IP-Address	03
Primary-DNS-Address	81
Primary-NBNS-Address	82
Secondary-DNS-Address	83
Secondary-NBNS-Address	84
Mobile-IPv4	89

two PPP link stations can agree to remove certain fields that add nothing to the functioning of the protocols under most circumstances. This compression technique is called the Van Jacobson (or VJ) protocol, after the man who developed and documented it [11], and is the only compression protocol currently selectable using option 2. Note that when VJ compression is being used, IP traffic is sent using three different protocol numbers depending on whether it carries TCP compressed (protocol number 0x002D), TCP uncompressed (0x002F), or regular IP data that were not eligible for VJ compression (0x0021, the standard IP protocol number).

The DNS (Domain Name Server) and NBNS (NetBIOS Name Server) options are likewise used to exchange addresses over the PPP link. There is some controversy over whether another dynamic discovery protocol such as BOOTP or DHCP should be used instead. Readers are advised to consult [6]. Finally, the most recent IPCP option, 08 Mobile IPv4, is designed to allow a mobile IP host to request a tunnel to its home agent; the option includes the IP address of the home agent.

7.8.2 IPX Control Protocol

After IP, the most common internetworking[2] protocol is probably Novell's Internetwork Packet eXchange (IPX). The IPX Control Protocol (IPXCP) defines a total of six options, all of which are in common use except option 04) (see Table 7.6). Of the remainder, options 01 (IPX-Network-Number) and 02 (IPX-Node-Number) contain addresses for the PPP link and the link station sending the IPXCP Configure-Request packet, respectively. Option 03 invokes one of several IPX header compression protocols similar to the VJ protocol

Table 7.6
IPXCP Options

IPX-Network-Number	01
IPX-Node-Number	02
IPX-Compression-Protocol	03
IPX-Routing-Protocol	04
IPX-Router-Name	05
IPX-Configuration-Complete	06

---

2. That is, excluding SNA, which we treat as a terminal–host protocol.

from IPCP. Option 05 allows a link station to send the name(s) of any local IPX servers. The last option, 06 IPX-Configuration-Complete, is an expedient included to terminate the IPXCP negotiations should convergence between the link stations prove elusive.

7.9 Transforming Protocols: Compression and Encryption

From NCPs we move on to look at two control protocols for configuring ancillary communications tasks that have also been standardized under the PPP umbrella, namely, compression and encryption. These are often referred to as *transforming* protocols [6] since the actual client data are altered or transformed. In contrast, though other network layer protocols may support transformation options, their main task is to encapsulate data as is, that is, without alteration.

Conceptually, encryption and compression can both be treated in terms of the encoders and decoders. The respective encoders actuate the data, modifying it either to reduce its size while retaining the information or to obscure its content, while the decoders reconstruct the original data by reversing the effects of the transformations. This is exactly the model we considered in Chapter 3, with error control coding. The encoder in all instances consists, logically if not actually, of a scheduler that decides what actuation to schedule and an actuator that executes it; and the encoder is basically an estimator that receives the actuated data and, since it knows the algorithm used by the scheduler, can reconstruct the original (preactuated) data. Note that for the algorithms to work correctly, if both compression and encryption are employed, then compression must be executed first and then the compressed data can be encrypted (Figure 7.9).

With both compression and encryption, the process is the same. The ULP packet is compressed or encrypted and the protocol code corresponding to the transformation protocol being executed (i.e., compression or encryption) becomes the new (or outer) protocol type code, whereas the original protocol type code for the affected packets, now called the inner protocol number, is inside the new packet. If both compression and encryption are invoked, then the process is repeated twice, first with the ULP packet being compressed, and receiving the outer protocol number 0x00FD or 0x00FB, and then this packet is passed to the encryption encoder, where it is encrypted and receives yet another outer protocol number, this time 0x0053.

When a PPP link station using encryption and/or compression receives a frame, a frame "router" (actually, workload manager) within the link station directs the packet to one of several destinations depending on the (outermost)

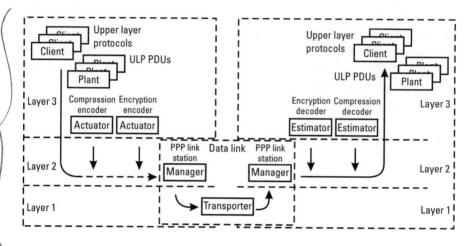

Figure 7.9 Encryption and compression in PPP.

protocol type code. Figure 7.10 illustrates this process limited to the encryption, compression, and ULP type codes. Other packets, for NCPs and LCP, would be similarly routed to the corresponding modules. If the (outermost) protocol type code indicates the packet has been encrypted or compressed it is sent to the appropriate module. After it has been "decoded" the frame is sent back to the frame router with its inner protocol type code now outermost. Note that if both encryption and compression were used then this process is repeated twice, once for each decoding, before the ULP packet is reconstructed and passed to the ULP state machine.

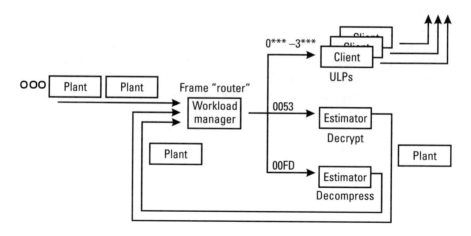

Figure 7.10 Processing received frames in a PPP link station.

Once the Compression Control Protocol (CCP) and/or the Encryption Control Protocol (ECP) have reached the opened state, the packets passed down from the ULPs are compressed and/or encrypted. The purpose of the ECP and CCP is to allow the two PPP link stations to negotiate which encryption and/or compression control protocols will be used on the link and with what parameters. Given that both encryption and compression can be expensive in terms of processor cycles and memory, costs that may exceed the capabilities of smaller or less powerful systems. With negotiation a less powerful link station can request a less onerous transformation algorithm or even refuse to use one at all. Of course, if it is encryption that is refused and security is essential, then the refused link station should tear down the LCP connection and terminate the PPP link.

Both ECP and CCP use the standard set of packets (the link configuration, link termination, and Code-Reject packets) augmented with two additional packets: Reset-Request and Reset-Acknowledge. The link stations conduct their configuration negotiation and link termination with the standard packets. The Reset-Request and Reset-Acknowledge packets are used to coordinate the encoders and decoders in ways particular to the respective protocols.

7.9.1 Compression and the Compression Control Protocol

Because it precedes the encryption process, we consider compression first. Many different algorithms are available for compressing data. The CCP, which uses protocol type code 0x8053, allows the PPP link stations to select which compression protocol will be used. This is done with the CCP option field.

The majority of these compression protocols use the Lempel-Ziv (LZ) algorithm [12] or some variant of it such as the Lempel-Ziv-Welch algorithm. LZ algorithms perform a real-time substitution of patterns in transmitted data with finite length code words, which are mapped back to the original patterns by the estimator at the destination.

Such compression is referred to as history based because it works by analyzing past traffic for patterns that are assumed to recur and using these patterns as shorthand to compress the data being transported. To some extent, history-based compression is like the story of the old comedians who have heard each other's material so often that they no longer repeat the jokes, but merely a shorthand numbering as in "How about number 42?" In the same way, the LZ encoder and decoder build up a set of shared references, called a *dictionary*, and use code word references to contents of the dictionary instead of the contents themselves. The encoder is responsible for adding words (patterns) to the dictionary.

Note that it is possible to use PPP compression in only one direction. That is, PPP compression can be used in one direction of a PPP link but not the other; for example, with asymmetric traffic flows, such as transaction processing with small messages going one way and large messages the opposite way, compression may be necessary in only one direction. To quote RFC-1962, "A different compression algorithm may be negotiated in each direction, for speed, cost, memory or other considerations, or only one direction may be compressed" [13]. Given that compression algorithms can require substantial CPU and memory resources, such a one-way compression may be advantageous.

As with the other control protocols, the CCP can be used to select and negotiate not just the compression protocol but also the protocol's parameters, to the extent these are negotiable. These parameters vary by the compression protocol used. For history-based protocols, one group of parameters relates to the dictionary. For example, among the parameters that describe the LZ implementation in V.42bis, negotiated in this case with LAPM XID frames, are the following:

- N1: Maximum size of code words;
- N2: Maximum size of the dictionary (in code words);
- C2: Current size of code words;
- C3: Threshold for changing size of code word;
- N4: Number of characters in the alphabet.

An excellent overview of the LZW algorithm as used in V.42bis compression can be found in [14].

Other parameters subject to negotiation include the granularity of the compression: per packet versus history based, and within the latter one or more than one history. For example, by maintaining separate histories for different clients of the data link (ULPs), greater compression can generally be achieved since the recurrence of patterns is likely to be higher the more coherence there is to the stream of data from which they are drawn. On the other hand, managing multiple histories imposes considerable additional overhead in terms of tagging compressed packets for one history or another, as well as managing multiple dictionaries, and so on.

This brings us to an important point. Because with history-based compression it is very important that the encoder and decoder maintain the same dictionary. A lost or corrupted frame can result in a loss of "synchronization" between the two sides of the PPP link and a subsequent propagation of errors similar to catastrophic error propagation in sequential decoding (Chapter 3). Compression, therefore, imposes a requirement for a more reliable transport service than PPP is designed to provide; some compression protocols require the usage of reliable PPP (RFC-1663), which essentially employs HDLC/ABM as an inner protocol within PPP to provide error recovery via retransmission, while others implement their own sequencing and retransmission within the compression protocol itself.

When synchronization is lost, a decoder may signal this by sending to the encoding link station a CCP Reset-Request; the latter responds by sending a Reset-Acknowledge packet, and restarts the dictionary construction from the beginning. To quote from the CCP RFC-1962, "Upon reception of a Reset-Request, the transmitting compressor [i.e., encoder] is reset to an initial state. This may include clearing a dictionary, resetting hash codes, or other mechanisms" [15]. Note that some compression protocols do not use these packets but rather some internal mechanism to indicate loss of synchronization with the encoder's dictionary.

7.9.2 Encryption and the Encryption Control Protocol

As indicated earlier, encryption and the ECP parallel in many ways compression and the CCP. ECP shares with CCP the same Reset-Request and Reset-Acknowledge packets. As with CCP, their role is to signal a loss of synchronization between decoder and encoder when history-based encryption is used. And, as with CCP, such protocols may either stipulate the use of reliable PPP or provide their own way of detecting and recovering from link faults that would otherwise disrupt the encoder/decoder synchrony.

We touched on some of the basic ideas behind encryption when we looked at LCP authentication protocols. There, however, it was not the actual data being encrypted. By encrypting the packets being transported, PPP encryption prevents so-called "man-in-the-middle" security breaches: A wiretapper or other collector of data exchanged would be unable to recover the original data without knowing the encryption mechanism. Designing good encryption is a very complicated task, and we refer the reader to [16] for more details. Most of the encryption mechanisms in use today are proprietary and not published, presumably for reasons of security. The only public encryption protocol that ECP can be used to select is the U.S. Bureau of Standards DES Encryption algorithm (ECP option 01). This is defined in RFC-1969.

7.10 PPP Multilink Protocol

All the data link protocols we have considered up to now have been designed to manage a single physical layer transporter. Two disadvantages of such single transporter implementations relate to their availability and bandwidth. If the transporter suffers a persistent fault, then its availability (and by implication its effective bandwidth) is reduced, even eliminated if the fault is fatal, until repair (bandwidth actuation) is effected. Likewise, if the performance of the transporter becomes unacceptable as the effective bandwidth of the physical layer transporter is exceeded by the workload (the traffic), then augmenting its capacity will typically involve replacing one or more elements—transmitter, channel, signal, and/or receiver, or all of these.

If a composite transporter is allowed then the situation changes. Instead of a single transporter, two or more transporters are bundled while preserving a single-system image to clients. Such a bundling is what we call a vertically, or single-stage, composite transporter, in distinction to a horizontally, or multi-stage, composite transporter, which is the result of concatenation of single-stage transporters. As we have indicated before, we will explore multistage transporters and the routing, switching, and tunneling mechanisms used to create them in the last section of this book, Part IV.

Figure 7.11 shows a vertically composite bundling of multiple transporters. There are several things to note. The first is the presence of the manager, a workload manager that is responsible for preserving the single-system image by mapping the RFSs to the virtual transporter to RFSs to one or more of the component transporters. This is workload actuation of kind, which we discussed in Chapter 1.

The second thing to note is that the task sets of the composite and the components are identical. This redundancy is the same principle behind massive parallelism in multiprocessor computers. The composite transporter has a nominal bandwidth that is an additive function of the nominal bandwidths of the component transporters, although due to management overhead it will be less than the sum of the components' bandwidth. Reliability is likewise additive. Two of the principal benefits of these multilink transporters are graceful growth and graceful degradation. If additional bandwidth is needed then a component transporter can be added to the bundle; and if a fault disables one

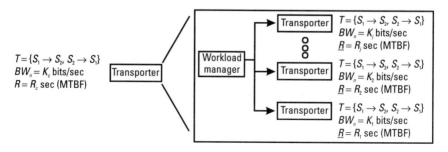

Figure 7.11 Multilink transporter.

link then the bundle still can transport data as long as at least one component transporter has not failed.

All of this brings us to the last PPP protocols we cover in this chapter, the Multilink Protocol and the Bandwidth Allocation Protocol and its associated Bandwidth Allocation Control Protocol. Together these three protocols allow PPP link stations to group multiple links and dynamically manage these. We note that the use of "bandwidth" here is consistent with our definition from Chapter 1: The Bandwidth Allocation Protocol is used to actuate the effective bandwidth (service rate) of the composite transporter; it incidentally actuates the composite's reliability as well, since this is also changed as the number of component links is increased or decreased.

7.10.1 Multilink Protocol

The PPP Multilink Protocol (MP) is defined in RFC-1990. To quote the RFC,

> The goal of multilink operation is to coordinate multiple independent links between a fixed pair of systems, providing a virtual link with greater bandwidth than any of the constituent members.... The bundled links can be different physical links, as in multiple async lines, but may also be instances of multiplexed link, such as ISDN, X.25, or Frame Relay. [17]

PPP MP is actually the second effort at defining multilink support for PPP. The first involved adapting the multilink mechanisms defined for LAPB in ISO 7776 [18]. This was eventually rejected for several reasons. First, ISO 7776 works only with LAPB, which as we saw in Chapter 6 is a "heavy" protocol that provides reliable transport with retransmission to prevent lost or missequenced packets. In contrast, while PPP MP allows the use of reliable PPP (RFC-1663) on the data links in a bundle, it is not required. The second reason that ISO 7776 was rejected is that it does not support fragmentation of ULP packets into multiple frames for concurrent transport by component data links. (We discuss fragmentation in detail later.)

7.10.1.1 MP Encapsulation

What the designers of MP eventually arrived at was a protocol that in many respects resembles the compression and encryption protocols. Whereas with standard PPP (non-MP) a ULP packet is encapsulated in a UI frame with the corresponding protocol code, with MP the ULP's protocol type code is moved inside the PPP frame and the outer protocol code is changed to 0x003D, which is MP's protocol code. This is just as we saw with compression and encryption. In addition, MP uses a special frame header that has fields for managing the segmentation and reassembly of ULP packets. This "inner" header immediately follows the standard PPP header and consists of either two or four octets, depending on whether 12- or 24-bit sequence numbers are used (this is one of the MP options subject to negotiation by the PPP link stations; see later discussion).

Figure 7.12 illustrates MP encapsulation and the two types of MP headers. We will discuss this header in more detail in the next section.

7.10.1.2 MP Fragmentation

One of the design objectives for MP was support of fragmentation. An obvious question is: "Why fragment with MP when standard PPP does not support fragmentation?" The answer has to do with exploiting fully the added bandwidth that comes with MP's multilink implementation. Assuming that ULPs prefer to send larger rather than smaller packets, MP fragmentation allows data links with MTUs less than the packet size to nonetheless carry the packets. To quote from RFC-1990, fragmentation "… offers the ability to split and recombine packets, thereby reducing latency, and potentially increase the effective maximum receive unit (MRU)" [19].

Fragmentation, however, opens the possibility of faults not possible with standard PPP, notably lost or out of sequence fragments. The original (prefragmented) ULP packet can only be reassembled if all the fragments arrive without being corrupted. In addition, there must be some way to reorder fragments that arrive out of order due, for example, to different bandwidths and latencies of the component data links. If a fragment is lost, however, then it may not

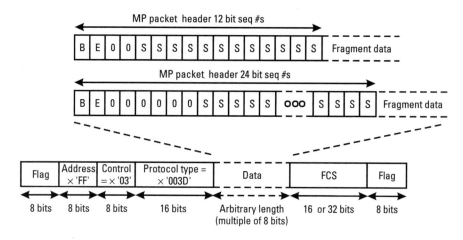

Figure 7.12 MP headers: 12- and 24-bit sequence numbers.

be possible for PPP/MP even to indicate to its ULP clients that a fault has occurred; the only fault detection mechanism in PPP is a frame's FCS and this will not detect a lost fragment.

To prevent this and detect lost or out of order fragments, MP augments PPP fault detection with two mechanisms in the MP header:

1. The Beginning-of-fragment (B) and End-of-fragment (E) bits: When an ULP packet is segmented into fragments, MP protocol machine sets the B bit in the MP header for the first fragment and the E bit in the MP header for the last segment.

2. The Sequence number field: MP uses sequence numbers for reordering fragmented packet segments, much like the sequence numbers we saw with SDLC and HDLC. By default these are 24 bits long, allowing over 16 million sequence numbers before the field wraps around. Some MP implementations support 12-bit sequence numbers, sacrificing some depth of the sequence number field in return for saving 12 bits per MP frame header.

Together the B/E bits and sequence numbers will correct out-of-sequence faults and detect lost fragment faults. The interested reader is advised to consult RFC-1990 or the general reference listed in [6] for more details.

7.10.1.3 LCP Negotiation of MP Parameters

Before two PPP link stations can open a multilink bundle, they must negotiate several options such as size of the sequence number field in the MP headers and, for that matter, multilink capability itself. This brings us to another reason why the MP is difficult to categorize in the PPP scheme, because it is neither an NCP nor does it have an NCP associated with it. At the same time, notwithstanding having a network protocol's protocol type code (0x003D), MP is not a true network protocol.

Whatever MP is, the configuration options for MP are negotiated with the LCP using options defined in RFC-1990 specifically to support bundling. These are:

- Option 11, Multilink Maximum Receive Reconstructed Unit (MRRU): This parameter serves two purposes. First, by using it the sending link station indicates it is capable of multilink operation. Second, it specifies the largest ULP packet that can be reassembled at the link station. This value is the equivalent to the MRU value in PPP, and, like it, is used to constrain the size of ULP packets.

- Option 12, Short Sequence Number Format: We have already discussed the fact that some MP implementations may choose to use short (12-bit) sequence numbers rather than the long (24-bit) sequence numbers. One downside to this is the potential for sequence number wraps on high-bandwidth MP transporters, with the potential ambiguity in reassembly this can cause.

- Option 13, Endpoint Discriminator: This is a unique system identifier that a PPP link station sends to its partner. The standard defines a number of candidate values, including IP address, Ethernet MAC address, magic number, or a locally assigned address.

Note that these options superseded LCP option 0C, "Multi-link Procedure," which was used to configure the ISO 7776 multilink capability.

7.10.1.4 MP, Compression, and Encryption

While basic MP encapsulation indicates the presence of the MP header by setting the protocol type code in the PPP header to the MP protocol type code, which is 0x003D, this may be complicated by the use of compression and/or encryption. Recall that in discussing compression and encryption we said that the choice of protocol type codes for the two transforming protocols depended on whether and how they were deployed in a PPP multilink. The difficulty is that while compression must precede encryption, there is no fixed order for encryption, compression, and PPP MP encapsulation. If compression and/or encryption precedes MP encapsulation it is said to be done at the bundle level. On the other hand, if either comes after MP encapsulation it is said to be done on the link level. PPP link stations need to know the sequence in order to properly parse received frames, hence the fact that both compression and encryption have two protocol type codes.

Taking these together, there are three valid encapsulation sequences:

1. A ULP packet may be compressed then encrypted then encapsulated in one (or more if fragmented) MP header(s) for transport. The outermost protocol code is MP's 0x003D, while the inner codes are 0x0053 and 0x00FD for the bundle-level encryption and compression, respectively, and finally the ULP's "native" protocol code.

2. A ULP packet may be compressed, then receive an MP encapsulation, with encryption done at the link-level. The sequence of protocol type codes is 0x0055 (link-level encryption)/0x003D (MP)/0x00FD (bundle-level compression)/"native" ULP.

3. Finally, if both compression and encryption are done at the link-level then the outermost protocol code is 0x0055 (link-level encryption)/ 0x00FB (link level compression)/0x003D (MP)/"native" ULP.

7.10.2 Bandwidth Allocation Protocol/Bandwidth Allocation Control Protocol

One criticism of RFC-1990 was that the standard does not adequately define how to manage the addition of a link to an MP bundle, nor how to handle removing one. For this reason, the IETF defined in RFC-2125 [20] a Bandwidth Allocation Protocol (BAP) and an associated Bandwidth Allocation control Protocol (BACP). With these the PPP link stations can dynamically manage the link bundle, facilitating the insertion and deletion of component links as conditions change. According to the RFC, the "BAP defines packets, parameters and negotiation procedures to allow two endpoints to negotiate gracefully adding and dropping links from a multilink bundle" [21]. For example, with BAP a standards-based mechanism exists for vendors to implement features such as bandwidth on demand (BOD) and dial on demand (DOD), which augment link bandwidth to accommodate surges in network traffic; or to implement transparent disaster recovery, allowing for example routers to use dial connections if frame relay circuits go down without disrupting any of the ULP clients of the PPP data link.

When two PPP link stations want to use the BAP to manage the links between them, they first exchange Link Discriminators, 16-bit numbers that uniquely identify the links in a multilink bundle, using LCP Option 0x17 (defined in RFC-2125). These Link Discriminators are necessary so that there is no ambiguity as to which links are the subject of a BAP request.

Next, the two PPP link stations must use the BACP to provide necessary configuration information. To date there is only one configuration option defined for BACP: option 01, Favored Peer. This is necessary to determine which PPP link station will win when both send the same BAP requests at the same time. BACP uses protocol type field 0xC02B.

Once the BACP has successfully reached an Opened state and negotiated the favored peer option, the BAP packets can be exchanged. The standard groups these into three categories:

1. Request permission to add a Link to a bundle (Call-Request);

2. Request that the peer add a link to a bundle via a callback (Callback-Request); and

3. Negotiate with the peer to drop a link from a bundle (this implies that the peer can refuse) (Link-Drop-Query-Request) [21].

The difference between a Call-Request and a Callback-Request is which link station is to originate the call:

- A Call-Request packet is sent if the implementation wishes to originate the call for the new link, and a Callback-Request packet is sent if the implementation wishes its peer to originate the call for the new link [21];
- The Link-Drop-Query-Request is provided to negotiate a graceful removal of a link. If a PPP link station *must* drop a link, it can circumvent negotiation by sending an LCP Terminate-Request packet.

Table 7.7 lists the currently defined packet types of BAP. Note that BAP packets use the protocol type code 0xC02D.

7.11 Summary

In this chapter we have covered the PPP protocol and its family of associated control protocols. First we reviewed the motivations and design rationale behind PPP and its modular architecture (courtesy of RFC-1547). We saw that PPP was itself designed to be a relatively simple protocol, with the only actual frame used being the *unnumbered information* (UI) frame. We placed PPP within the management continuum alongside SDLC and HDLC: Although connection oriented like these, PPP is a best effort protocol lacking

Table 7.7
BAP Packet Type Codes

Call-Request	01
Call-Response	02
Callback-Request	03
Callback-Response	04
Link-Drop-Query-Request	05
Link-Drop-Query-Response	06
Call-Status-Indication	07
Call-Status-Response	08

retransmission capabilities; likewise, flow control was omitted in order to keep the protocol processing to a minimum.

Instead, we saw that the complexity and sophistication of PPP's management of the data link is realized in the ancillary control protocols. The first is the Link Control Protocol, which we saw replaces the various management frames used for connection actuation and configuration management in SDLC and HDLC. Following the LCP came the Authentication and Link Quality Monitoring protocols, selection of which is also effected via LCP options. Next we discussed the role played by network control protocols, and illustrated it by examining the IP and IPX network control protocols.

After the NCP exchanges occur, there may be negotiated the so-called transformation protocols, including compression and encryption. We discussed these in both a single-link context and with regard to multilink operation, the last topic of this chapter. We saw that the PPP Multilink Protocol allows two or more individual data links to be aggregated and treated as one with respect to upper layer protocols; and that the Bandwidth Allocation Protocol allows the management of additions and deletions to such composites, realizing dynamic bandwidth to meet changing requirements for capacity and/or reliability.

References

[1] Perkins, D. (Ed.), "Requirements for an Internet Standard Point-to-Point Protocol," RFC-1547, originally prepared June 1989

[2] *A Nonstandard for the Transmission of IP Datagrams over Serial Lines: SLIP,* RFC-1055, June 1988.

[3] Simpson, W. (Ed.), *The Point-to-Point Protocol,* RFC-1661, July 1994.

[4] RFC-1661, pp. I, 26, 34, 36, 43.

[5] RFC-1547, pp. 6, 8, 11, 13.

[6] Carlson, J., *PPP Design and Debugging,* Reading, MA: Addison-Wesley, 1998.

[7] Rivest, R., *The MD5 Message-Digest Algorithm,* RFC-1321, 1992; see also Cheswick, W., and S. Bellovin, *Firewalls and Internet Security,* Reading, MA: Addison-Wesley, 1994, especially Chap. 13.

[8] Simpson, W. (Ed.), *PPP Link Quality Monitoring,* RFC-1989, 1996; obsoletes RFC-1333.

[9] McGregor, G. (Ed.), *The PPP Internet Protocol Control Protocol (IPCP),* RFC-1332, 1992; obsoletes RFC-1172.

[10] draft-ietf-pppext.ipcp.mip.02.txt

[11] RFC-1332, p. 9; see also V. Jacobson, *Compressing TCP/IP Headers,* RFC-1144, 1990.

[12] Lempel, A., and J. Ziv, "A Universal Algorithm for Sequential Data Compression," *IEEE Trans. Info. Theory,* Vol. IT-23, No. 3, May 1977.

[13] Rand, D., *The PPP Compression Control Protocol (CCP),* RFC-1962, 1996, p. 2.

[14] Black, U., *The V Series Recommendations,* New York: McGraw-Hill, 1991, pp. 117-128.

[15] RFC-1962, p. 4

[16] Cheswick and Bellovin, *Firewalls and Internet Security.*

[17] Sklower, K., et al., "The PPP Multilink Protocol (MP)," RFC-1990, 1996, p. 3; obsoletes RFC-1717.

[18] International Organisation for Standardization, *HDLC- Description of the X.25 LAPB-Compatible DTE Data Link Procedures,* International Standard 7776, 1988.

[19] RFC-1990, p. 3.

[20] Richards, C., et al., *The PPP Bandwidth Allocation Protocol (BAP)/The PPP Bandwidth Allocation Control Protocol (BACP),* RFC-2125, 1997.

[21] RFC-2125, p. 7.

8

Data Link Management: Local-Area Networks

8.1 Introduction

In this chapter we conclude our study of data link management by examining local-area networks (LANs) and their protocols. LANs are data links that rely on peer mechanisms to manage the execution of a shared channel by means of closed-loop workload scheduling. In contrast to PPP and HDLC/ABM, which are also peer protocols, LAN protocols are not limited to point-to-point topologies. Our focus here is on the principal protocols of the IEEE 802 family of LAN standards: the IEEE 802.3 Carrier Sense Multiple Access/Collision Detection and IEEE 802.5 Token Ring protocols, along with the IEEE 802.2 Logical Link Control protocols, LLC1 and LLC2. Note that we explored the physical layer signaling and interfaces for these LANs in Chapter 4; here our primary focus is on the workload and bandwidth management mechanisms above the physical layer.

The chapter begins with a landmark effort known as THE ALOHA SYSTEM, an early packet-radio network that pioneered distributed management. Next came the first commercial LAN, Ethernet, which was developed by Xerox's fabled Palo Alto Research Center and introduced carrier sensing as a means of closed-loop workload scheduling. From there we consider the work of the 802 committee, especially its division of the data link layer into two sublayers, the Medium Access Control (MAC) and the Logical Link Control (LLC). We explain what prompted this division as the IEEE 802 committee sought to satisfy very divergent demands from the TCP/IP and SNA communities and the ramifications of the MAC address space with its 48-bit, globally significant addresses.

We then turn to the IEEE 802 protocols themselves. First we explore the 802.3 protocol and how it uses carrier sensing with collision detection to realize random access to the channel. From there we move on to consider the other method used to implement distributed workload management, namely, sequential allocation via token passing, which is at the heart of the 802.5 Token Ring protocol. As we will see, the 802.5 architecture is enormously more complicated in its management than 802.3's, in large part because of the mechanisms included to manage the token—to handle, for example, faults such as lost or duplicate tokens.

Finally, we will briefly discuss the Logical Link Control protocols LLC1 and LLC2 and their relationships to each other and to HDLC/ABM. The two protocols are very different: LLC1 is a connectionless, best effort protocol that is almost devoid of management apart from support for upper layer protocol multiplexing; LLC2 is a connection-oriented, reliable protocol that is almost identical to LAPB and its extensive management services.

8.2   Peer Management, THE ALOHA SYSTEM, and Ethernet

Recall that with serial protocols the progression from the Normal Response Mode/Unbalanced Normal Class of Procedure (NRM/UNC) (i.e., SDLC) to the Asynchronous Response Mode/unbalanced Asynchronous Class of procedure (ARM/UAC) to the Asynchronous Balanced Mode/Balanced Asynchronous Class of Procedure (ABM/BAC) entailed devolving ever more control authority to secondary link stations. At each step we move closer to peer management and away from master/slave. The change from Normal to Asynchronous eliminates polling and allows a secondary link station (SLS) to send at will; and the change from Unbalanced to Balanced eliminates the distinction between link stations altogether and allows either side to manage the mode of the data link connection.

But these changes come at the expense of support for multipoint topologies: Only NRM/UNC supports more than two stations on a data link. This is because all of the data link protocols we have considered so far rely on open-loop scheduling of workload (traffic); in none of them does a workload scheduler monitor (measure and/or estimate) the current state of the channel. If any link station could send data at will (so-called random access) then the inevitable result would be continuous collisions. Only NRM/UNC's primary link station, with its ability to strictly allocate the channel to an SLS or to inhibit an SLS from transmitting, can prevent collisions. The PLS maintains an open-loop estimate of the state of the channel based on its previous scheduling decisions, and based on this, it determines if the channel is available.

Yet allowing even two stations autonomy to transmit at their own discretion (i.e., self-schedule) would seem to create an impossible situation since the open-loop estimate could not be maintained reliably—in effect, we would have a multiple writers problem. So how do ARM/UAC and ABM/BAC, which are also open loop yet allow autonomy, solve this problem? The answer is that most if not all implementations of these protocols require FDX channels, which with their concurrent multitasking are effectively two simplex channels; likewise, PPP is limited to FDX channels.[1] Consequently, these protocols are arguably less examples of peer management than of coexisting master/slave management mechanisms controlling decoupled components of a logically composite transporter.

Closed-loop scheduling and peer management were not realized earlier for several reasons. First, until the ARPANET was developed computer communications were limited to terminal–host networks, in which master/slave control was the rule. It was the management philosophy of the Internet, which can be expressed succinctly as "distrust authority," that elevated peer management to its current importance. Second, closed-loop scheduling requires a level of sophistication that could not be implemented economically before the advent of VLSI circuits.

In the rest of this chapter we look at the milestones in the development of peer management, starting with the ALOHA network; through to Ethernet, the first LAN with its use of closed-loop scheduling; and finally the 802 family of LAN protocols.

8.2.1 THE ALOHA SYSTEM

In 1968 the University of Hawaii was using an expensive network of leased telephone lines connecting a number of terminals on distant islands to a mainframe computer in Honolulu. As the costs of this network grew to consume a large part of their budget, a research program was started by a team led by Norman Abramson to replace these telephone lines with an experimental UHF packet-radio network. This network, called THE ALOHA SYSTEM, employed two different frequency bands to provide 24-Kbps communications links to and from the central site. A centralized management architecture using polling was rejected partly for cost reasons and complexity and also because

---

1. Although it is at least theoretically possible to realize ARM/UAC and ABM/BAC protocols over HDX channels by exploiting the physical layer circuits such as request to send to signal between DTEs, this is rare, to say the least.

greater efficiencies could be obtained by exploiting the bursty nature of terminal traffic. Indeed, by using what was termed *random access radio channel multiplexing,* the resulting network was able to support hundreds of terminal users simultaneously who shared the mainframe while enjoying better response times than before.

To do this Abramson and his team devised a simple way for each site to manage its own transmission (workload). A remote site was allowed to broadcast at any time (hence the "random access") on the return channel, sending data in packets up to 704 bits long, including a 32-bit CRC. If two or more sites transmitted concurrently then obviously the interference would corrupt their packets, which would be detected at the central site by comparing the CRC of an incoming packet to the CRC it carries. If no fault was detected then the central site would send to the remote site a positive acknowledgment. Conversely, if the remote site did not receive an acknowledgment within a finite interval of time it would infer that retransmission was necessary (Figure 8.1). In terms of our data link taxonomy, THE ALOHA SYSTEM used a connectionless protocol that relied on positive acknowledgments with retransmission to achieve reliable transport.

Recall that in Chapter 1 we distinguished between preventive and corrective maintenance. The crucial step that Abramson took with THE ALOHA SYSTEM was to focus not on preventing collisions by workload scheduling but rather on correcting the collision's consequences, and to do so by exploiting mechanisms already present, namely, CRC fault detection and retransmission. What Abramson, one of the pioneers of error control coding theory, did harkens back to the remark in Chapter 3 that it was wasteful to build too good a channel since it was more economical to use an error control code. Of course, a collision is not, strictly speaking, a fault: there is no deterioration of the server in question, namely, the physical layer transporter (channel, transmitter, and receiver); rather, RFSs from two or more clients have interfered with each

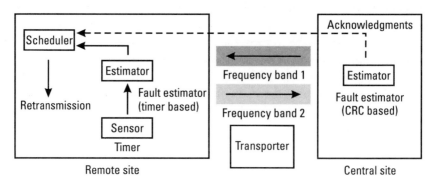

Figure 8.1 ALOHA retransmission management.

other, causing all to fail to be executed correctly (nominally). Nonetheless, it is *effectively* a fault in that the transporter cannot transport or execute any of the requested transport tasks. We will refer to these as *workload faults*.

We should note that a second-generation version, called *slotted ALOHA*, combined random access and centralized scheduling in the form of a master clock that fixed a periodic sequence of "slots"; a remote site that wanted to transmit waited until the beginning of a slot rather than transmitting at will. This resulted in packet collisions overlapping completely rather than partially and approximately doubled the attainable utilization of the shared radio channel [1]. Together, ALOHA and slotted ALOHA represent an absolutely pivotal landmark in the development of distributed protocols, most notably the crucial protocol called Ethernet.

8.2.2 Local-Area Networks: Ethernet

We now come to Ethernet, the first true LAN. Work on Ethernet was started at the Xerox Palo Alto Research Center (PARC) in 1972 [2], where a high-speed communications mechanism was sought to connect the workstations being developed. The original Ethernet used baseband signals over a coaxial cable (called the *bus*) to connect up to 256 stations at speeds up to 2.94 Mbps. In 1980 DEC, Intel, and Xerox collaborated to produce an updated specification, referred to as DIX, which increased the speed to 10 Mbps and increased the number of stations enormously by moving from 8- to 48-bit station addresses. In both cases the Ethernet LAN was single tasking. The channel executed only one transport task at a time; if two or more stations transmitted at once the result was a collision, that is, a workload fault.

From the central innovation of THE ALOHA SYSTEM—allowing random access and employing corrective maintenance to "repair" the consequences when this results in collisions—it was a short but critical step to Ethernet's principle of random access via carrier sensing with collision detection. (As we saw in Chapter 2, baseband signals do not use a modulated carrier; but notwithstanding this, the protocol became known as carrier sense multiple access/collision detection.) That is, Ethernet's architects took the random access of THE ALOHA SYSTEM and augmented it by having each station monitor the coaxial channel before transmitting. If the channel was idle the station was free to start transmitting; otherwise the station waited until the channel was free. In contrast, with THE ALOHA SYSTEM a remote site transmitted at will; there was no management (actuation) of its workload/traffic generation.

This is closed-loop scheduling; the actuation of the arrival of traffic (RFSs) to the transporter (the coaxial channel) is scheduled by each station based on feedback about the state of the channel. However, if a collision occurred because two or more stations attempted to transmit at the same time then the stations involved would detect the collision also by monitoring the channel, stop transmitting, and wait a random amount of time before attempting to retransmit. Monitoring the transporter (channel) to detect if it was already in use amounts to preventive maintenance. Although a station's collision detection mechanism(s) can determine if a workload fault has occurred, it is clearly preferable to avoid these (if possible without too much effort).

Figure 8.2 shows the formats of the experimental (original) Ethernet frame and the DIX Ethernet frame. Both start with a field called the *preamble*, originally 1 bit but revised in the DIX standard to be a 64-bit-long bit pattern, which ensured that receivers could accurately estimate the clock of the sender—to "train" the receiver's PLL. Next are the addresses for the destination and source stations, 8 bits in the original and 48 bits in DIX. After the addresses, in the original Ethernet frame came the data field, which could be up to 500 bytes long, and a 16-bit CRC. Note that, unlike the serial protocols we considered in Chapters 6 and 7, there is no trailing flag; the absence of a carrier (i.e., waveform) indicates the end of a frame. In addition, there is no control field since Ethernet used only a single frame type. Without connections to be managed, flow control, or other management tasks such as we saw with SDLC and HDLC, no management frames were needed. In terms of our data link taxonomy, both the original and DIX Ethernets constituted connectionless, best effort protocols with fault detection but with fault recovery limited to workload faults, that is, collisions.

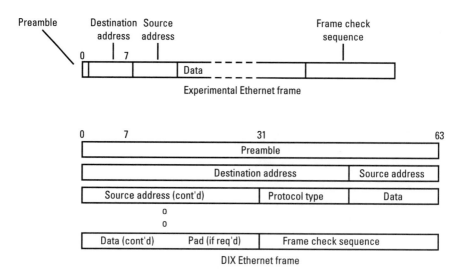

Figure 8.2 Experimental and DIX Ethernet frames.

The DIX frame differed in other ways from the original Ethernet frame. The maximum data field was increased to 1500 bytes. On the other hand, because small frames can cause difficulties with collision detection, the architects of DIX Ethernet established a minimum frame size. If the data field is too small then the sender inserts a pad field to extend the frame length. The CRC was correspondingly increased to 32 bits. The DIX frame, finally, has another field not present in the original, namely, a 2-byte protocol type code. We already discussed the introduction into Ethernet's L2 header of a protocol type code in Chapter 6 with respect to efforts to incorporate multiprotocol support into HDLC and later with PPP. As we said then, the importance of this field cannot be overstated since it is what allowed the development of multiprotocol data links—Ethernets, for example, that can carry IP, IPX, and AppleTalk.

8.2.2.1 LANs and Serial Protocols

As we have remarked several times, a local-area network is, in fact, not a network but rather a data link. And when the term *LAN* is used in distinction to *serial* data link protocols such as SDLC, HDLC, and PPP, the potential for confusion is compounded by the fact that most LAN protocols, including Ethernet, token ring, and FDDI, are serial, meaning 1 bit is transmitted at a time; the opposite of "serial" is not LAN but rather "parallel."

8.3 The IEEE 802 Standards

Shortly after the DIX standard was released, the Institute of Electrical and Electronics Engineers (IEEE) formed the 802 committee to standardize LAN protocols. Almost immediately, however, a conflict broke out between the Ethernet community, which wanted the 802 committee simply to ratify the DIX standard, and IBM, which insisted that LANs be reliable rather than best effort. This was because the design of IBM's Systems Network Architecture (SNA) was explicitly predicated on reliable data link protocols such as SDLC (see Chapter 11 for more details on SNA). The Ethernet community, coming largely from the TCP/IP world with its opposing management philosophy, was equally unwilling to bend. The result was that the two camps had staked out irreconcilable positions.

The only solution was a compromise that essentially allowed the two sides to claim victory. The 802 committee split the data link layer into two sublayers: the Medium Access Control (MAC) sublayer and the Logical Link Control (LLC) sublayer. Within the LLC two principal LLC protocols were defined, one called LLC1, which was best effort and connectionless like Ethernet, and one called LLC2, which was reliable and connection oriented like HDLC or LAPB. Figure 8.3 shows the IEEE 802 family of standards and their relationships.

8.3.1 IEEE 802 MAC Addressing

As we said earlier, though the original Ethernet specification used 8-bit addresses, the DIX standard increased this to 48 bits. The 802 committee adopted this, although they also made provision for the use of 16-bit addresses as an option for smaller networks that did not want to incur the overhead of the larger addresses. However, few vendors supported the smaller addresses and 48 bits became the rule.

The use of 48 bits for addresses enabled the IEEE to maintain a global address space: No two 802 devices should have the same MAC address. As we discussed earlier, addresses are overhead and most data link protocols prior to Ethernet and its 802 successors attempted to minimize the overhead by having only locally significant addresses and, with SDLC, even using only one address in the frame header. The fact that LANs used globally significant addresses meant that it was possible for the first time to concatenate at layer 2, that is, without layer 3 addressing at all. Such L2 concatenation is called *bridging* or, if implemented in hardware, *switching*; we discuss this in more detail in Part IV.

However, because it was anticipated that some protocols would embed MAC addresses in upper layer configurations and it would be undesirable to require these to be changed if, for example, a network interface card (NIC) had to be replaced, the 802 committee included the ability for network administrators to override the global MAC addresses and instead define their own *local* addresses. In addition, provision was made for defining multicast and broadcast addresses, although it was left to the individual MAC protocols to specify how these were utilized.

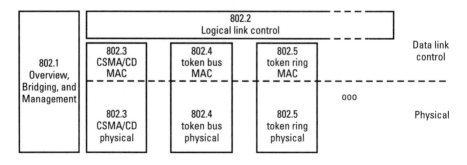

Figure 8.3 IEEE 802 family of standards.

8.3.2 IEEE 802.3 MAC Protocol: CSMA/CD

The results of the 802.3 committee's work was the MAC protocol officially called Carrier Sense Multiple Access/Collision Detection, no longer Ethernet. In terms of the 802.3 MAC frame header (Figure 8.4), perhaps the most significant change was that the DIX protocol type code was omitted and replaced with a 2-byte length field. (As we see later, the 802 committee did retain support for multiple upper layer protocols in the form of 1-byte *service access point* identifiers that effectively function the same way and are located instead in the LLC protocol header.) The only other change to the DIX header was that the preamble was reduced to 7 bytes and a new field, called the *start frame delimiter* (SFD), was included in place of the eighth preamble byte. The preamble pattern is a repetition of the bit pattern 10101010, whereas the SFD is the bit pattern 10101011.

8.3.2.1 Estimating the Channel State: Carrier Sensing

Recall that in Chapter 4 we looked closely at the physical layer of an Ethernet/802.3 station, including the modularization of management/decision mechanisms between the station proper and the multistation access unit (MAU), also called the transceiver (see Figure 4.9). A transceiver that is powered on and connected to the channel continuously measures the signals on the channel both to monitor the state of the channel for collisions and because the signal (waveforms) may be part of a frame that is addressed to a station that is attached to it. The transceiver contains an estimator that, based on the measurements it receives from the receiver circuits, continuously decides among three choices: (1) the bus is idle, (2) a collision (or other channel anomaly) has occurred, and (3) a valid waveform has been received,

As we saw in Chapter 4, the 802.3 standard calls for the transceiver to send valid waveform information to the station's physical layer signaling (PLS) layer across the *Data In* circuits of the AUI cable, while its estimates of the channel's state are sent via the *Control In* circuits of the AUI cable. It may do

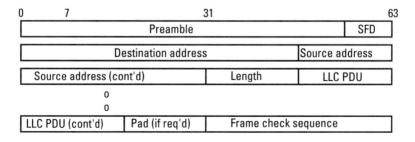

Figure 8.4 IEEE 802.3 frame format.

this by means of three different waveforms: (1) the CS0 waveform, which carries a *signal_quality_error* (SQE) message; (2) the IDL waveform, which carries a *mau_available* message; and (3) the CS1 waveform, which carries the *mau_unavailable* message, though this is optional to the 802.3 standard.

The scheduler within the transceiver decides what waveforms should be transmitted (actuated) on to the shared channel. However, it is the scheduler within each station's MAC layer that manages collision recovery via retransmission. The channel state estimate is sent to this scheduler, which in turn makes its own decision based on the state of the channel and whether it has data to send and/or has sent data already. If it has data to send and if the bus is idle, then it schedules transmission; on the other hand, if a collision or valid waveform is detected, then the scheduler waits until the channel is idle. If it has sent data and if a valid waveform is detected, then it continues to schedule transmission until the frame is transmitted or the maximum frame size is reached. If a collision is detected then the scheduler stops scheduling additional transmission of waveforms and enters collision recovery. Finally, if the scheduler neither has data to send nor has sent data then it ignores the information completely.

Figure 8.5 shows the division of responsibility between the transceiver and the MAC sublayer, as well as other tasks of the latter including serializing and deserializing data between the LLC and PLS layers, channel fault detection via CRCs, and so on. Note that, unlike THE ALOHA SYSTEM, neither Ethernet nor IEEE 802.3 uses the CRC for collision detection. Indeed, since collision detection in Ethernet/802.3 occurs within the transceiver at the waveform layer, CRCs are unnecessary.

8.3.2.2 Collision Recovery

Any station on the bus may detect a collision; it is not the sole responsibility of a transmitting station. When a station detects a collision the 802.3 standard specifies

> [t]o ensure that all parties to the collision have properly detected it, any station that detects a collision invokes a *collision consensus enforcement procedure* that briefly jams the channel. Each transmitter involved in the collision then schedules its packet for retransmission at some later time. [3]

By requiring that any station which detects a collision briefly jam the channel, the standard increases the overall probability of successful detection.

When a collision has occurred and been detected, the transmitting stations must take additional actions above and beyond those listed above. The standard states

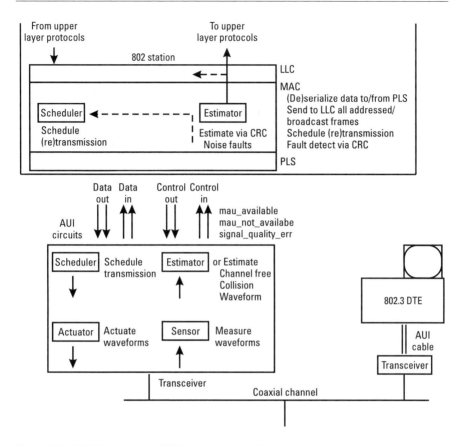

Figure 8.5 Collision recovery: MAC versus transceiver tasks.

[w]hen a transmission attempt has terminated due to a collision, it is retried by the transmitting CSMA/CD/CD sublayer until either it is successful or a maximum number attempts (attemptLimit) have been made and all have terminated due to collisions.... The scheduling of the retransmissions is determined by a controlled randomization process called "truncated binary exponential backoff." At the end of enforcing a collision (jamming), the CSMA/CD sublayer delays before attempting to retransmit the frame. The delay is an integer multiple of slot time. The number of slot times to delay before the nth retransmission attempt is chosen as a uniformly distributed random integer r in the range:

$$0 <= r <= 2^k$$

where

$$k = \min(n, 10) \text{ [3]}$$

Note that the slot time is fixed to be larger than the sum of physical layer round-trip propagation time and the maximum jam time. This last parameter is set such that all stations on the LAN can detect a jam.

8.3.3   IEEE 802.5 MAC Protocol: Token Ring

Just as IBM lobbied the 802 committee to get a reliable connection-oriented data link protocol, so, too, it pressed the case for an alternative to CSMA/CD's random access that would provide a more deterministic and predictable service. Because CSMA/CD stations could theoretically be delayed for arbitrarily long times by the exponential backoff process if two or more colliding stations chose the same random numbers, many in the industrial control community were unhappy with the prospect of its use in closed-loop control of real-time systems. By making common cause with large companies such as General Motors that were interested in adopting an 802-compliant LAN for use in industrial (i.e., factory) networks, IBM was able to get the 802 committee to sanction an alternative that relied on sequential, token-passing mechanisms rather than random access.

The result was that not one but two token-passing MAC protocols were defined, the 802.4 Token Bus and the 802.5 Token Ring. Since the latter has been much more widely adopted, especially in IBM's customer base, we concentrate here on the 802.5 standard and its management mechanisms. In fact, 802.4 Token Bus's principal distinction was that, at least initially, it was the only 802 MAC protocol that had a broadband (carrier-modulated) physical layer. This was again a concession to its intended operating environment, factories with high-powered electrical equipment, against which it was felt carrier-modulated signaling would prove more robust.

8.3.3.1   Sequential Allocation via Token Passing

The Token-Ring MAC protocol defined in the 802.5 standard is at the opposite end of the spectrum to CSMA/CD in terms of design philosophy and implementation complexity. Although it, too, is a connectionless protocol, 802.5 Token Ring has perhaps the most elaborate management design of any protocol we have discussed up to now. As we see later, the Token-Ring protocol uses over two dozen special frames to aid in the management of the ring and its ring stations. (Rather than link stations, the standard speaks of *ring* stations.)

The focus of much of this in management is the token, which the 802.5 standard defines as "a symbol of authority that is passed between stations using a token access method to indicate which station is currently in control of the medium" [4]. In other words, a token indicates that the channel is free; the token functions exactly like one of Dijkstra's semaphores used to protect critical regions in an operating system (see, for example, [5, 6]). When there is a token on the channel (ring) it passes each station in turn, indicating that the channel is free and can be scheduled and that the station can have its task executed by the channel. The station responds by changing/removing the token from the channel, thereby indicating that no other station can have its task executed at that moment.

A second reason for the token-ring protocol's complexity is that the ring is, in fact, a logical fiction that must be created and maintained by various management mechanisms. Recall from Chapter 4 that while a Token Ring, topologically, is a circular/looped concatenation of communications channels, physically, the ring is actually realized as a star, with each ring station connected by a pair of lobe cables to a MAU. Each channel is called a *lobe*; the concatenation of lobes is a ring (Figure 8.6). To facilitate knitting together the individual stations and their lobes, the 802.5 standard defines a number of management servers (described later) and corresponding management frames to be exchanged between these and the ring stations.

As we see, another important aspect of the Token-Ring protocol is its support for prioritization. The 802.5 standard defines eight levels of priority that can be used to expedite access (workload scheduling) for important traffic such as real-time or multimedia data; in effect this amounts to class of service support.

Last, we should mention a small but notable difference between the 802.3 and 802.5 standards, namely, the order in which bits are transmitted on the channel: 802.3 bytes are transmitted least significant bit first, whereas 802.5 bytes are transmitted most significant bit first. This includes MAC addresses, which when interpreted per 802.3 are said to be in canonical format while 802.5 addresses are said to be in noncanonical format. This does not generally

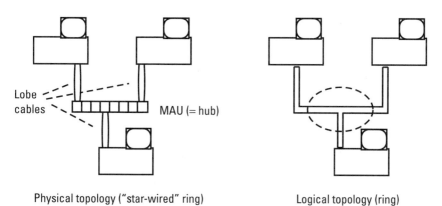

Physical topology ("star-wired" ring)   Logical topology (ring)

Figure 8.6 Token-Ring: physical versus logical topologies.

cause difficulties except for two circumstances: (1) when MAC addresses are embedded in upper layer protocol addresses and software and (2) when translational bridging between 802.3 and 802.5 LANs is desired (see Part IV).

8.3.3.2   IEEE 802.5 MAC Frame

The format of an 802.5 frame is shown in Figure 8.7. The frame begins with a starting delimiter that uses the two 802.5 nondata symbols J and K (see Chapter 4) to distinguish it from ordinary data. Next comes a field called Access Control (AC), which is at the heart of Token-Ring workload management. After this comes the Frame Control (FC) field, which like the control field in SDLC and HDLC is used to indicate the frame type—whether the frame is carrying LLC data (an LLC frame) or management data (a MAC frame). The frame's destination and source MAC (noncanonical) address fields are next, 48 bits each.

An optional element of the token-ring frame that follows these is called the Routing Information Field (RIF). The RIF is used in source-route bridging (L2 concatenation), which we discuss in more detail in Part IV. The data field, like the data field in an 802.3 frame, carries an LLC PDU (see later discussion). Note that, unlike 802.3, 802.5 does not define a maximum data field or frame size; instead, each station maintains what is called the Token Holding Timer, which in conjunction with the data rate (originally 4 Mbps, later 16 Mbps with higher speeds under discussion) of the ring determines how large a frame can be. This is followed by a 32-bit Frame Check Sequence Field, which is calculated using the same CRC generating polynomial as that used in 802.3.

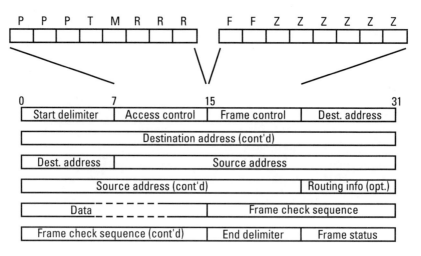

Figure 8.7 IEEE 802.5 frame format.

The 802.5 frame, however, does differ markedly from the 802.3 frame in its remaining fields. Unlike 802.3, 802.5 frames require a trailing flag, here called the Ending Delimiter. Like the Starting Delimiter, the Ending Delimiter includes the nondata J and K symbols as well as two subfields called the Intermediate Frame Bit (the I bit) and the Error-Detected Bit (the E bit) to indicate, respectively, that the frame is part of a multiple frame transmission and/or that an error such as CRC failure or nondata symbol violation has been detected.

The last field in an 802.5 frame is called the Frame Status (FS) field, which contains two important subfields called the Address Recognized (A) bits and the Frame Copied (C) bits. A station that receives a frame addressed to it sets the A bits (for redundancy there are two bits for each A and C) as it forwards the frame downstream; if in addition the station copies the frame to its local buffer it sets the C bits. As we see, the 802.5 architects used the A and C bits to realize MAC layer management; however, as we see in Part IV, they also created a thorny dilemma when it comes to concatenation of Token-Ring LANs.

As can be seen in Figure 8.7, the AC field contains four subfields, two of which are concerned with managing the Token-Ring priority mechanisms: a 3-bit priority subfield, which carries the currently available priority of the token, and a 3-bit reservation field, which is used by ring stations to raise the next available token's priority level. That is, a token circulating on the ring will always have a priority level ranging from 0 to 7. A station that receives the token checks the priority level it carries in the Access Control field and compares it with its own priority to see if it can claim the token and start transmitting a frame or must instead wait for the token to eventually circulate with a lower priority. The reservation subfield, on the other hand, allows a station with high-priority data to indicate this and thus "jump the queue" in front of lower priority ring stations that would otherwise get to use the Token Ring first if they were "ahead" of it on the ring. With the reservation bits set, these lower priority stations cannot claim a token because they could not meet or beat the required priority. Obviously, stations are limited in their ability to claim higher priorities at their own discretion.

The remaining two fields of the AC frame are the Token (T) and Monitor (M) bits. A token proper is a special structure that consists of just the Starting Delimiter field, an AC field with the token bit equal to 0, and an Ending Delimiter field. After a ring station receives the bits of a Starting Delimiter it will examine the AC field to see if the T bit is 0. If it is, then this is a token, and the station then checks the priority field to determine if it can claim the token; if the T bit is 1, however, then this is a frame and the station will check the destination address to see if the frame is addressed to itself. In addition, a ring station must check the source address. If it placed the frame on the ring, then it is responsible for removing it as well, which it does by resetting the T bit to 0 and thus recreating the token.

Clearly, the correct operation of a Token Ring is contingent on the stations adhering to the protocol concerning the claiming and release of the token; if a station fails to obey the protocol and does not release the token then it could conceivably continue to send frames and monopolize the ring indefinitely. Anticipating this, the 802.5 architects created a special management mechanism called the active monitor, a ring station that has been elected as part of the 802.5 protocol to act as a manager of the ring with responsibility for managing the token. This is where the monitor bit comes in. The station running the active monitor uses the M bit to detect token faults and recover from them, for example, by discarding a frame that is continuously circulating on the ring. We discuss the role of the active monitor and other 802.5 management servers in more detail later.

8.3.3.3 Early Token Release

As originally defined, no station on the Token-Ring LAN could begin to transmit until it received a free token of usable priority, and a token was only put back onto the LAN when the previous sending station had removed its frame. However, this was subsequently modified with what is known as early token release (ETR). Using ETR, a station that has finished transmitting its frame's end delimiter but has not yet received the frame's header will issue a token on the ring, enabling another station to begin transmitting its frame before the first frame has been removed. This exploits the fact that on large Token-Ring LANs the physical dimensions of the ring may easily accommodate two or more frames, especially if the frames are relatively small. Put another way, while the token-ring LAN was originally single tasking, with ETR it now supports concurrent multitasking.

8.3.3.4 MAC Versus LLC Frames

Before continuing with 802.5 management we should discuss in more detail the FC field and the two types of 802.5 frames, MAC frames and LLC frames. The two high-order bits of the FC field indicate the type of frame—00 for a MAC frame and 01 for an LLC frame (note that 10 and 11 are reserved). MAC frames are concerned with management. LLC frames carry upper layer PDUs encapsulated in LLC frames. The data link carries upper layer protocol data in LLC frames; these originate in and are delivered to the LLC layer. If the 802.5 frame is carrying an LLC PDU then three of the remaining 6 bits of the FC field may be used to carry priority information on an LLC–LLC level (not to be confused with the priority carried in the AC field).

Management data is carried in MAC frames in the form of 802.5-defined records called vectors which, in some cases, have additional subvectors. These data are exchanged largely by ring stations with those ring stations that are hosting the 802.5 management servers (see later discussion). This is "in-band" management signaling and the management data all stay within the MAC layer (Figure 8.8). As with other management data, this is overhead, that is, it displaces user data and uses transport bandwidth that could otherwise carry user data.

Table 8.1 lists the 25 MAC frame types defined in the 802.5 standard. The majority of these have the same FC value x'00' and are only differentiated by the contents of the frame, namely, the vector and (possibly) subvectors. Note that the vectors corresponding to a number of the MAC frames in Table 8.1 are only partly specified; the asterisks are in fact placeholders that are filled in accordance with the function (i.e., management) class of the ring station that sends or receives the MAC frame (more later).

The 802.5 standard defines four such function classes: an ordinary ring station is class 0; a Configuration Report Server is class 4; a Ring Parameter Server is class 5; and a Ring Error Monitor is class 6. Thus, for example, the vector carried in a Request_Station_Address MAC frame that is sent by an ordinary station is 0x000E. If the same frame is sent by the Ring Error Monitor the vector is 0x060E.

8.3.3.5 Ring Management

Every ring station contains a component called the monitor, capable of detecting and recovering from various LAN anomalies. However, only one monitor,

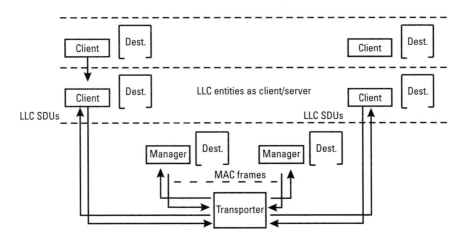

Figure 8.8 IEEE 802.5 MAC and LLC frames.

Table 8.1
IEEE 802.5 MAC Frames and Associated Vectors

MAC Frame	FC	Vector	MAC Frame	FC	Vector
Beacon	02	0002	Request station attachments	00	0*10
Claim token	03	0003	Report new active monitor	00	4025
Ring purge	04	0004	Report SUA change	00	4026
Active monitor present	05	0005	Request initialization	00	5020
Standby monitor present	06	0006	Report neighbor notification incomplete	00	6027
Duplicate address test	01	0007	Report active monitor error	00	6028
Lobe media test	00	0008	Report error	00	6029
Remove ring station	01	040B	Response	00	*000
Change parameters	00	040C	Report station addresses	00	*022
Initialize ring station	00	050D	Report station state	00	*023
Request station addresses	00	0*0E	Report station attachments	00	*024
Request station state	00	0*0F	—	—	—

called the active monitor, is allowed to exercise these responsibilities at any given time; all the monitors in other stations are said to be in standby mode. The role of the active monitor according to the 802.5 standard is "to recover from various error situations such as absence of validly formed frames or tokens on the ring, a persistently circulating priority token, or a persistently circulating frame" [4].

To recover from these anomalies the active monitor broadcasts the Ring_Purge MAC frame. This frame serves two purposes. First, when it returns to the active monitor it confirms that the ring is whole and operating. Second, it directs all other ring stations to reset their state machines and timers.

An active monitor will signal its presence on a ring by periodically sending out the Active_Monitor_Present (AMP) MAC frame. (Standby monitors may likewise send out Standby_Monitor_Present frames under certain circumstances.) Normally the active monitor will be the first station on the ring but what if this station powers down or is otherwise lost? That is why every other station on a token-ring LAN monitors the traffic for these periodic AMP frames. If they fail to come within a certain interval (by default 18 seconds) then it is assumed that there is no active monitor on the ring and the task of electing a new active monitor begins.

The process by which it is determined which station will be the active monitor is called *token claiming*. This consists of broadcasting several Claim_Token MAC frames. Though every 802.5 station is capable of becoming the active monitor, the default configuration option is for a station not to participate in the election process unless it was the first to detect the absence of AMP frames. If two or more stations do seek election, then the station with the lowest MAC address wins. The new active monitor signals the election is complete by broadcasting a Ring_Purge frame prior to issuing a new token.

Above and beyond the management executed by monitors, the 802.5 standard discusses three additional management entities: the Ring Parameter Server, the Ring Error Monitor, and the Configuration Report Server. Like the active monitor, these are not statically bound to any given station but rather are elected among these stations configured to host them. The purpose of these servers is to assist in ring management related to actuations such as adding, removing, or moving a station's position on the logical ring, as well as reporting faults and other anomalous conditions.

The Ring Parameter Server (RPS) assists when a station wants to join the logical ring (Figure 8.9). A station will send a Request_Initialization MAC address to the RPS functional address in order to get the operational parameters for the ring; these include the local ring number (used in source-route bridging; see Chapter 14). The RPS, if one is present, will respond by sending an Initialization MAC frame with the ring parameters. If no RPS is present then the station can still join the ring but will use default values for the ring parameters.

In contrast to the RPS, the Ring Error Monitor (REM) provides a centralized repository to which stations *send* data. A ring station is instrumented to monitor the ring and the station itself [4]. Eleven different counters (state variables) are maintained as part of this monitoring; these record such events as line errors (either invalid waveforms or FCS detected), internal errors within a station, Access Control Errors, and token errors. These measurements are sent to the Ring Error Monitor using the REM functional address by means of the error MAC frame (Figure 8.10).

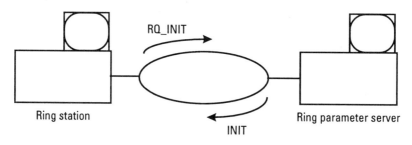

Figure 8.9 Ring Parameter Server.

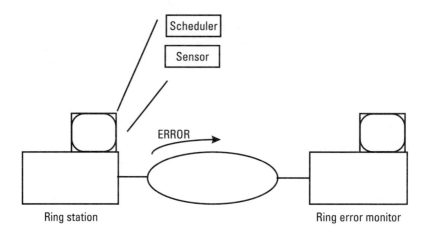

Figure 8.10 Ring Error Monitor.

The presence of the REM on the ring is a twofold convenience:

1. Without the REM, stations on the ring would have to store measurements indefinitely. This could require unacceptable storage capacity at individual ring stations.

2. Without the REM, the systems management applications would have to interrogate each ring station to obtain statistics about the ring's condition.

Finally, the Configuration Report Server (CRS) "can alter the configuration of the ring by requesting stations to remove themselves from the ring. The CRS can also query ring stations for various status information" [4]. Four MAC frames are exchanged between the CRS and individual ring stations: Remove_Ring_Station, Change_Parameters, Report_New_Active_Monitor, and Report_SUA_Change. We discuss the last of these frames and the concept of a stored upstream address (SUA) next.

8.3.3.6 More on Ring Operation

As we discussed in Chapter 4, a ring station is by default bypassed from the physical ring by means of a relay in the Token-Ring MAU that is only actuated when a station is powered on and sends a request by means of a phantom current carried on the lobe cable; until this, the lobe cables are, in effect, looped back—much as we saw with the V.28 serial interface. The first stage in joining a Token-Ring LAN is that, while the lobe cables are in loopback, a station will first execute a loopback test to determine the condition of its lobe cables by sending to itself Lobe_Media_Test MAC frames followed by Duplicate_Address_Test MAC frames. Only if these are successfully transported by the lobe cables does the station send the MAU the phantom current to physically join the ring.

The logical ring, however, must still be joined. Once a station has joined the physical ring there is a finite interval of time, measured by what is called its attachment timer T(attach), during which it expects to see an AMP, SMP, or Ring_Purge MAC frame. If the timer expires before any of these frames arrive, then it assumes that there is no active monitor on the ring, either because it is the first station on the ring or because of some anomaly. In either case, the station will attempt to become the active monitor by means of the token claiming process mentioned earlier, the duration of which is limited by the T(claim_token) timer.

Assume, however, that an active monitor is present. Then the inserting ring station will determine if any other ring station is using its MAC address by broadcasting a Duplicate_Address_Test MAC frame. If no conflict is detected the next step is to find the address of the station on the ring immediately preceding it. That is, the 802.5 fault detection and recovery mechanisms require that a station on the logical ring know the MAC address of its nearest upstream neighbor.

The process of learning the upstream neighbor address (UNA) involves the AMP frames sent out by the active monitor. These frames carry as a subvector the UNA of the station the active monitor last determined to be its upstream neighbor. The first downstream station that receives an AMP frame will copy the source MAC address from the frame header and store it as its UNA. It, in turn, will broadcast an SMP frame with this address as a subvector. The first downstream station that receives this SMP frame will then copy the source MAC address as its UNA and itself issue an SMP frame. This process continues until the upstream neighbor of the active monitor sends an SMP frame, at which point the neighbor notification process concludes.

A station's UNA is used in fault determination and recovery. When a ring station estimates that a severe fault such as a broken ring has occurred, it broadcasts a Beacon MAC frame, including its UNA to aid in fault determination. A station that receives a Beacon frame removes itself from the ring via the phantom current signaling and performs self-testing, including the lobe-loopback described earlier. If no internal fault is detected then the station reinserts itself into the ring. To prevent a beaconing station from disabling a Token-Ring LAN, the protocol requires a beaconing station to remove itself and perform self-test when its nearest upstream neighbor has performed a self-test and returned to the ring.

8.3.4 Logical Link Control Protocols

Last we want to touch on the Logical Link Control layer and its protocols. Like Ethernet and Token Ring, the two LLC protocols defined as part of the 802.2 standard are at opposite ends of the management spectrum. The management of LLC1 is so minimal that it would almost constitute a pass-through layer were it not for the presence of the SAP fields within the LLC headers. LLC2, on the other hand, is essentially a repackaging of HDLC/ABM. The result is to superimpose a point-to-point connection orientation on top of the inherently broadcast connectionless Token Ring. Note that both LLC 1 and LLC 2 share a common frame format (Figure 8.11).

The first two fields are the destination and Source Service Access Points, each 8 bits long. These are similar to the DIX protocol type field but obviously smaller. In fact, because 2 bits are used to indicate whether the SAP address is global or local and individual or group, there are effectively only 6 bits to represent upper layer protocols. The inadequacy of this led to the definition of the SubNet Access Protocol (SNAP) field, which provides an additional 5 bytes to expand the protocol type code. If a SNAP field is used then both the DSAP and SSAP fields are set to 10101010.

After the SNAP fields comes the control field. Because LLC2 uses 7-bit sequence numbers some of its frames have 8-bit control fields while others need 16 bits. After this comes the data field (unless a SNAP field is included, in which case it follows the control field and precedes the data). As we mentioned earlier, the size of the data field depends on the MAC protocol. Whereas 802.3 limits the field to 1500 bytes, 802.5 has no fixed maximum. Instead, this is determined by the ring speed and the token-holding timer.

8.4 Summary

In this chapter we have discussed the workload and bandwidth management mechanisms employed in the most important elements of the 802 family of protocols, namely, the 802.3 CSMA/CD and 802.5 Token Ring Medium Access Control protocols; and, briefly, the 802.2 Logical Link Control protocols LLC1 and LLC2. We also explored why the 802 committee divided the

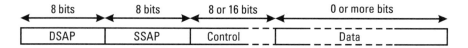

Figure 8.11 LLC header.

data link layer into two sublayers to meet the conflicting requirements of the Internet community, on the one hand, and IBM and its allies, on the other.

The chapter began by reviewing the innovative random access workload management used in Abramson's ALOHA SYSTEM project. From this we moved on to the Ethernet protocol developed by Xerox PARC, the first protocol to use true closed-loop scheduling. We discussed how Ethernet and later the 802.3 CSMA/CD protocol realized fault recovery using peer management mechanisms. We then moved on to the more complex Token-Ring process and explained how token passing is actually a type of open-loop scheduling, the token constituting a semaphore indicating whether the LAN is free or busy. Indeed, a measure of the protocol's complexity can be seen in the large number of management frames it requires. Along with these frames, we saw that the 802.5 protocol defines extensive mechanisms for monitoring and controlling a Token-Ring LAN.

References

[1] Abramson, N., "Packet Switching with Satellites," *1973 National Computer Conf., AFIPS Conf. Proc.,* Vol. 42, 1973, p. 698.

[2] Shoch, J., et al, "Evolution of the Ethernet Local Computer Network," *IEEE Computer Magazine,* August 1982.

[3] *Carrier Sense Multiple Access with Collision Detection (CSMA/CD) Access Method and Physical Layer Specification,* IEEE 802.3, New York: Institute of Electrical and Electronics Engineers, 1992, p. 47.

[4] *Token Ring Access Method and Physical Layer Specification,* IEEE 802.5, New York: Institute of Electrical and Electronics Engineers, 1992, p. 18.

[5] Lorin, H., and H. M. Deital, *Operating Systems,* Reading, MA: Addison-Wesley, 1981, p. 270.

[6] Hoare, C. A. R., *Communicating Sequential Processes,* Upper Saddle River, NJ: Prentice Hall, 1985, p. 278.

Part III:
Management in End-to-End Protocols

9

End-to-End Management: Basics

9.1 Introduction

We now continue up the encapsulation hierarchy to look at upper layer protocols and their role in managing end-to-end communications. As with data link protocols, there is a difference of opinion as to how much management should be incorporated into end-to-end protocols. And, as before, the decision is between simplicity/efficiency, on the one hand, and complexity/robustness, on the other. However, the management trade-off with end-to-end protocols is complicated by several factors, not least being that end-to-end management issues extend past the scope of a single data link: Multiple data link concatenations introduce new fault scenarios, where relays may fail or become congested, necessitating mechanisms of fault detection/correction, flow control, and/or additional management to support distributed transaction processing, for example.

Another complication comes from the fact that the encapsulation hierarchy can be arbitrarily deep, the only constraint being the overhead incurred by adding layers. At one extreme is the X.25 Packet Layer Protocol, a monolithic end-to-end protocol in which all the management tasks, from connection management to flow control to error control, are included in the single layer. At the other extreme is the seven-layer OSI model, which divides responsibility for managing end-to-end communications into five upper layers. Rather than follow either the OSI, TCP/IP, or any other model of protocol layering, in this chapter we derive from first principles the tasks of end-to-end management and then discuss their modularization—that is, how many layers there are above the data link.

As before, we decompose end-to-end management into bandwidth and workload management; and these into the corresponding measurement, estimation, scheduling, and actuation tasks. After defining what end-to-end management is, we discuss why it is necessary in real-world communications networks. We also explain why end-to-end management should be decoupled from concatenation. Next, we present a management taxonomy for classifying end-to-end protocols: connections versus connectionless operations, flow control, fault detection and recovery, and segmentation; whether these are executed at each stage (relay) or only the endpoints; and in how many layers. We also examine the locus of control in end-to-end protocols, that is, whether the management mechanisms are peer or master/slave.

In the remaining chapters of Part III we apply this framework to the analysis of the two most important families of protocol architectures, namely, TCP/IP and SNA (including APPN). Each of these represents different approaches to the management of end-to-end communications and each has strengths and weaknesses. We discuss their respective management mechanisms and how these have been modularized within the various protocol architectures. Finally, we look at the question of managing high-speed transporters, with their large delay bandwidth products, and the implications for the types of management mechanisms required.

9.2 End-to-End Management and Protocols

9.2.1 What Is End-to-End Management?

Although the meaning of end-to-end management may seem obvious, as we will see in this chapter it needs some elaboration. For example, what does end-to-end management mean if the transporter consists of a single data link? The answer depends on the management executed by the data link protocol versus the management desired by client(s). If the data link protocol is "light," that is, no retransmission, no flow control, and so on, and reliable service is desired then an end-to-end protocol must provide this additional management. Seemingly the situation is clearer if the transporter is made up of multiple data links, that is, if it is multistage. In this case the role of an end-to-end protocol may include management tasks that no data link protocol can effect, for example, end-to-end retransmission to recover from relay faults. But if concatenation is achieved via MAC bridging then the LLC protocols in fact span end to end (see later discussion). So as we said, end-to-end management is not so simple to define.

First we should ask what do we mean by an end system? At the physical layer an end system is a DTE and the channel interface module is a DCE. At the data link layer all the link stations constitute end systems. However, once we get above the data link layer we again distinguish between types of systems: End systems are clients of the transport service, whereas intermediate systems are components of the transport server. As we noted in Chapter 1, different terminologies are in use that complicate the discussion. Synonymous with end system are DTE (ITU/CCITT) and "host" (IP); unless otherwise specified, an end system is both a client (i.e., source) and a destination. Likewise, synonymous with intermediate system, are DCE (ITU/CCITT) and "relay" (IP).

Next we must ask what we mean by an *end-to-end* protocol? It turns out that there are two ways to answer this. The first is that *end-to-end* means end system to end system (host to host in IP parlance), to the exclusion of those protocol layers that are in both end systems and intermediate systems. With this definition, the lowest end-to-end layer in the OSI model is the transport (layer 4) [1]. Figure 9.1 illustrates the distinction, where the end system-only definition includes the Application, Presentation, Session, and Transport protocol layers, whereas the more expansive End System/Intermediate System definition includes the Network protocol layer as well.

On the other hand, between the transport layer and the data link layer there is another layer, namely, the network layer in the OSI model, that is inarguably end to end in its span and which is present in both end systems and

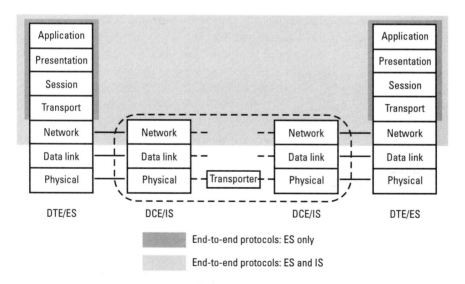

Figure 9.1 Protocol stacks (OSI model): DTEs and DCEs.

intermediate systems (except layer 2 intermediate systems, i.e., MAC bridges and switches). Again excepting L2 intermediate systems, the network layer is absolutely essential because it manages the global address space, without which end systems on different data links (other than 802 LANs) could not communicate with each other. For this reason, in this book we will follow the second definition and consider any protocol above the data link layer to be an end-to-end protocol.

With either definition, however, the management scope of an end-to-end protocol is global, whereas the principal defining characteristic of a data link protocol is that the scope of its management is a single stage or channel (Figure 9.2). (Even this statement needs to be qualified, since the "single channel" in question is in fact composed of two or more channels concatenated at the physical layer to produce the appearance of a single channel. The only exception to this would be two DTEs directly connected over a channel, but in most cases this is not possible, as we discussed in Chapter 4, because of the asymmetry between DTE and DCE interfaces. Hence with serial data links the minimal configuration involves two channels, whereas an actual connection over a public telephone or data network may involve tens, hundreds, or even more channels as the signals are routed over the internal switch-to-switch circuits. Likewise, with Token-Ring LANs, each station is connected with a pair of channels (the lobe cables) to the MAU, so a LAN with k stations consists of at least $2k$ channels. And the various types of Ethernet media all involve multiple channels: the shared bus as well as the AUI and MDI cables, and so on. Therefore, when we speak of data link protocols managing a "single channel" it should be clear that this is a convenient fiction.)

In contrast, an end-to-end protocol may span a single stage (channel or data link) or it may span multiple stages (generally speaking, data links) concatenated together—this is why we speak of the "global" scope of an end-to-end protocol. And it is this that drives the need for implementing an

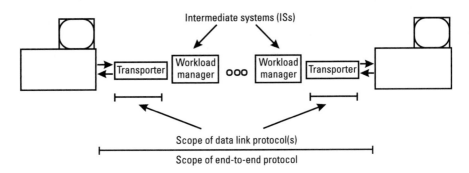

Figure 9.2 Span of data link versus end-to-end management.

end-to-end protocol, to augment the management provided by the data link protocol(s), including these tasks:

1. To define a global address space;
2. To abstract data links and isolate details from upper layer clients;
3. To manage end-to-end faults and flows; and
4. To provide additional multiplexing of upper layer clients.

We explore these tasks in the rest of this section.

9.2.1.1 Global Addressing

The first difficulty that comes with end-to-end protocols and their global scope is that the addressing that was adequate for a single data link is likely to be inadequate when multiple data links are connected. The reason for this is that data link addresses generally have only local significance. This is why an early "advanced course" on distributed systems stressed that "[e]nd-to-end protocols (EEP), datagram or VC, are required to provide the unique address space needed…" [2].

With all protocols, and especially data link protocols, a major concern has been to minimize the protocol overhead (header size) by keeping the address field(s) as small as possible. With the exception of the 802 MAC addresses, which were defined for LANs with relatively high-bandwidth channels and in any case are a comparatively recent development, most data link addressing is not sufficient to identify unambiguously a set of end systems/hosts that are on different data links. For example, SDLC only included one address field, for the secondary link station(s). If two SDLC links are concatenated, the addressing fails if for no other reason than that we now have two primary link stations which both assume the implicit address. And PPP carried minimization to the extreme with the configuration option of eliminating address fields altogether.

When we consider end-to-end protocols, therefore, managing a global address space is the first task. To do this, an end-to-end protocol must map global addresses to the addresses used by the data links in the network. These, after all, realize the end-to-end transport task(s). This is similar to employing network address translation (NAT) or similar interconnecting gateways that operate at an upper protocol layer; even with these, however, it is necessary to define a mapping of a globally unique address space to, for example, the "locally" unique network addresses in the different component networks. An example of this in SNA is called the *boundary function* (see Chapter 11).

For now, however, look at a simpler scenario: the management of a single global address space mapped to component data link address space(s). Consider the composite transporter composed of two data links and a layer 3 relay (i.e., a router) shown in Figure 9.3. There are two end systems, each with global (layer 3) and local (layer 2) addresses. Client end system A has the layer 3 address A^{L3} and the layer 2 address A^{L2}; the destination end system B has the layer 3 address B^{L3} and the layer 2 address B^{L2}. Just as the (local) addresses define the task set of the data link, so the global addresses define the task set of the composite transporter. In Figure 9.3 the task set of the composite transporter (i.e., the network) is $A^{L3} \rightarrow B^{L3}$.

Note, however, that neither component data link has the transport task $A^{L2} \rightarrow B^{L2}$ in its task set. Therefore, the only way to realize the transport task $A^{L3} \rightarrow B^{L3}$ (i.e., the end-to-end transport task) is with a schedule of two layer 2 transport tasks: $A^{L2} \rightarrow C^{L2}$ and $D^{L2} \rightarrow B^{L2}$, where C^{L2} and D^{L2} are the layer 2 addresses of the relay corresponding to its link stations on the first and second data links, respectively. The relay here acts as the proxy destination for the client A and the proxy client to the second data link.[1] We should also remark here that end-to-end protocols vary according to whether an intermediate system needs global addresses for each of its local (i.e., data link) addresses or if it suffices to have a single global address for the intermediate system. IP is an example of the former approach, and SNA is an example of the latter.

If the layer 2 addressing used on each transporter (data link) is only locally significant then, in the absence of the global addresses A^{L3} and B^{L3}, the relay would have no way of intelligently forwarding the data to the destination

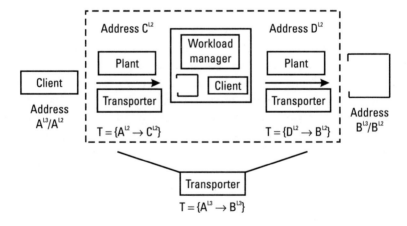

Figure 9.3 Address aliasing: L3 and L2 addresses.

---

1. Unless source routing is employed (see Chapter 13), these proxies should be transparent to the client(s) and destination(s)—otherwise, the goal of location transparency is sacrificed.

end system B. A station with destination address B^{L2} could be on one, both, or neither data link. On the other hand, even if globally significant layer 2 addresses are used then the relay still has the dilemma when a frame with destination B^{L2} arrives at the relay's link station (either C^{L2} or D^{L2}), namely, that it can only employ the relatively unsophisticated forwarding technique of broadcasting the frame out all its other interfaces, a process otherwise known as *flooding*. To reduce the need for such inefficient discovery, transparent bridging relies on a cache-based learning mechanism (see Chapter 14 for more details).

This is because the IEEE 802 addressing, while globally significant, defines a *flat* address space. A flat address space has no order in the assignment of addresses. In other words, there is no spatial correlation between addresses and the location of the addressed system. On the other hand, a *hierarchical* address space implies a clustering or correlation of addresses and locations. Human beings naturally gravitate toward hierarchical addressing, including mail (house number, street, city, state, country) and phone (country code, area code, branch exchange, line number). We will discuss the implications of hierarchical address spaces with respect to routing in Part IV.

9.2.1.2 Abstracting Data Links

Just as data communications would be much easier from an addressing perspective if no data link had to be connected to another, so too would other aspects of management be simpler if there were only a single type of data link. Once we admit multiple types of data links. However, we are confronted by the need to isolate implementation details beyond just the data link's addressing structure. This is a logical continuation of how the data link layer shielded its clients from the details of the physical layer and of how the DCE shielded the DTE from the channel and signals used. For example, an HDLC link station does not need to know if its DTE is connected to a DCE via a V.35 or RS-232 interface connector or whether the channel waveforms use phase modulation, frequency modulation, or digital signaling.

These are all instances of abstraction. The process of global addresses aliasing local addresses is, in effect, just such an abstraction of data link(s). The task of the network layer/end-to-end protocol workload manager is to map the end-to-end RFS to the L2 RFSs that are executable by the L2 transporters (data links) (Figure 9.4).

In practice, this means encapsulating end-to-end PDUs in L2 PDUs, which are then sent down to the physical layer, where the bits are sent across the DTE–DCE interface. In general, the manager in a layer will take the $L(k+1)$ SDU in the $L(k)$ PDU. The plant at the L2/L1 interface sends bits: The bits of the L2 PDU are sent to the L1 manager, which transforms them to the interface standard waveforms and then to the DCE. The plant sent by

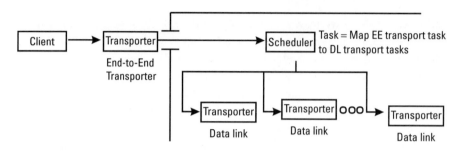

Figure 9.4 Layered abstraction of embedded transporter.

DTEs to DCEs are standardized waveforms (V.28, V.35, and so on). Finally, the plant carried by channels is waveforms that depend on the channel and signaling used (Figure 9.5).

Thus, when a data link (L2) protocol machine, also known as a link station, receives a ULP SDU from a ULP client, it encapsulates this ULP SDU in an L2 PDU and then assumes its role as client of the physical layer (L1) transporter. The real or ultimate client is the upper layer protocol client but the

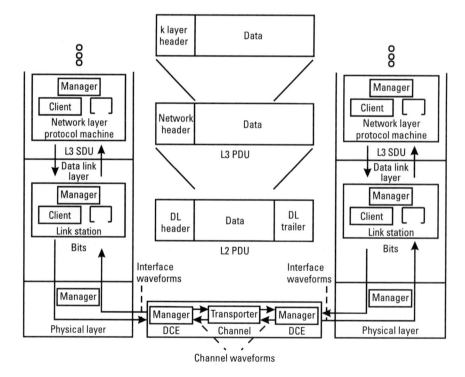

Figure 9.5 Transporter as composite of link stations and physical transporter.

physical layer (L1) knows nothing of this; the transporter only deals with the data link (L2) protocol machine, that is, the proxy client.

Splitting and Blocking

At each layer interface, the encapsulation relationship of layer $k + 1$ SDUs and layer k PDUs can be one-to-one, one-to-many, or many-to-one relationships. We are specifically looking at the layer 3/layer 2 interface, where a single L3 SDU can be encapsulated in one L2 PDU (frame); an L3 SDU can be split or fragmented and its segments each encapsulated in its own frame; or two or more L3 SDUs can be encapsulated or blocked together in one L2 PDU.

The choice of whether to split/fragment an L3 SDU or to block multiple L3 SDUs together is dictated by the relative sizes of the packets and frames. As we saw in Part II there is a wide of range of frame sizes among the data link protocols in common use. Note that not all data link protocols support fragmentation (splitting) and/or blocking of packets to fit into frame sizes—PPP, for example, allows only one packet (L3 PDU) to be encapsulated in a PPP frame.

There are thus five scenarios in which an L3 transport RFS (task) is mapped to one or more L2 transport RFSs—the actual RFSs being L3 and L2 PDUs, respectively, and the mapping of encapsulation of the L3 PDUs in the L2 PDUs:

1. One L3 SDU (RFS_e) in one L2 PDU (RFS_i): The transporter is single-stage, and no segmentation of the L3 SDU is used.

2. One L3 SDU (RFS_e) in two or more L2 PDUs (RFS_i). There are two subcases:

   2a. The L2 PDUs (tasks) are different in kind—heterogeneous tasks;

   2b. The L2 PDUs (tasks) are different in degree—homogeneous tasks.

3. Two or more L3 SDUs (RFS_e) in one L2 PDU (RFS_i): The transporter blocks two or more plant for transport within a single frame.

4. Two or more L3 SDUs (RFS_e) in two or more L2 PDUs (heterogeneous): The transporter is multistage but nonetheless blocks two or more L3 SDUs.

Subcase 2a corresponds to a multistage transporter consisting of two or more data links, which we discussed earlier. Conceptually it is, in effect, a workload actuation of kind: The end-to-end RFS is transformed into data link RFSs. Subcase 2b occurs with the segmentation of a large L3 SDU for transport by a single data link.

9.2.1.3 End-to-End Reliability and Flow Control

As we saw in our discussion of data link protocols, there is no consensus over the appropriate level of management with respect to faults and flow control at the data link layer. At one extreme are advocates of what are essentially "management-light" layer 2 protocols—protocols without fault recovery, flow control, connections, and so on, while at the other extreme are advocates of "management-heavy" protocols with all these mechanisms.

The same is true as we climb the protocol stack to end-to-end protocols. Taking reliability first, we are confronted by the same issues as at the data link layer: Should the end-to-end protocol attempt to detect faults? To correct faults via forward and/or backward error control? What about clients that are sending data such as real-time voice or video, in which the occasional lost packet is less important than timely delivery? And what end-to-end reliability is needed if the data link layer already employs reliability mechanisms?

Just as with data link protocols, there has been an evolution in the thinking about what management an end-to-end protocol should effect. One definition of end-to-end control as "... a technique for ensuring that information transferred between two terminals [DTEs] is not lost or corrupted" [3] reflects the thinking that management should attempt to provide additional fault recovery by end-to-end retransmission. In addition, it may negotiate end-to-end connections and/or other parameters and it may effect end-to-end flow control and other management tasks we discussed apropos of data link protocols in Chapter 5.

Let's address the last question first. Clearly, if a single data link is all that is being considered and the data link protocol managing that link includes fault recovery via retransmission then there is little benefit to including similar mechanisms in an end-to-end protocol above the data link layer. However, as we saw in Part II the newer data link protocols—for example, PPP or LLC1—are designed to defer fault recovery to an upper layer protocol. Clearly, if the network is to offer a reliable transportation service then the required mechanisms must be included in the end-to-end protocol. And if an end-to-end protocol is to run over many types of data link protocols this argues for inclusion of at least optional reliability mechanisms in end-to-end protocols.

The same arguments apply to flow control. We saw earlier that some data link protocols such as SDLC and HDLC employ sliding window mechanisms to throttle a sending link station. Limiting ourselves to stand-alone data links, this should be adequate to the task. However, other data link protocols that are more "lightweight"—again, we note PPP and LLC1—do not include any flow control at all. As with reliability, if such management is to be offered over these data link protocols then it must be included in an end-to-end protocol, at least optionally.

We should note that, in addition to flow control, many researchers have investigated the use of so-called *rate* control, by which the transporter and/or destination can limit the rate at which the client can send data; this is, in control theoretic terms, derivative control while basic flow control is proportional control (see, for example, [4]).

Transporter Versus Relay Faults

The other circumstance that can necessitate the inclusion of (additional) reliability and/or flow control in an end-to-end protocol is the management of multistage (composite) transporters—that is, networks composed of two or more data links along with relay(s) to concatenate them. With multistage transporters we must consider relay faults—when a relay fails to correctly forward data from one transporter to another.

The inclusion of relays means that we must consider such faults as the relay corrupting a PDU, misforwarding it, or simply discarding it if buffers are not available. Consider two reliable transporters concatenated by a relay that itself has finite bandwidth and finite reliability. Either the finite bandwidth or the finite reliability of the relay may result in data not being forwarded. For example, if the relay's buffers fill up then it may, depending on the protocols and traffic management, throw away PDUs. Likewise, errors in the software and/or hardware of the relays can corrupt the packets.

And the point is that an L2 protocol cannot correct a relay fault unless it involves an L2 relay (i.e., MAC bridge). That is to say, L2 (data link) retransmission protocols do nothing to repair a faulty or saturated L3 relay. The link stations were designed to manage transporter faults—that is, noise faults (faults of traditional channels, amenable to FEC)—as well as flow control between two link stations on a single data link. Multistage transporters require management in an end-to-end protocol (which includes LLC2 when concatenation is via MAC bridges).

Other Composite Transporter Faults: Out-of-Order and Multiple Packets

Another contingency that must be anticipated when we consider composite transporters is out-of-order delivery: The sequence in which several packets arrive at their destination is different than the sequence in which they left the client. For such out-of-order delivery to occur, the topology of the composite transporter must admit an alternative schedule of component transporters (i.e., there must be an alternative route); and there must be a change in the scheduling (routing) between subsequent PDUs. (These PDUs are also known as "birds in flight.")

For example, in Figure 9.6 we can see that some packets follow the A-B-E-F schedule while others may follow A-B-C-E-F. If packets 1 and 3 follow the former path and packet 2 follows the latter, then it is likely that out-of-sequence delivery will occur due to timing mismatches between the two paths.

Likewise, sequence numbers or similar mechanisms are necessary to deal with multiple copies of the same packet arriving at the destination. For example, let's assume that a simple end-to-end protocol with retransmission is running in Figure 9.6. If packet 2 is delayed in relay C for a sufficiently long time to cause retransmission by node A but is then forwarded nonetheless from C to E and from E to F, then F will receive two copies. This is unacceptable to the majority of clients seeking to have data transported—imagine if the packet carried financial transactions that might be duplicated, resulting in spurious debits or credits. Thus, we see the need for additional sophistication with composite transporters to avoid such faults.

We should note that amidst this catalog of the faults of multistage transporters, that in fact there are many virtues to such realizations that make them the best way to design highly available transporters. Multistage transporters that have alternative routings are, in fact, generally more reliable than noncomposite (single-stage) transporters. Indeed, the advantage of such a network is that single points of failure can be eliminated. We discuss this in more detail in Part IV.

Data Link and End-to-End Management: Fault Recovery and Flow Control

As we said earlier, with a single-stage transporter retransmission at both layer 2 and layer 3 is redundant: If retransmission is used at layer 2 then the plant is transported if transport is possible. However, employing retransmission at layer 2 means a client that does not want reliable transport is getting it anyway (unless the layer 2 protocol is itself subsetted) (Table 9.1).

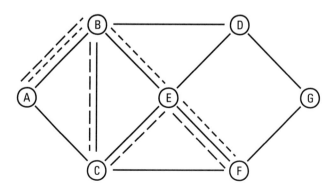

Figure 9.6 Out-of-sequence delivery.

Table 9.1
Retransmission Versus Stages

Layer 2 Retransmission?	Transporter	
	Single Stage	Multiple Stages
No	Layer 3 retransmission useful	Layer 3 retransmission useful/necessary
Yes	Layer 3 retransmission not useful	Layer 3 retransmission useful/necessary for relay faults

If every data link is managed by a connection-oriented protocol with flow and error control then is an end-to-end protocol necessary? Consider the scenario of a client connected to one transporter that is concatenated by some relay to another transporter to which is connected the destination. An L2 PDU containing an L3 PDU is successfully transported by the first transporter. However, assume the second transporter suffers a fault that is nonrecoverable. Then the client never knows it because the first transporter has already confirmed the successful transport of the L2 PDU.

9.2.1.4 Multiplexing Upper Layer Clients

One question that we see come up in discussions of various upper layer protocol architectures is protocol multiplexing. At each layer k is only one protocol allowed or can there be two or more protocols that the layer $k-1$ protocols must be able to encapsulate? In addition, there is the question of "openness." Beyond basic multiplexing does the upper layer protocol, at any or all of its layers, allow arbitrary types of client protocols to be transported? SNA is an example of an architecture that allows multiple upper layer protocols but is nonetheless not open—only protocols that are part of the SNA family, that is, those that conform to SNA formats, are allowed. TCP/IP, on the other hand, *is* open: IP as well as TCP and UDP can all encapsulate many different types of protocols and over the years dozens have been defined both for experimental and for production purposes.

The key to enabling such openness is the use of protocol type fields. Recall from Chapter 5 that with data link protocols the enabling element for such multiprotocol transport is the presence of a protocol type field in the data link header. When a destination link station receives a frame it uses the protocol type field to send the payload to the correct upper layer protocol machine for further processing; without the protocol type field the payload is just a set of bits. Generalizing this, for an encapsulating (carrier) layer $k - 1$ protocol to encapsulate more than one type of payload layer K protocol it is necessary for either the carrier protocol or the payload protocol to include a type field. Just as at the data link layer, including a protocol type field in the layer $k - 1$ protocol header allows a receiving protocol machine to disambiguate the payload.

Note that the same results are achieved by requiring that the payload protocols be "self-identifying," that is, contain a unique protocol type code at the beginning of each packet. Such an approach, however, is much less "open" in that it requires conformance to an *a priori* format, entailing modifications to existing protocols and thus sacrificing what we might term "transparency." SNA achieves its multiplexing precisely in this manner. Though SDLC lacks any protocol type field support, at the Path Control Layer (layer 3) the packet header contains a field called the Format Identifier that identifies the type of packet being encapsulated and with which the receiving Path Control protocol machine (called the Path Control Element) can determine the correct destination for processing each packet that arrives. We discuss SNA's formats and protocols in more detail in Chapter 11.

9.2.2 Additional Management

9.2.2.1 Dialog Management

After the end-to-end management tasks we have discussed so far comes the question of whether there should be a workload management mechanism that schedules the execution of the end-to-end transportation of data—a process sometimes called dialog management. This amounts to treating the end-to-end transporter in effect as a (virtual) channel. To the extent that the end-to-end transporter is point to point, this is only an issue if it is half-duplex; otherwise, with full-duplex transporters, just as we saw with PPP and HDLC ABM, there is no need for this level of workload management. For example, TCP is full duplex and management of the TCP connection is symmetric between both sides. Likewise, if the end-to-end transporter is point to multipoint (multicast) with only one client sending data then such management is unnecessary; the single sender schedules itself.

With half-duplex end-to-end transporters, however, the two clients cannot both send and receive concurrently. End-to-end transport in SNA is largely half-duplex. This is exactly the workload management problem we examined at great length with data link protocols, and not surprisingly the same mechanisms are employed higher in the protocol stack as at the data link layer. If control is symmetric (i.e., distributed) then either tokens or contention-based random access are employed. On the other hand, if control is master/slave then some sort of polling is employed. As we see in Chapter 11, SNA employs various techniques in its different upper layer protocols.

9.2.2.2 Transaction Support

A major reason for implementing data communications systems is to distribute programs and their execution (processing) among two or more computers, to exploit concurrency and/or increase reliability. When the interactions between the distributed application components are structured and well defined, this arrangement is often referred to as transaction processing or on-line transaction processing (OLTP) and includes the so-called "client/server" model as a special case. OLTP examples include point-of-sale (POS) networks and airline reservation systems.

Such distributed processing may require additional management to execute correctly in the face of the faults and other phenomena that can occur when two or more separate computers cooperate (as opposed to multiprocessor systems under a single executive operating system). For example, consider the transaction processing system shown in Figure 9.7, where we have a client program A that is sending (via a transporter, i.e., computer network) two RFSs requesting transactions T_1 and T_2, respectively, to server programs B and C executing in separate computers. It may be that transactions T_1 and T_2 are completely decoupled and their respective executions do not have any effect on the other. On the other hand, T_1 and T_2 may be transactions that must be performed together or not at all—this is an instance of what is called *atomic execution*. It may even be that T_1 and T_2 are identical transactions being executed on replicated data being maintained on both server computers for added

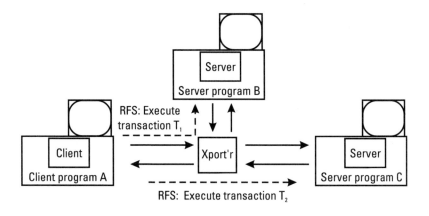

Figure 9.7 Distributed transaction programs.

reliability. In this case there may be a requirement to keep each copy of the replicated data from ever becoming out of synchronization with each other. This is known as consistency.

Note that atomicity may be applied to the transportation of data itself. Consider an application component (program) that wants to send a transaction request or a case in which the amount of data is too large for a single packet to carry. The obvious solution is to divide the message and carry its fragments in two or more packets. However, it may be that the sending program does not want partial delivery of its message—either all of the transaction details are to be sent to the destination program or none. This transport indivisibility is sometimes called *quarantining* (see, for example, Section 7.3 of X.200, *Reference Model of Open Systems Interconnection*).

Such atomicity introduces a new unit of recovery and retransmission, namely, a set of packets that in the OSI model are called dialog units and activities while in SNA they are referred to as chains. This can be as simple to implement as 2 bits in a header indicating whether the packet is the first, last, or both. If a single packet in a set of packets is corrupted, then either it is resent (selective retransmission) or the set is resent as a unit. Packet ensembles may also group related messages and treat them as a unit. Examples of such mechanisms are examined later in the OSI Session layer and SNA's Data Flow Control layer discussions.

Consistency is even stronger than atomic delivery in that it effects the synchronization of execution among components of a distributed application. Replicated data—for example, in distributed databases—should never be allowed to become inconsistent. To prevent this, distributed database management systems (DBMSs) often implement a two-phase commitment process, in which database transactions are sent to each DBMS but are not executed until these have all confirmed to the requesting application that they have received the transactions. In addition, there is generally an ability to "roll back" transactions if one of the component systems fails to execute correctly, in which case the remaining systems undo their respective transactions and the whole set of systems as well as their respective data are returned to the *status quo ante*.

Here we can see that distributed transaction processing places additional responsibility on application developers, even if there are management mechanisms implemented to ease their task. The heart of synchronization is the definition of synchronization points, otherwise known as checkpoints, by the sending client; these demarcate recovery boundaries to which rollbacks will occur if a fault happens. Checkpoints are, in fact, the delimiters of atomic transactions and hence define the beginning and end of packet aggregations.

Beyond atomicity and consistency, two additional transaction properties are called *isolation* and *durability*. Isolation means that the resources affected by a transaction are "locked," that is, the local operating systems prevent any other application from accessing resources (including data) while a transaction is being executed. Finally, durability means that, at each stage of executing a multistage transaction (a composite RFS), the results of each stage's execution are secure once that stage executes; in other words, subsequent transactions cannot tamper with the results of earlier transactions.

These mechanisms are sometimes grouped together under the rubric *ACID*, which stands for Atomicity, Consistency, Isolation, and Durability. These are four properties of distributed transactions that are desirable in terms of robustness. What they have in common is that they all provide transaction support in the form of workload management mechanisms to prevent certain types of faults associated with incomplete transactions. Note that some transaction support protocols implement one or more of these, for example, atomicity, without implementing the balance. Regardless of how many of these are implemented, such transaction support in effect defines mechanisms of interprocess communication that extend past the confines of a single computer.

9.2.2.3 Presentation Services

The last upper layer of management that we touch on is presentation services. Given the range and diversity of computer systems and software, it may not be surprising that computers seeking to exchange data and programs are frequently confounded by the absence of a uniform representation. We discussed in earlier chapters one notable division in computers, namely, between the world of (mainly IBM) mainframes with their 8-bit EBCDIC character set and the rest of the world, which uses the 7-bit ASCII character set. Another division is between big-Endian and little-Endian computers, which refers to the convention according to which computers store the bits or bytes—the most or least significant bit/byte first, respectively.

To overcome the communication difficulties posed by these and similar incompatibilities, some communications architects have defined additional services that a program can invoke to translate the data it seeks to send into a network *lingua franca*. To quote Tanenbaum again,

> The key to the whole problem of representing, encoding, transmitting, and decoding data structures is to have a way of describing the data structures that is flexible enough to be useful in a wide variety of applications, yet standard enough that everyone can agree on what it means. [1]

9.2.3   End-to-End Management: Implementation

9.2.3.1   How Much Management?

Having defined these end-to-end management tasks (global addressing, data link abstraction, etc.), we need to discuss their implementation. The first implementation question is how much management? Note that not all of the management tasks we discussed in the previous sections need to be implemented—this is what differentiates "light" and "heavy" end-to-end protocols. For example, a light end-to-end protocol may omit fault detection and recovery and leave this up to clients to implement if desired. Other management tasks may not even be required—a case in point being global addressing, which may be unnecessary if the L2 addresses are globally unique.

This question of how much management should be implemented within an end-to-end protocol amounts to how "smart" to make the transporter. By analogy, consider automatic versus manual transmissions in cars. With a manual transmission, the scheduling of gear actuation is by the driver, who must engage the clutch and select the gear desired; in contrast, an automatic transmission includes a scheduler that executes this task, freeing the driver from shifting. Of course, to automotive purists automatic transmissions are anathema precisely because they dislike this delegation of control.

Such drivers are like the advocates of minimal management in upper layer protocols. Huitema [5] cites three reasons why clients may prefer to have minimal management incorporated into the transporter:

1. Very simple exchanges (a single query and reply) are short in duration—requiring connection setup and teardown would bring little benefit and impose significant overhead.

2. Some applications are intended to run in limited memory, and the programming logic necessary to interact with connection management would be excessive.

3. Some applications, notably management and secure programs, do not want to delegate any management to the transporter because its actions may mask important information concerning faults and other problems.

On the other hand, some clients *will* desire to delegate management, and not wish any but the simplest (i.e., high-level) interfaces to a transporter. Take, for example, reliable transportation (via retransmission, alternative routing, and so on). Clients that want this generally do not want to be concerned with the details of *how* this is effected; they just want to be ensured their data will get to the destination or at least they will be informed if the transport fails. The workload manager, which is the proxy client for the actual client, is responsible for managing the retransmission and other management tasks and hides all of this from its clients.

9.2.3.2 How Many Layers?

This brings us to the second implementation question, namely, how end-to-end management tasks are modularized, specifically meaning how many layers of end-to-end protocols are needed? By way of illustration, recall the IEEE 802 architecture. We saw in Chapter 8 that the division of the management tasks of the data link layer by the 802 architects resulted in data link management that is realized in two sublayers: the Medium Access Control layer provides a basic connectionless transport, whereas the Logical Link Control layer added further management to provide either connectionless (LLC1) or connection-oriented (LLC2) service. And LLC1 is itself a subset of the (data link) management tasks that LLC2 implements.

When it comes to protocols for the management of end-to-end transportation the issues are the same as with data link protocols. For example, some clients desire highly reliable end-to-end transport while others are more concerned with latency and/or jitter. Do we implement a reliable end-to-end protocol and force all clients to use it? Or do we implement the basic transport management and offer additional management on top of this as an option?

The challenge of providing some clients with low-level access while affording others a high-level abstraction is most easily met with a modular implementation of the management. That is to say, if the end-to-end management tasks are implemented monolithically (in a single layer) then all clients are going to get the same level of abstraction. This is not the only deficiency of monolithic end-to-end protocols. As Piscatello and Chapin comment, surveying the range of protocol management issues from data link to application layer,

> [a monolithic] protocol to deal with all these functions would be inefficient, inflexible, and inordinately complex.... Imagine, for a moment, what the state machine would look like. [6]

A modular design, on the other hand, allows the tailoring of the management "offloading" to the needs of several classes of clients. Because the difference between unreliable and reliable protocols is management, a reliable transport service may be realized by intervening with one or more managers, which then become the proxy client(s) of an (unreliable) end-to-end transporter. That is to say, direct access to the transporter provides connectionless service for clients content with best effort delivery but greater reliability service requires that a client use connection-oriented service provided by a manager that is the client of the best effort transporter.

This illustrates the power of modularization: Rather than constraining all clients to a common level of service and management (a "one size fits all" approach), modularization enables different levels of management to be made available to meet the varied needs and sophistications of different clients. And this is why such modular designs are employed at the data link layer—in HDLC's base and towers, PPP's core protocol and ancillary/control protocols, and the 802 division of the data link layer into MAC and LLC layers.

On the other hand, it is possible to have too many layers. Each additional layer introduces additional overhead, and as the developers of layered or object-oriented software have discovered this can kill performance. Commenting on the seven layers of the OSI Reference Model (ISO 7498/ITU X.200), Tanenbaum writes:

> They are also difficult to implement and inefficient in operation. One problem is that some function, such as addressing, flow control, and error control, reappear again and again in each layer ... repeating [error control] over and over in each of the lower layers is unnecessary and inefficient. [1]

Piscatello and Chapin, looking at just the top three (end-to-end) layers, echo these comments:

> Even though the OSI upper layers are formally organized in a hierarchy, from an examination of the information passed through the presentation layer it is obvious that a purely "clinical" interpretation of the interactions among application, presentation, and session entities would be extremely inefficient. Well-behaved implementations of the OSI upper layers frequently lump association, presentation, and session connection management together.... [6]

So why do so many in the networking community believe that seven is the "right" number of layers? It is mainly a matter of conventional wisdom. However, as Tanenbaum has written, there is much to question in this:

> Most discussions of the seven-layer model give the impression that the number and contents of the layers eventually chosen were the only way, or at least the obvious way.... This is far from true. The session layer has little use in most applications, and the presentation layer is nearly empty. In fact, the British proposal to ISO only had 5 layers, not 7. [1]

What are the alternatives to the OSI Reference Model? At the other end of the layering spectrum is the X.25 Packet Level Protocol (PLP), which is a monolithic end-to-end protocol. PLP management tasks include connection management, retransmission, and flow control. Some may question how robust and/or efficient X.25 networks can be given this monolithic end-to-end protocol. However, it is a fact that even today many public data networks in the world still rely on just X.25's PLP to manage the end-to-end transportation of client data.

Between these two extremes we have the principal protocol families that dominate networking at present: TCP/IP and SNA. As we have frequently mentioned, TCP/IP divides end-to-end management into a best effort/connectionless IP and ensures end-to-end reliability in TCP providing additional management for those clients that desire it; a third protocol, UDP, is a best effort transport protocol that basically provides additional multiplexing for upper layer clients but otherwise no greater management than native IP.

As for SNA, the traditional layering model has three layers above the Data Link Control protocol. These are the Path Control layer, which is responsible for managing end-to-end connections and data transfer across what is known as the path control network (PCN). The Transmission Control and Data Flow Control layers, in turn, provide additional dialog and transaction support, notably for the aggregation of large messages and recovery from their faults.

Figure 9.8 shows the protocol stacks for X.25, TCP/IP, SNA, and OSI. Note that the layer boundaries for TCP/IP and SNA do not align exactly with those of the OSI Reference Model (and by extension X.25). The functionality of the IP layer is somewhat less than the network layer, whereas the TCP transport protocol includes some of the missing layer 3 functionality as well as the Transport layer management tasks. With SNA the reverse is true, that is, the Path Control layer is more functionally rich that the OSI layer 3 protocols, whereas the Transmission and Data Flow Control extend up into the Session layer.

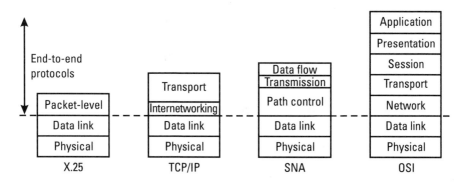

Figure 9.8 Protocol stacks: X.25, TCP/IP, SNA, and OSI.

Finally we note that some experimental high-speed protocols have returned a monolithic end-to-end protocol, sometimes called a transfer protocol. XTP (the eXpress Transfer Protocol) is typical. It aggregates the network and transport layers into a single layer to increase the efficiency of the implemented protocol machines. Defending this departure from the OSI model, XTP's advocates write:

> It should be remembered that the OSI Reference Model represents only one effort in constructing a framework for computer communication. The OSI Reference Model is not sacred; it is under on-going review, and other (successful) network architectures exist. [7]

9.2.4 End-to-End Transport Versus Concatenation

We should take a moment to explain why we are distinguishing between end-to-end management and the management of concatenation; and why we are deferring consideration of the latter until Part IV. After all, as we indicated in Chapter 1, the traditional model of networking places in layer 3, the network layer in the OSI model, the responsibility for concatenation and its "knitting" of data links together. In fact, this association of concatenation with the network layer was natural because, prior to the development of globally unique L2 addresses such as we have with the IEEE 802 MAC layer, concatenation *did* have to take place at layer 3. It is clear that if the composite transporter is realized, for example, from PPP data links or other data links without globally unique addresses, there is no way for the two end systems to communicate unless layer 3 addresses and relaying are employed.

But it remains that concatenation is a completely distinct task from end-to-end management, notwithstanding that the OSI Reference Model places the responsibility for routing in the network layer. This can be seen in innovations such as bridging and frame relay, which allow composite transporters to be constructed without any L3 concatenation at all. That there are so many ways to concatenate without involving end-to-end protocols is another indication that the two subjects are decoupled and hence need to be treated separately.

9.3 Taxonomy of End-to-End Protocols

Before presenting our management taxonomy for end-to-end protocols, let's step back and look at the "big picture." We are considering the management of an end-to-end transporter S, which may or may not be composite; with task set

$\{T_j\}$, where these are transport actuations, of course; and that serves a set of clients C_i to send data to destinations D_k.

With this end-to-end transporter we again confront the two basic "facts of life" concerning all servers: finite reliability and finite bandwidth. The composition of multiple data links can increase the task set, reliability, and/or bandwidth of the transporter, but in no case can it result in infinite reliability and/or bandwidth. Thus all the management issues we discussed apropos of data links in Part II are relevant to the composite or end-to-end transporter. However, because we are no longer dealing with the management of a single server (i.e., a data link) but rather with a (possibly) composite server, the management "degrees of freedom" increase several fold for each of our management tasks. As we saw in discussing concatenation, it is precisely the presence of intermediaries that greatly complicates management of end-to-end transporters.

That said, let us move on to our management taxonomy for end-to-end protocols. This taxonomy is, with suitable modifications, the same as the one we first presented in Chapter 5 for classifying and understanding data link protocols. This should not be surprising since, as we have stressed several times, the recursive nature of the *transporter* definition means that a transporter plus management—such as a data link protocol machine—is still just a transporter. End-to-end management, therefore, should not be fundamentally different than the management we discussed in Part II. The principal difference is that with each criterion we break it down into two and sometimes three subcategories to define the participants in the management:

1. Client and destination (end system-to-end system management);

2. Client and intermediate system ("portal" management); and

3. Intermediate and intermediate system (stage-by-stage management).

With this in mind, here are our taxonomic criteria:

- Does the end-to-end protocol require that a connection be set up prior to transportation of data?

- Does the end-to-end protocol implement any fault management?

- Does it attempt to correct faults via retransmission (backward error control) and/or forward error control? What about routing around faults?

- Does the end-to-end protocol have any mechanism for throttling the transmission of clients? Does the end-to-end protocol implement flow control and rate control to prevent overwhelming the transporter and/or the destination?

- Does the end-to-end protocol use packet (L3 PDU) sizes related or unrelated to the size of the frame (L2 PDUs)? That is, does the end-to-end protocol support fragmentation?

- Does the end-to-end protocol support any ancillary upper layer management such as dialog, transaction, or presentation management?

In addition to these questions, to characterize end-to-end protocols we look at workload scheduling of end-to-end transport tasks—stopping short of discussing the "how" of the concatenation mechanisms because this is being deferred until Part IV—and its relation to the debate over virtual circuits and datagrams. Finally, we touch on modularization: In how many layers is an end-to-end management task realized: in a single layer or in multiple layers? If multiple layers, is the protocol stack "top heavy" or "bottom heavy," that is, in which layer is the management concentrated?

9.3.1 Connection-Oriented Versus Connectionless End-to-End Protocols

The debate over connections that figured so prominently at the data link layer persists with upper layer protocols. Indeed, at each layer in a multilayer protocol stack the questions of connection-oriented versus connectionless operation need to be answered anew. As with data link protocols, with end-to-end protocols the question of whether a protocol is connection oriented or connectionless depends at its most basic on whether the client of the end-to-end transporter can send data without prior arrangement. However, as we saw in our earlier discussion of connection-oriented versus connectionless data link protocols, there are many other dimensions to the question of which approach is superior, notably the management overhead and the workload (traffic) patterns exhibited. Each approach has been adopted by end-to-end protocol architects and each has advantages over the other.

With a few exceptions, most protocol stacks concede the need for supporting a connection-oriented mode, if only as one of several modes of service available to clients. However, now the question arises as to from which entity the client end system must receive permission: the destination end system, the intermediate system to which the client is directly attached via a data link, *all* the intermediate systems that will forward the client's data to the destination end system, or some combination of the three?

On the other hand, if the client is allowed to send without any such permission being secured then the protocol is said to be connectionless. Note that the fact a protocol is connectionless does not imply that a client can send data at will. It may still be required to wait for allocation (scheduling) of the transporter, as for example is the case with the 802 MAC protocols, all of which are connectionless but which nonetheless have workload scheduling mechanisms to allocate the transporter.

In the discussion of connectionless versus connection orientation at the Data Link layer, we neglected to ask if there is a negotiation with the transporter. The answer depends on the transporter—for example, with serial channels a dial-up (i.e., *switched*) telephone connection must be established before any data can be sent to the destination (including data link connection requests). Put another way, if the data link protocol is connection oriented then the connection request (e.g., an SNRM frame) cannot be sent until the channel is "up"; on the other hand, even a connectionless data link protocol requires a channel "connection" if the channel is a dial line.

Let's return to connection-oriented end-to-end protocols. The simplest form of connection orientation, as we indicated earlier, occurs if the client end system must request permission from only the destination end system, not any of the intermediate systems. This is the case, for example, with TCP/IP: The IP intermediate systems (routers) will forward traffic without any connection setup, and in fact the destination end system will receive IP packets without any prior arrangement. However, permission for a TCP connection is granted by the destination end system alone, and if this is denied then the client end system will not send any TCP data.

We should note that the original ARPANET model had a two-tiered management structure that ran a reliable protocol over a reliable protocol. The latter, commonly called the Network Control Protocol, was connection oriented with retransmission and other reliability mechanisms. This was replaced by the Internet Protocol (IP) as part of a major design shift. Because it was assumed that some component networks (such as packet radio) would be very unreliable, it was felt that a reliable NCP was the wrong approach. We cover this in detail in the next chapter.

9.3.1.1 Datagram/Virtual Circuit Versus Connection-Oriented/Connectionless

Despite our intention to defer consideration of the internal details of concatenation until Part IV, we should take a moment and make clear that the question of connection-oriented versus connectionless end-to-end protocols is separate from another long-running debate in data communication, namely, that over datagrams versus virtual circuits. Unfortunately, the two issues have become conflated, and there is a considerable amount of confusion on the topic. It is incorrectly assumed by many that a connection-oriented end-to-end protocol necessarily runs over a network that uses virtual circuits, and that a connectionless end-to-end protocol necessarily runs over a network that uses datagram.

To understand why this is incorrect, we need to discuss virtual circuits and datagrams. In its simplest form, a virtual circuit is a sequence of transporters that is followed by all the packets (end-to-end PDUs) sent from client to destination. All the scheduling (routing) decisions are made for the first packet, after which the remaining packets are forwarded on the basis of the virtual circuit, not on an individual level. Part of the original rationale for virtual circuit architectures back in the 1970s was that all packets sent subsequent to the call setup packet need only carry a virtual circuit identifier and hence could omit destination as well as source addresses, since these would not be necessary to forwarding, and thus save on header overhead.

The alternative to virtual circuit protocols is datagram protocols. The essence of datagram service is that each packet is routed independently of its predecessors or successors. Of course, because of this each packet must contain addressing information sufficient to allow the relays to forward the packet to its destination. At the very least this means the globally unique destination address must be included in the header, although almost always the source address is included if for no other reason than security and authentication.

Part of the traditional argument for datagrams over virtual circuits is robustness. Because each datagram packet is relayed independently, if a component transporter fails the adaptation to this is simply to forward subsequent packets along a different path (that is, use a different schedule of transporters). With virtual circuits, on the other hand, because the concatenation follows a (semi)static or *persistent* schedule, a failure of a component transporter necessitates setting up the virtual circuit all over again.

We return to virtual circuits and datagrams in Part IV. For now, let's summarize their relationship to connection-oriented and connectionless operation by saying that it is possible to run a connection-oriented end-to-end protocol over a datagram network (this, in fact, is how TCP/IP is structured) or a connectionless end-to-end protocol over a virtual circuit network (IP over X.25 is an example).

9.3.2 Reliable Versus Best Effort End-to-End Protocols

The next taxonomic criterion is an end-to-end protocol's fault management—if there is any. Recall that a server with finite reliability is subject to two broad classes of faults: transient and persistent. These require very different solutions. As we saw in our discussion of data link protocols, recovery from transient faults generally involves retransmission—what we have referred to as temporal replication or redundancy. A persistent fault, on the other hand, requires bypassing the faulty transporter, which is possible only if there are alternative routes (transporters). This spatial redundancy and the replication it enables are, of course, principal reasons for implementing composite transporters. We discuss these in more detail later. First, however, we review the tasks of fault management.

9.3.2.1 Scope of End-to-End Fault Detection: Header Versus Plant

As we saw in earlier chapters, fault management is typically decomposed into two or sometimes three tasks: fault detection, fault isolation (an estimation task that may be aggregated with fault detection), and fault recovery. We discussed in Chapter 3 various mechanisms of error detection and correction—from simple parity checks to sophisticated block and convolutional codes—all of which function by introducing redundant information into the transported data; this redundancy is then used by the decoder to detect and possibly recover from the fault. And the larger the block of data being "covered," the larger the overhead in terms of redundant information that must be included.

It was to reduce such overhead that some protocol designers came to make a distinction between faults that affect client's data and faults that affect the management fields in the PDU headers. In part this is based on the fact that, while faults that corrupt the data being transported are obviously deleterious, those faults that alter the management information in an end-to-end PDU's header are more troublesome since they may result in misrouting of PDUs that must ultimately be discarded anyway—itself a waste of network bandwidth. We saw earlier that this led to the decision by the designers of broadband ISDN/ATM to limit fault detection to the cell headers (Figure 9.9.).

Seeking to minimize bandwidth costs, this is also why the architects of IP Version 4 ("standard" IP) included a checksum that only covers the header to detect faults that corrupt the various fields—such as addresses—that are used to forward IP packets. Notwithstanding the fact that IP is a best effort protocol,

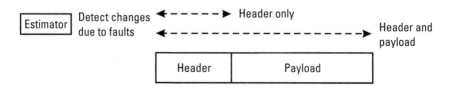

Figure 9.9 Fault detection coverage: header versus header and payload.

it was felt that the advantages of detecting faults in the packet headers, and thus preventing the wasteful forwarding of corrupted packets, merited this limited level of fault detection. Detecting faults in client data is only possible if TCP encapsulation is employed prior to IP transport.

However, as we see in the next chapter, with IP Version 6 (IP Next Generation) even the header checksum is omitted. In part, this was because there are other ways to detect faults beyond checksums and similar "codes." Discussing the decision to omit header checksums in IPv6 protocol, Huitema [8] points out that faults that corrupt one or more fields in the IPv6 packet header can be detected by ordinary protocol processing. For example, if the version field is altered the receiving router will discard it. Similarly, if the destination IP address is affected then the packet will be discarded at the destination end system that receives it, and so on.

9.3.2.2 Fault Management: End Versus Intermediate Systems

The principal rationale for eliminating the header checksum in IPv6 packet header was to minimize the processing overhead associated with forwarding packets at L3 intermediate systems (IP routers). Any fault detection (let alone recovery) that requires each stage in a multistage transporter to make decisions will significantly slow forwarding rates, particularly with high-speed data links. Many router vendors, attempting to increase the forwarding rates of products, were already failing to perform the header checksum calculation on incoming IPv4 packets; so it was not difficult for the IPv6 protocol architects to eliminate the header checksum.

A seeming contradiction is apparent if we contrast this move toward eliminating fault detection and, as we see in a moment, flow control and other management in intermediate systems—that is, stage-by-stage management—with the contrary philosophy in manufacturing. After all, two of the most prominent examples of multistage servers are assembly lines and computer networks. Yet the conventional wisdom concerning maintenance of the two is completely opposite. With assembly lines, the emphasis is on repairing at each stage while with computer networks the practice is increasingly becoming to repair only on an end-to-end basis.

Why this discrepancy? In part, the focus of the maintenance actuators in the two cases is different. With assembly line servers, the focus is on repairing the servers since few mechanical faults are transient and an unrepaired server will likely lead to a cascade of faulty products (see, for example, [9]). With computer networks, the focus is on repairing the plant since it is an accepted fact of life that any transporter will have faults that are transient and hence self-limiting.

9.3.2.3 Layered Communication and Multiechelon Maintenance

When some faults are easier to diagnose and/or repair than others, it is convenient to implement maintenance with a hierarchy of servers (bandwidth managers): the first echelon, which fixes the most tractable problems; the second, which fixes the problems that cannot be fixed at the first echelon; and so on, until repair no longer is cost effective or practicable. As we have remarked before, the fault management mechanisms of an end-to-end protocol are part of just such a multiechelon maintenance system, where the physical layer's demodulation and/or decoding, as well as any fault management in the underlying data link protocols, constitute the lower levels of the hierarchy.

As with fault management in the data link layer, where we saw that protocols ran the spectrum from the management-intensive SDLC and LLC2 to the management-light PPP and LLC1, such is also true for end-to-end protocols: the fault management spectrum runs from the fault management-light SNA to the fault management-intensive TCP. How much fault management is executed in the data link protocols will determine how much fault management must be executed in an end-to-end protocol to meet client requirements.

SNA assumes that underlying the data link protocols assures reliable delivery; and that its upper layer (end-to-end) protocols need be concerned with fault management only to the extent of routing around failed links. That is, SNA's philosophy is to rely on fault detection and recovery via retransmission at the data link layer so that upper layer protocols can assume the reliability of the transporter is "black and white": Either the data link transporter delivers the data correctly (to within the ability of the CRC to estimate the occurrence of faults) or it is declared to have suffered a fatal fault and will no longer be used by the upper layer protocols. This is why

> [t]here are few timers in SNA, but the principle is to use them only when absolutely necessary and at as low a layer as is possible. For example, data link controls use an inactivity timer to recognize outages and as an acknowledgement timer to protect against lost messages or acknowledgments. The use of DLC timers is critical to SNA: because SNA requires a reliable delivery by data link control, it can avoid high-layer timers. [10]

The TCP/IP approach is completely opposite. Because IP was designed to be an "internetwork" protocol able to run over many different types of networks (L2 and L3 protocols) ranging from presumably reliable X.25 to decidedly unreliable packet radio, it is TCP that contains the principal fault detection and recovery mechanisms, including timers for deciding when to retransmit lost PDUs (called segments). And, just as IP is itself merely a best effort end-to-end protocol with only limited fault detection capability (omitted entirely in IPv6), so are the IP-related data link protocols, notably LLC1 and PPP, also management-light.

Each approach (traditional SNA and TCP/IP) represents a set of trade-offs. However, as we saw in Part II and as we see in the remaining chapters of Part III, the direction in which protocols are going is to limit management in the lower layer protocols in order to facilitate high-speed networking and forwarding. More intensive fault recovery is implemented in end systems where the clients require the highest level of reliability and are willing to accept the higher overhead that comes with it. This, again, reflects the advantage of a modularized protocol architecture.

9.3.2.4 Spatial Replication/Routing Around Faults

All the fault management that we have discussed so far is limited to recovery from transient faults. Recovery from faults that are persistent and fatal cannot be effected with either retransmission (backward error control) or redundancy (forward error correction). Attempting to resend data over a broken channel (for example, the fiber optic cable cut by a backhoe operator) will not result in any greater success. Likewise, forward error correction cannot improve the situation since neither the redundant information nor the client's data ever gets to the destination.

In fact, short of repair, the only way to recover from a failed component is to use alternative routing (schedules) to bypass it. This is why many large customers insist on multiple disjoint telephone lines into their facilities. Such disjoint connectivity is a form of spatial replication. What this means is that a given end-to-end transport task may be realized by two or more schedules of component transport tasks (i.e., transport tasks of component transporters). A fault that disables a component transporter will leave the task set of the composite unaffected if there is an alternative path.

In graph theory, this is referred to as multiple connectivity. If the graph corresponding to a composite transporter is k connected then that means there are k distinct schedules that realize any of its end-to-end transport tasks. Multiple connectivity means that, for a given transport task, the topology of the composite transporter admits an alternate path (= schedule of transporters). Then fault recovery at the level of the composite transporter becomes the solution.

Such spatial redundancy is complementary to the temporal redundancy of reliable end-to-end protocols. If the adaptation time is within the retransmission time window of the sending workload manager then it is even possible that the client and destination may be shielded from any knowledge of the faults. With best effort end-to-end protocols, spatial redundancy will allow recovery of the end-to-end transport tasks lost when the component transporters failed.

The client data that were sent during the failure will not be recovered, but communication between the client and the destination will be restored. This is why we remarked in Chapter 1 that fault management transcended its traditional definition and a prime example is routing as a means of recovering from faults.

The key is that the schedule to realize a given task in the composite transporter's (end-to-end) task set will have to be changed if the current schedule includes a task in the task set of the failed components. Some composite transporters effect this bypassing (adaptation) with a routing protocol; others rely on a central scheduler (SNA). As with our discussion of connection-oriented versus connectionless operation, we want to defer until Part IV delving more deeply into how workload management mechanisms effect this rerouting around faults.

9.3.3 Flow Control in End-to-End Protocols

Just as with fault management, end-to-end flow control is similar to flow control at the data link layer. But, as with fault management, the issues that arise at this higher level are more complicated in that, beyond the end systems, the impact of the various flow control workload management mechanisms on the intermediate systems (relays) must now be considered. Above all, two facts must be kept in mind: that the longer an intermediate system holds an end-to-end PDU, the greater the total latency; and that as enqueued PDUs approach the storage limits of a relay, relay performance will suffer to the point that it may be unable to receive any new PDUs, let alone forward them.

Thus we should distinguish between an end-to-end protocol that implements flow control that is confined to end systems and an end-to-end protocol that implements flow control which is executed stage-by-stage—meaning intermediate systems, if any, must also execute this throttling. Many end-to-end protocols do not support stage-by-stage flow control for the same reason that we saw many do not support stage-by-stage fault management: to keep the intermediate systems unencumbered by any decision making apart from that involved in concatenation.

However, such a strategy raises the importance of flow control as admission control in the portals. That is, those intermediate systems that constitute the entry points into a composite end-to-end transporter must execute workload management—flow and/or rate control—lest the intermediate systems become overwhelmed. Various approaches to admission control—credit based, rate based, and so on—have been investigated, particularly with respect to ATM, and we discuss some of the results in later chapters.

We also see the need to separate monitoring from control. Even if the intermediate systems do not actuate the traffic directly, it is possible that the intermediate systems are instrumented to send state information to the clients and/or destinations of the end-to-end transporter. This may include estimates of round-trip delays (RTDs) as well as other parameters such as utilizations and queue capacities. Frame relay, for example, does not have stage-by-stage flow control but nonetheless it includes Forward Explicit Congestion Notification (FECN) and Backward Explicit Congestion Notification (BECN) bits for signaling by the intermediate systems (Frame Relay switches) regarding their respective states.

9.3.4 Segmentation of End-to-End PDUs

Another aspect of end-to-end protocols that can have a major impact on relay performance is whether segmentation of end-to-end PDUs is supported. As we saw in Part II, different data link protocols have a wide range of maximum transmission unit (MTU) sizes—that is, the largest frames (L2 PDUs) that can be sent. Given these L2 MTU size constraints, an end-to-end protocol has two choices. The protocol can either support some segmentation mechanism allowing large end-to-end PDUs to be split into sizes that can be transported across the transporters involved and later recombined at the destination; or the protocol can keep its MTU size no greater than the smallest L2 MTU size of any network or data link that transports its PDUs.

Segmentation can adversely impact relay performance in several ways. First of all, if a large end-to-end PDU has already been segmented by an upstream intermediate system and the protocol requires that the segmented pieces (fragments) be reassembled before being forwarded then the forwarding of all the pieces must await the arrival of the slowest piece. And, of course, the fragments that arrive prior to the last one must be stored awaiting reassembly. This results in delay and the need for large memories.

This also creates the potential for a situation known in operating systems as a deadly embrace, where competing RFSs each have exclusive use of some resources (in this case storage), creating a shortage of resources sufficient for any to finish. (This is similar to what happens with congested city traffic when cars that are too far into an intersection block cross traffic from moving when the light changes.) In this case, if the storage of a relay is completely full because it is holding the fragments of assorted end-to-end PDUs, none of which can be reassembled because there is no room for the relay to receive any more fragments, then the result is known as *reassembly lockup*. In fact, just such a situation led to a "meltdown" of the early ARPANET.

To prevent such lockups, some end-to-end protocols employ some form of resource reservation. A notable example is SNA, where the activation of end-to-end connections (known variously as virtual routes or sessions, depending on the level in the SNA stack) is allowed only if sufficient storage is available in the intermediate systems through which these virtual circuits will flow. To monitor this, SNA has defined a set of messages between these intermediate systems and a central manager known as the Systems Services Control Point (SSCP), which makes the decision on connection activation. We discuss this in more detail in Chapter 11.

However, a change in thinking occurred as experience in the 1980s was gained with high-performance networks and the routers used to build them. Segmentation and reassembly were identified as being impediments to attaining performance rates of more than 100,000 packets per second (pps). In a way, segmentation constitutes a second or outer level of packetizing, above and beyond the multiplexing that occurs at the data link layer, and this is pure overhead. The conventional wisdom now is that any segmentation necessary should be the responsibility of the end systems, not intermediate systems.

Because of this, the decision was made when designing IPv6 to eliminate the segmentation of IP packets. IPv6 networks are expected to support a minimum MTU size of 536 octets. If end systems wish to send larger packets than this ensured minimum, then they are expected to employ MTU path discovery to determine the largest MTU size that they can use. We cover this and other aspects of IP in Chapter 10.

9.3.5 Workload Management: Scheduling the End-to-End Transporter

Whether an end-to-end transporter implements flow control or fault management, there remains the issue of its tasking and the associated workload management. (We limit consideration to so-called unicast transportation—that is, point to point between two end systems; multicast transport—point to multipoint—is considered in the next chapter.) Does the end-to-end transporter allow concurrent sending and receiving or must the two end systems take turns? These correspond to full-duplex (FDX) and half-duplex (HDX) transporters, respectively.

As we saw with our examination of data link protocols such as HDLC/ABM and PPP, with FDX transporters there is no need for global workload management because either side can send at will. Put another way, FDX end-to-end protocols implement peer management mechanisms. Some of the most important end-to-end protocols such as TCP are full duplex.

On the other hand, many end-to-end protocols are half-duplex. SNA and similar terminal-oriented protocols are half-duplex in their workload scheduling; half-duplex "flip-flop" operation means there is a strict alternation between the two end systems, and half-duplex contention means that the two end systems "bid" for the right to use the transporter. As with HDX serial data link transporters, HDX end-to-end protocols may rely on either peer or master/slave management mechanisms. With master/slave end-to-end protocols, one end system is the master and is responsible for "polling" the slave end system; only when polled can the latter send data. This is the management model employed in SNA, for example.

9.4 Summary

In this chapter we have covered management in the remaining layers of the protocol stack, the so-called upper layer protocols. We identified four basic tasks executed by all end-to-end protocols, namely, the management of a global address space, the abstraction of the underlying (data link) transporters, the implementation of end-to-end reliability and flow control (if any), and the multiplexing of upper layer clients, particularly multiple upper layer protocols.

We then discussed how such end-to-end management should be implemented. We saw that some advocates of end-to-end protocols argue for putting all management functions in a single layer, such as with the X.25 protocol stack. Others have opted for two layers of end-to-end protocols, notably the developers of the TCP/IP protocol stack. SNA's original architects chose a five-layer stack, with three additional layers of end-to-end protocols above the data link layer. Last, we saw that the Open Systems Interconnect Reference Model with its seven-layer protocol stack presents the most elaborate decomposition but one widely criticized for its inefficiency and overhead.

Finally, we stressed throughout the chapter that end-to-end management is distinct from the concatenation. Although with composite transporters there must be some workload management mechanism(s) to map the end-to-end transport tasks to the tasks executable by the component data links, we saw that this is independent of the nature of the upper layer protocols carried by the component transporters. Most notably, we saw that although a virtual circuit implies that routing/forwarding decisions are made on the basis of circuit IDs and that all the packets follow the same schedule, and a datagram implies that every packet contains sufficient information (for example, addresses), this is distinct from the general protocol question of connection-oriented versus connectionless operation.

References

[1] Tanenbaum, A. S., *Computer Networks,* 2nd ed., Upper Saddle River, NJ: Prentice Hall, 1988.

[2] Lampson, B. W., M. Paul, H. J. Siegert (Eds.), *Distributed Systems: Architecture and Implementation—An Advanced Course,* New York: Springer-Verlag, 1981, p. 116.

[3] *Van Nostrand Reinhold Dictionary of Information Technology,* 3rd ed., New York: Van Nostrand Reinhold, 1989, p. 193.

[4] Dorf, R. C., *Modern Control Systems,* 3rd ed., Reading, MA: Addison-Wesley, 1981, p. 380.

[5] Huitema, C., *Routing in the Internet,* Upper Saddle River, NJ: Prentice Hall, 1995, p. 57.

[6] Piscatello, D. M., and A. L. Chapin, *Open Systems Networking,* Reading, MA: Addison-Wesley, 1993.

[7] Strayer, W. T., B. J. Dempsey, and A. C. Weaver, *XTP: The Xpress Transfer Protocol,* Reading, MA: Addison Wesley, 1992, p. 18.

[8] Huitema, C., *IPv6,* Upper Saddle River, NJ: Prentice Hall, 1996, p. 36.

[9] Womack, J. P., D. T. Jones, and D. Roos, *The Machine That Changed the World,* New York: Harper Dellacourt, 1990.

[10] Pozefsky, D., D. Pitt, and J. Gray, "IBM's Systems Network Architecture," in *Computer Network Architectures and Protocols,* 2nd ed., C. Sunshine, Ed., New York: Plenum, 1989, p. 504.

10

End-to-End Management: The IP Protocol Suite

10.1 Introduction

Having examined in Chapter 9 end-to-end management in the abstract, this chapter will focus on TCP/IP or, more precisely, the IP protocol suite: the Internet Protocol (IP), the Transmission Control Protocol (TCP), and the User Datagram Protocol (UDP); and an ancillary protocol to fault and configuration management, the Internet Control Message Protocol (ICMP). For each protocol we determine the MESA tasks involved and map them onto our workload management/bandwidth management model.

The chapter begins by discussing TCP/IP's antecedents in the protocols of the ARPANET, the predecessor of today's Internet. TCP/IP's approach to end-to-end management came directly from lessons learned with the ARPANET protocols, the most important being the idea of internetting by means of abstracting or "enveloping" underlying networks and protocols. This led to TCP/IP's two-tiered modularization of end-to-end management with a host-level end-to-end protocol (TCP) running on top of an internet hop-by-hop protocol (IP).

From this we segue to IP itself, and consider both the currently deployed IP Version 4 (IPv4) and the next-generation IP Version 6 (IPv6). These are end-to-end protocols with minimal management: no connections, no retransmission, and no flow control. IPv4 does little more than define a global address space and provide a uniform encapsulation to abstract the underlying transporter details; IPv6, however, is more streamlined—its architects having abandoned, for example, such IPv4 features as fragmentation in order to increase routing performance. Last we consider a new feature in IPv6, called *flows*, which approximates certain aspects of virtual circuits for the purpose of improving the transport of real-time traffic such as multimedia.

Next we examine ICMP. Both ICMP for IPv4 and ICMP for IPv6 are used by the principal protocols (IP, TCP, and UDP) to send error messages for fault management/bandwidth monitoring. Finally, we cover TCP and UDP, the transport protocols of the IP suite. It is TCP that provides most of the end-to-end management in the IP protocol suite. Within the context of end-to-end connections, TCP uses retransmission to achieve high reliability and flow control to keep from overwhelming the destination end systems. UDP, on the other hand, adds relatively little to the (minimal) end-to-end management of IP. It is connectionless and makes no effort at providing additional reliability or flow control beyond that of the underlying transporters that carry the IP datagrams. In fact, we see that UDP's principal management task, which it shares with TCP, is to provide multiplexing of multiple upper layer protocol clients.

10.2   The ARPANET and Its Protocols

The roots of today's Internet and the IP protocol suite lie in the ARPANET, so-called because its development was sponsored by the Advanced Research Projects Agency of the U. S. Department of Defense. However, contrary to net folklore, the ARPANET was *not* built to survive a nuclear war. Rather, the ARPANET was a testbed of two related ideas: remote computing and resource sharing; and computer–computer communications, as distinct from computer–terminal data entry, to make this sharing possible. As we see, the most significant contribution of the ARPANET was the emphasis on peer management in its protocols; these avoided wherever possible any hint of master/slave or centralized control.

10.2.1   Elements of the ARPANET

The ARPANET was the first packet-switched network in the world when it came on-line with four nodes in late 1969. Crucial to the ARPANET design was that the end systems were to communicate via intermediate systems, called interface message processors (IMPs), which executed the actual packet switching. Although IBM and other mainframe manufacturers had previously employed specialized processors to off-load communications from the CPU (for example, the IBM 27XX channel processors), the concept of the IMP, which we would now call a router, proved enormously important not least because it effectively modularized the transporter, isolating implementation details from the end systems.

Hosts sent and received PDUs in the form of *messages*, which were up to 8096 bits long. IMPs exchanged *packets* that had a maximum transmission unit (MTU) size of 1008 bits. Allowing hosts to send and receive larger MTUs meant that the I/O burden was lower on systems that were designed primarily to support batch processing. Packet switching, on the other hand, stressed the serial reuse of communications channels; and the smaller the MTU exchanged by IMPs, the finer the "time-slice" quantum and the greater granularity of control.

Hosts and IMPs were connected at speeds up to 200,000 Kbps with custom-built cables at up to 2000 feet long, whereas IMPs were connected to each other over wide-area (telephone) serial channels at rates up to 56 Kbps; and over satellite channels to Hawaii and Norway at 56 and 10 Kbps, respectively. (We should note that in the ARPANET the term *links* also referred to logical end-to-end connections.) Both the host–IMP and IMP–IMP communications channels were limited to point-to-point topologies. Although multi-dropped data links had been employed for more than a decade (for example, in the airline industry's SABRE technology deployed in the late 1950s), these were all based on centralized (master/slave) management. With the technology of the time, the ARPANET developers could not realize multidropped data links without sacrificing peer management.[1]

10.2.2 ARPANET Protocols

The ARPANET had three principal protocols: the host–IMP protocol, also known as 1822; the host–host protocol; and the IMP–IMP protocol, which included both stage-by-stage (IMP–IMP) and end-to-end (source IMP–destination IMP) management (Figure 10.1).

10.2.2.1 BBN 1822

The host–IMP protocol was generally known in the ARPANET community as 1822. The reason for this was that much of the original work on the ARPANET was performed by Bolt, Beranek, and Newman (BBN); and

---

1. In fact, it is significant that the principal developer of Ethernet, the first peer-managed multidropped data link, Robert Metcalfe, had been one of the early developers of ARPANET hardware and software. He built the MIT host–IMP interface circuits.

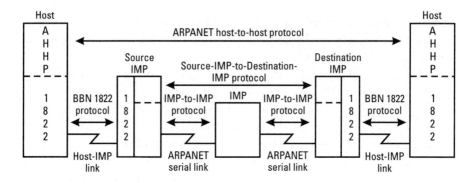

Figure 10.1 ARPANET protocols.

the host–IMP protocol was first described in BBN report 1822, titled *Specification for the Interconnection of a Host and an IMP*. BBN 1822 defined what later became known as the host access protocol (HAP).[2]

Host PDUs were up to 8063 bits in length, not including a 32-bit header (originally called a leader). The first 8 bits of the header specified control information. The next 8 bits specified an address—the destination address if the leader is from host to IMP, the source address if the leader is from IMP to host. The third field is an 8-bit link identifier; as we said earlier, links are logical end-to-end connections. Finally, the last 8 bits of the original 1822 leader were reserved.

The 1822 messages for the host-to-IMP and IMP-to-host exchanges included connection management messages as well as notifications of errors and other anomalous conditions. In both the host-to-IMP and IMP-to-host communications, client data (from and to the host, respectively) were carried in regular messages (message type 0). All other messages were concerned with the management of the hosts, the IMPs, and the transportation of data. One of these was the Ready For Next Message (RFNM) message, which was used by IMP to control the flow traffic from the host; this, of course, is workload actuation. In addition, messages were defined that allowed host and/or IMP to inform its partner that it was going down and to specify the reason (for example, scheduled maintenance). Finally, a set of IMP-to-host messages was defined that carried fault information concerning the destination host, IMP, and so on.

---

2. Not to be confused with the host access protocol specified in RFC-907, which "specifies the network-access level communication between a host and a packet-switched satellite network" [1].

10.2.2.2 The IMP–IMP Protocol

Once data had been handed off from a host to its IMP, it was up to the IMP-to-IMP protocol to ensure that the data were successfully transported to the destination host. The overriding design goal was to embed as much management as possible within the subnet so as to give hosts the simplest possible interface. In particular this meant that, with the exception of nonrecoverable faults such as lost destinations, the ARPANET hosts did not participate in fault recovery.

There were two parts to the IMP–IMP protocol: (1) management of the transportation of data across a WAN line between directly connected IMPs; and (2) management of the transportation of data between source–IMP and destination–IMP. These were substantially decoupled and treated as point-to-point (between any two pairs of IMPs) and end-to-end protocols (between source–IMP and destination–IMP pairs). Thus management in the ARPANET's IMP–IMP protocol encompassed both the data link and network layers.

In terms of our data link taxonomy, the point-to-point management of data transport between IMPs was of intermediate complexity: It was connectionless and had no provision for parameter negotiation, but did include sufficient management to detect and recover from faults. Thus, unlike the data link protocols we examined in Part II there was no need for a multiplicity of frame types defined for purposes of connection establishment, workload scheduling, fault recovery, and so on. And because IMP–IMP connections were limited to point-to-point topologies there was no need to include local addresses.

Reliability in the IMP–IMP protocol relied on a Cyclic Redundancy Code similar to the CRC/FCS field used today with the data link protocols we considered in earlier chapters. After an IMP calculated the CRC for an arriving frame/packet and compared this with the CRC sent, if there was no discrepancy then it was assumed that no fault had occurred, the packet was processed, and a positive acknowledgment was sent to the sending IMP. These acknowledgments were piggybacked on reverse-flow packets much the same way as we saw with SDLC using fields in the IMP–IMP header. On the other hand, if the two CRCs did not match then a fault was estimated to have occurred, in which case the receiving IMP simply discarded the IMP–IMP packet. The sending IMP relied on timers to initiate retransmission; in other words, no negative acknowledgments were used in the protocol.

The end-to-end management part of the IMP–IMP protocol was considerably more elaborate than its point-to-point (data link) management and had three main tasks:

1. Fragmentation and reassembly of host messages longer than 1008 bits;

2. Flow control between source–IMP and destination–IMP; and

3. Fault detection and recovery, including out-of-order and duplicate packets.

To realize this management the protocol defined eight different packet types sent between source and destination IMPs (Figure 10.2). Host-to-host data were sent in Regular (type 0) packets. Additional packets were defined for the purpose of subnet (end-to-end) control. These packets were strictly internal to the subnet, meaning that they were never passed to or otherwise visible to hosts; in addition, they were treated as data by the intermediate IMPs—they had visibility only to the source and destination IMPs. The overhead of these control packets was significant, comprising during one monitoring experiment in 1974 nearly half the traffic [2].

The IMP–IMP protocol also included no fewer than four packets dedicated to managing the buffers of destination IMPs, reflecting the fact that before the era of inexpensive memory (some of the earliest IMPs had as little as 12 kbytes and consequently were severely constrained in their buffering of packets) storage management was a major concern. While intermediate IMPs merely forwarded packets as quickly as link bandwidth allowed, destination IMPs were constrained by the need to reassemble the packets of a multiple packet message before they could forward it to the destination host and thus free up its memory.

Fragmentation and reassembly meant that end-to-end flow control was vital to prevent overwhelming the limited buffer storage of the destination IMP. The source IMP-to-destination IMP protocol employed feedback for flow control: The destination IMP sent an RFNM to the source IMP to indicate that it could accept another packet. This flow control implied coupling of

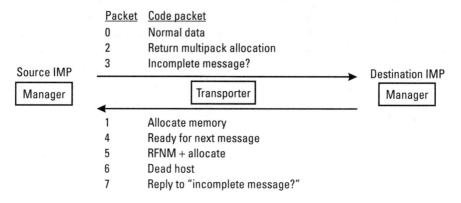

Figure 10.2 IMP–IMP protocol packet types.

the IMP–IMP and host–IMP (1822) protocols. When an RFNM IMP–IMP packet arrived at a source IMP the source IMP sent an 1822 RFNM message as outlined in BBN 1822:

> … when all packets arrive at the destination, they are reassembled to form the original message and passed to the destination Host. The destination IMP returns a positive acknowledgment for receipt of the message to the source IMP, which in turn passes this acknowledgment to the source Host. This acknowledgment is called a Ready For Next Message (RFNM) and identifies the message being acknowledged by name. [3]

A source IMP would therefore reserve storage by sending to the destination IMP a request allocation packet (type 1) to ensure that buffers were available. The destination IMP could ignore this request or it could respond with an allocation packet. A source IMP could return memory it was allocated by sending a Return Multipacket Allocation packet to a destination IMP that had previously allocated its buffers.

In addition, because the hosts and the host–host protocol assumed that the subnet was fault free, out-of-order delivery was unacceptable to the ARPANET designers two of whom wrote

> [t]he task of the ARPA Network source-to-destination transmission procedure is to reorder packets and messages at their destination, to cull duplicates, and after all the packets of a message have arrived, pass the message on to the destination Host and return an end-to-end acknowledgment called a Ready For Next Message (or RFNM) to the source. [4]

That is to say, the end-to-end management of the IMP–IMP protocol was designed to detect and recover from those faults that were introduced by the dynamic routing and datagram forwarding of the ARPANET. Finally, there were also IMP–IMP packets with which the destination IMP could also send explicit fault estimates. For example, if a destination host was not up then the IMP sent a Dead Host message.

10.2.2.3 The ARPANET Host-to-Host Protocol

The last component of the original ARPANET protocol suite was the ARPANET Host–Host Protocol (AHHP) [5]. The AHHP consisted of a set of control messages exchanged between hosts to manage the flow of data. Three AHHP control messages were used to start and terminate connections between source and destination hosts. An additional three were used to actuate (reserving and returning) the buffers of the destination host; there was likewise a requirement for memory management similar to what we just discussed with regard to the IMP–IMP protocol. Finally, to aid in fault detection (bandwidth estimation) the AHHP included two messages that effected an end-to-end "loopback."

10.3 Internetting

Notwithstanding the success of the ARPANET and its various protocols, by the early 1970s many had concluded a new approach was required. A particular concern was that, while the management mechanisms of the IMP-to-IMP and source IMP-to-destination IMP protocols resulted in the ARPANET being a very reliable transporter of data, the fact that the AHHP was predicated on a reliable transport service constituted a serious design flaw, one that limited the scope of the ARPANET in terms of potential communications technologies.

And this was precisely what ARPA was funding in the early 1970s, most notably an experimental packet-radio network developed by SRI International for use in the San Francisco Bay area. Such networks were inherently unreliable, and "opening" up the ARPANET to these networks had major implications in terms of the operation of the ARPANET protocols. There was, however, an alternative to requiring that new networks implement the IMP-to-IMP and source IMP-to-destination IMP protocols: develop a new protocol for internetworking or "internetting" these various networks, one that was not predicated on any given level of reliability or service. Cerf wrote about this some 20 years ago:

> The successful implementation of packet radio and packet satellite technology raised the question of interconnecting ARPANET with other types of packet networks. A possible to solution to this problem was proposed by Cerf and Kahn [CERF 74] in the form of an internetwork protocol and a set of gateways to connect the different networks. [6]

It was this protocol that eventually gave rise to TCP/IP.

10.3.1 The Transmission Control Program

What Cerf and Kahn discussed was the creation of what they called the Transmission Control Program (*sic*) to replace the Network Control Program of the ARPANET. The aim was to implement a "simple but very powerful and flexible protocol which provides for variation in individual network packet sizes, transmission failures, sequencing, flow control, and the creation of process-to-process associations" [7].

The challenge of constructing such a protocol was to accommodate different networks that might vary in terms of:

1. Addressing;
2. MTU size;
3. Performance (delay, throughput, and so on);
4. End-to-end reliability, and;
5. Instrumentation (such as state information) (Figure 10.3).

To concatenate networks of varying capabilities, Cerf and Kahn placed gateways to prepend the appropriate local headers (for example, the ARPANET host–IMP leader). What this meant was that the ARPANET and its protocols were not so much superseded as subsumed. The ARPANET remained an operational network but now would be only one of many networks to be "internetted" together with the new protocol. The philosophy behind this was one of abstracting, or encapsulating, the details of the component networks (transporters). To quote from an exposition of the idea:

> It should be noted that this approach, known as *encapsulation*, has some distinct advantages in the interconnection of networks. It is never necessary to build a "translation" device mapping one network protocol into another. The Internet layer provides a common language for communication between hosts and gateways, and can be treated as simple data by each network. [8]

A crucial change that went hand in hand with this was the division (i.e., modularization of management) of Cerf and Kahn's monolithic end-to-end protocol into several pieces. Credit for this is due in large part to the efforts of the late Jon Postel, who trenchantly argued that:

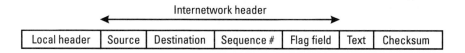

Figure 10.3 Original TCP internetwork packet (from [7]).

We are screwing up in our design of internet protocols by violating the principle of layering. Specifically we are trying to use TCP to do two things: serve as a host level end-to-end protocol, and to serve as an internet packaging and routing protocol. These two things should be provided in a layered and modular way. [9]

Postel argued for a basic design principle that said reliability mechanisms should, if employed at all, be used at the highest level of the protocol stack. And, as we saw in the last chapter, there are compelling reasons for such an architecture. First, end-to-end transport introduces the possibility of what we have called relay faults, from which stage-by-stage bandwidth (fault) management cannot recover. Second, eliminating retransmission and flow control from the intermediate systems enhances their bandwidth by simplifying their forwarding processes. In addition, not all clients desire reliable transport. Timeliness is much more important, for example, with multimedia or other real-time applications.

Postel prevailed in this argument and the subsequent developments owe much to his approach. The consequence was, to quote from Cerf's *Final Report of the Stanford University TCP Project,* that, while "[o]riginally designed as a monolithic internet protocol, the internet aspects were separated into a distinct protocol layer in early 1979 with the publication of version 4 of TCP" [10].

The final result of this effort was the TCP/IP architecture, or as it is also known the IP protocol suite with its four core protocols (Figure 10.4):

1. IP, which defines an end-to-end transporter that is connectionless and which offers its clients only best effort delivery;

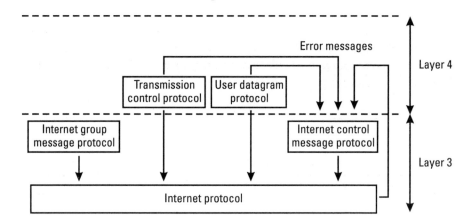

Figure 10.4 IP protocol suite.

2. ICMP, an ancillary protocol that complements IP by providing a delivery mechanism for bandwidth monitoring (measurement and/or estimation) information principally concerning faults that may occur in an IP network;

3. TCP, a connection-oriented/reliable protocol; and

4. UDP, a connectionless/best effort protocol.

10.3.2 End and Intermediate Systems in TCP/IP

The TCP/IP architecture also erased much of the distinction between end and intermediate systems. Recall that in the ARPANET hosts were presented with a very reliable abstraction by the subnet, so much so that they communicated with IMPs using different protocols than the IMPs used among themselves. With TCP/IP, in contrast, IP is used between end systems and intermediate systems, between intermediate systems, and between end systems directly if they share a transporter (data link). In fact, since every IP system, end or intermediate, must implement IP and ICMP, only two differences remain between the two types of systems (Figure 10.5):

1. An intermediate system is by definition multihomed *and* must be able to concatenate the IP networks (transporters) to which it is connected. An end system, single homed or multihomed, cannot concatenate.

2. An end system by definition has one or more upper layer protocol clients/destinations (otherwise there are no data for the IP transporter to transport). An intermediate system, on the other hand, need not implement any upper layer protocol, although typically

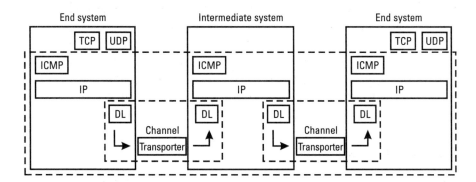

Figure 10.5 IP protocol stacks: end versus intermediate systems.

both TCP and UDP are included to support such applications as Telnet and SNMP.

10.4 The Internet Protocol

As we said earlier, effecting basic end-to-end transport in a TCP/IP network is the responsibility of the Internet Protocol. Whereas the IMP–IMP protocol did not extend past the subnet and was completely hidden from the ARPANET hosts, IP was designed to be a true end-to-end protocol by extending all the way out to hosts. The current version is IP Version 4 (IPv4), and the next-generation protocol, not widely deployed yet, is IP Version 6 (IPv6). Although the most conspicuous difference between the two versions is that IPv6 defines a larger address to help keep up with the growth of the Internet, there are others but we postpone discussing these until we have first explored IPv4.

Another aspect of IP's design is worth stressing at the outset: Like the original ARPANET, the goal with IPv4 was to have any-to-any connectivity using peer management mechanisms. The task set of an IP transporter is the any-to-any permutation of the IP addresses. Although this may seem common-place, many protocols such as SNA, designed primarily to support transaction processing and terminal handling, do not support communication between so-called outboard devices and systems, only between remote systems and the mainframes at the center of the network.

Within the constraint of peer control the designers of IP set out to create a protocol that was, if not management free, then at least very management light. IP's promise of "best effort" delivery meant that its approach to faults, buffer exhaustion, and other anomalous conditions was to simply discard the affected packet(s). In fact, from a management perspective it is scarcely an exaggeration to say that the Internet Protocol was designed in the negative, that is, more for what it would not do than what it would. As much as possible, it was the opposite of the IMP–IMP protocol. To quote from the original RFC:

> The internet protocol does not provide a reliable communication facility. There are no acknowledgments either end-to-end or hop-by-hop. There is no error control for data, only a header checksum. There are no retransmissions. There is no flow control. [11]

10.4.1 The IP Address Space

So what *does* IP do? According to its architects, "[t]he key feature of IP is the Internet address, an address scheme independent of the addresses used in the particular networks used to create the Internet" [8]. Of course, as we said in Chapter 9, one of the principal reasons for introducing an end-to-end protocol is precisely to define a global address space. To say that IP does little more than just this should not be seen as trivializing the crucial importance of the task: by providing a global address space that aliases the local address space(s) of the transporters (data links and networks) which it is internetting, IP achieved a decoupling of the Internet from these component networks. While IP addresses may be bound to local addresses by static definitions, a number of mechanisms have been defined such as the Address Resolution Protocol for automating this; we discuss these more in later chapters.

When the architects of IP were deciding the size of the addresses to be used, they faced the usual dilemma of overhead versus flexibility. What they chose was to use 32-bit addresses, which many at the time opposed as an unnecessary extravagance. The IPv4 address is conventionally expressed in what is known as dotted decimal notation: *a.b.c.d.*, where each of these are 8-bit groups converted into their decimal equivalent between 0 and 255. We will on occasion abbreviate these as A^{IP}, where the superscript indicates the protocol.

The ARPANET had employed a hierarchical address space, albeit of very limited size. In their first attempt [12] at the problem, the IP architects followed this by partitioning the 32 bits into an 8-bit network field (defining the network number) and a 24-bit host field. This meant that there could be a maximum of 256 networks each with up to 4 million hosts. This was not as unreasonable as it might appear today. By 1975, the ARPANET had approximately 50 IMPs and 70 hosts [4], and in any case the Internet itself was intended only to connect large providers such as Transpac, Telenet, and, of course, the ARPANET.

What changed all this and disrupted their calculations was the explosive growth of computer connectivity that came with the arrival of LANs and inexpensive modems with sophisticated error correction. Because of this new demand, the IP addressing scheme was modified [13] to allow many more networks by using additional bits for the network field for a portion of the total address space. This resulted in the division into three classes of addresses: (1) those IP addresses with $a = [1,126]$ were class A addresses, which retained the 24-host/8-network partitioning of the 32-bit addresses; (2) those addresses with $a = [128,191]$ were class B addresses, which used a 16/16 partitioning to provide networks each with up to 65,000 hosts; and (3) those addresses with $a = [192,224]$ were class C addresses, which is an 8/24 partitioning to provide networks each with up to 255 hosts. This later came to be called *classful* IP addressing, in contradistinction to *classless* addressing (see Chapter 13).

A further division of the 32-bit IP address came in 1985 with the introduction of subnetting, which basically consisted of taking bits from the host field of a classful address to define a third level of hierarchy. Subnetting allowed the definition of two or more subnets within a regular IP network number. However, subnetting complicated the routing of IP packets since it entailed the distribution of additional information called subnet *masks*. We discuss this in Part IV.

10.4.2 IPv4 Header

Because IP was a datagram protocol this meant that every IP packet had to contain both the source and destination IP addresses, constituting an overhead of 64 bits per packet. Beyond this, the IP packet contained additional fields to support IP's limited management mechanisms, notably fragmentation and reassembly; multiplexing upper layer protocols; and detection of faults, header and otherwise. However, unlike the IMP–IMP protocol, there is only one type of IP packet, so no field was needed to indicate packet type. The format of IPv4 packet headers is shown in Figure 10.6.

Let's start with fragmentation and reassembly. Fragmentation is an instance of workload management of degree: An RFS to transport a payload of k bytes is transformed into two or more RFSs to transport smaller payloads. And it was considered key to IP: "The internet protocol implements two basic functions: addressing and fragmentation" [13].

Although IP was designed to be simpler than the IMP–IMP protocol, this is one area where its management is more complicated. In the IMP–IMP protocol fragmentation and reassembly was performed, if at all, only at those IMPs at the edge of the subnet, that is, those directly attached to hosts. Intermediate IMPs did not have to concern themselves with it, in part because the

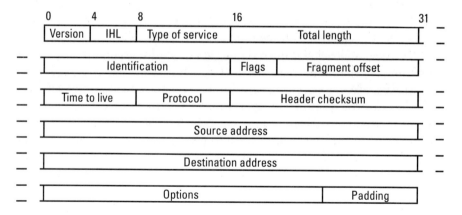

Figure 10.6 IPv4 packet header.

homogeneity of the ARPANET WAN links ensured that all could support IMP–IMP PDUs 1008 bits long. On the other hand, because IPv4 was intended to be an open protocol the design decision was made to require every IP system be able to support fragmentation of IP packets (reassembly was only executed by the destination host) since it could not be assumed a priori that any given network could transport a packet of arbitrary size.

The consequence of this complexity is that no fewer than three fields in the IPv4 header are involved in fragmentation and reassembly. First off, when an IP packet is created it is given a unique packet number, carried in the *Identification* field (16 bits). When an IP intermediate system must fragment a packet, all fragments are given the same 16-bit number to aid in their reassembly at the destination intermediate system. Next, each IP header contains a *Fragment Offset* field (13 bits) which indicates the position of this fragment in the datagram; this is necessary because IP does not guarantee in-order deliver. Finally, within the Flags field (3 bits), the third bit indicates whether this packet is the last fragment; without this bit, the destination host cannot determine if it has received all fragments that constitute the original packet. We should also note that another bit of the Flags field allows an IP system to prevent any intermediate systems from fragmenting a packet; this *do not fragment* bit is used, for example, to prevent an executable program from arriving partially at a destination and causing a mishap.

Next is the question of multiplexing and supporting multiple upper layer protocols. Toward this end every IPv4 header includes a Protocol field (8 bits) that identifies the upper layer protocol of the PDU encapsulated in the IP payload. As we noted before, ICMP (and IGMP), TCP, and UDP are all clients of IP and each has its own protocol number (ICMP 1, IGMP 2, TCP 6, UDP 17) carried in the Protocol field of their respective IP packets. Over the years dozens of additional upper layer protocols, mostly experimental or of limited use, have been defined and been assigned IP protocol numbers (see the latest version of the *Assigned Numbers* RFC for a list).

The remaining fields in the IPv4 are all, in one way or another, concerned with fault detection. The most obvious example is the Header Checksum (16 bits) field, which provides simple fault detection for the fields in the header and is calculated by relatively primitive means of taking the one's complement arithmetic sum of the header 2 bytes at a time and then the one's complement of that (as opposed to the more sophisticated mechanisms discussed in Chapter 3). The rationale for excluding the payload from the checksum span was both convenience and philosophy: convenience, since a header-only checksum would take less computation in intermediate systems, a nontrivial factor as this was implemented in software; and philosophy, since the IP design predicate was that clients desiring reliable transport would implement the necessary management mechanisms in their upper layer protocols. In fact, just as with ATM and its header-only checksum, IP's header checksum was intended merely to keep corrupted packets from being forwarded and consuming network bandwidth.

The prospect of PDUs being forwarded endlessly like so many *Flying Dutchmen,* looping around and wasting resources along the way, is a major concern in any network, and this is particularly true when, as in IP, there is no central mechanism to coordinate routing. Such faults may be caused by corrupted headers but a much more likely source is inconsistencies in the routing tables of IP intermediate systems. It was to detect this type of fault and to prevent the effects from lasting indefinitely that the IP architects included in IPv4 headers the *Time to Live* (TTL) field (8 bits), which was originally intended to record the lifetime of a packet, to be decremented as it transited the IP networks by each IP router as it was forwarded. But because time stamps were unreliable, this was quickly converted into a hop count, measuring not the time elapsed but the number of networks transited. In either case the effect was the same: When the TTL field reached zero the packet was not forwarded but instead discarded. A fault message could then be sent using ICMP (see later discussion). Fault recovery, of course, requires other means such as routing updates to correct the inconsistent tables and retransmission to transport the discarded packet(s).

The last type of fault that was considered by IPv4's architects was the truncation, for example by some software or hardware error, of an IP packet as it is being forwarded. A truncated or otherwise incomplete packet would not be detected by the header checksum. To detect these faults IPv4 includes a *Total Length* field (16 bits). These data are also used to detect reassembly faults when used in conjunction with the Fragment Offset field, to determine if the received fragments contain all of the original IP packet.

We should not leave our discussion of IPv4 without mentioning one of its more intriguing features, the *Type of Service* (TOS) field (8 bits). This was originally included to allow end systems to select different routes according to the delay, throughput, reliability, and precedence desired for an IP packet. However, no router vendor ever implemented support for multiple types of service; typically, this field was never even examined and was implicitly assumed to be set to 0 in all IP packets. There were a number of reasons for this disuse of what seems a very handy feature, but perhaps the most compelling was that its use would have required that the IP routers maintain distinct routing tables for each TOS supported, which would have consumed significant resources. Because there was little demand (no major applications were ever written that used different types of service), the TOS routing was never widely implemented.

10.4.3 IPv6

Notwithstanding the seeming extravagance of 32-bit addresses, by the early 1990s the limited quantity of network numbers provided by IPv4's addressing was rapidly being exhausted by the explosive growth of the Internet fueled by traffic for the World Wide Web. In addition, experience had shown some IPv4 features to be ill suited to high-speed networking, while others, such as TOS, were deemed not worth the benefit (if any) that they brought. Consequently, the Internet Architecture Board started to draft a successor protocol, to be known as IPv6 since IP Version 5 had already been used to designate ST, a real-time streaming protocol now largely abandoned. In this section we outline the major features of IPv6 and how it differs from IPv4. Despite the changes, as we see IPv6 is a much less radical revision than IPv4 was relative to the ARPANET IMP–IMP protocol.

The first and most obvious difference between the two versions of IP is the size of addresses. As a result of the plunging prices of memory and communications bandwidth, the IPv6 architects felt free to "future-proof" the new protocol by using 128-bit addresses. This provides an address space so much beyond any conceivable demand that the likelihood of having to change addresses again is remote for the foreseeable future.

Another factor influencing IPv6's design was to eliminate fields and corresponding management mechanisms that had never been embraced or had fallen into disuse. As we can see from Figure 10.7, the IPv6 header omits the

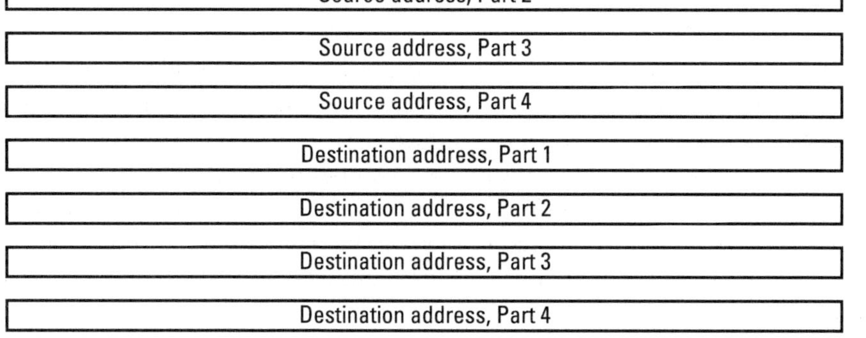

0	4	8	16	31
Version	Priority		Flow label	
Payload length			Next header	Hop limit
Source address, Part 1				
Source address, Part 2				
Source address, Part 3				
Source address, Part 4				
Destination address, Part 1				
Destination address, Part 2				
Destination address, Part 3				
Destination address, Part 4				

Figure 10.7 IPv6 packet header.

header checksum; in part this was because of the improved reliability of IP networks and relays, in part because the checksum failed to detect many types of faults, but mainly the header checksum was eliminated from IPv6 because router vendors had, in the interest of improved performance, ceased calculating it.

At the same time, IPv6 preserved the option of including additional management information by means of an Option field. We do not discuss these here but such extensibility recalls PPP's mechanism for incorporating header fields to support various control protocols.

IPv6 does, however, retain the TTL field from IPv4, now renamed the *Hop Limit* (8 bits). As with its predecessor, the purpose is one of fault detection: A routing fault that might otherwise cause a network meltdown can be detected if the Hop Limit is exceeded.

10.4.3.1　No Fragmentation

Another casualty of the "need for speed" that came to dominate the race for the fastest IP routers and switches was support for fragmentation. Fragmentation disrupted the smooth processing of IP packets because a router had to decide, based on the packet's next hop, if an additional set of tasks (creating the fragments) had to be executed, in addition to which, it then had to check if the *do not fragment* bit had been set by the client. Consequently, IPv6 routers do not fragment IP packets that are too large to be transported by the next stage transporter. When such PDUs are received, the router simply discards them; in addition, the router may send an ICMPv6 message (see later discussion) back to the IP client (source end system) indicating that the packet was too big and also information about the size of the largest packet that *could* be transported. Alternatively, every IPv6 network by definition must be able to transport packets that are up to 576 bytes in length. IPv6 packets no larger than this will never be discarded.

10.4.3.2　Priority and Flows

Despite the fact that IPv4's TOS mechanism was never really implemented, many in the IP community still felt that there was a need to provide clients with different rates of service according to their needs. For example, as we saw earlier, multimedia and other real-time traffic do not benefit from retransmission and multimedia decoders (estimators), in particular, experience high fault rates when there is jitter in the arrival of packets. File transfer, on the other hand, is largely immune to response time and jitter and is concerned mainly with bulk transport at the least cost.

To meet these needs the IPv6 architects included two fields in the IPv6 header, *Class* (originally called *Priority* and with 4 bits, now 8 bits in the revised IPv6 standards) and *Flow Label* (originally 24 bits, now 20 bits in the revised IPv6 standards). *Priority/Class* is largely equivalent to IPv4's *TOS* field in intent, but awaits further definition; currently the field has a default value of all 0's. Whether *Priority/Class* will be more successful than its IPv4 predecessor is uncertain.

The purpose of Flow label is to support a new forwarding mechanism, namely, flows. A flow is defined in the IPv6 RFCs as "a sequence of packets sent from a particular source to a particular (unicast or multicast) destination..." [14]. The idea behind flows is that conventional datagram forwarding was never designed with a view to supporting traffic like multimedia in which thousands, millions, or even more packets might be sent all between the same source and destination(s); indeed, IPv4 and the IMP–IMP protocol were originally designed to support remote computing, which might entail exchanging at most a few dozen packets for file transfer. Seeking ways to speed IP routing, researchers came up with the idea of "tagging" these packets with a common identifier that would then be used by intermediate systems to determine the next hop. These identifiers are carried in the *Flow label* field in IPv6 headers; note that each packet in a given flow will carry the same flow label value.

If this sounds similar to virtual circuit forwarding, it should: Flows are *de facto* virtual circuits, albeit without the static definitions of earlier virtual circuit networks, notably SNA and X.25. We saw in Chapter 9 that virtual circuit architectures allow forwarding decision making (scheduling) to be greatly simplified: Rather than calculating the next hop according to some optimality criterion for each packet, forwarding decisions were made based on a virtual circuit ID.

Such virtual circuit concatenation is exactly how flows work: A flow is set up by the initial packet and the remaining packets follow its path as long as the topology of the composite transporter (the IP network) does not change. And while some deny that flows are in any way similar to virtual circuits, the fact remains that, once a flow has been established, all subsequent packets are forwarded using their flow IDs, not their IP destination addresses. As we will see in the next chapter, a similar mechanism has been employed by IBM's Advanced Peer to Peer Networking protocol for more than a decade. We should also note the close relationship of flows and the ReSerVation Protocol (RSVP) (see, for example, [15]).

10.5   The Internet Control Message Protocol

As we just saw, both IPv4 and IPv6 are instrumented to detect faults in a packet and/or the IP network itself. A question confronting IP's architects was what should be done with the management data (measurements and/or estimates) that were produced. One option, of course, was to simply discard these along with the affected packet(s). Although this would not have contradicted IP's goal of best effort delivery, it nonetheless would have squandered hard-won information and, more importantly, done nothing to remedy the cause of the fault. Instead it was decided that "[e]rrors may be reported via the Internet Control Message Protocol (ICMP) which is implemented in the internet protocol module" [11]. That is to say, guided by the principles of modular design, IP's architects were led to the definition of a separate protocol—ICMP—to handle many of the configuration and fault management tasks that other, more monolithic designs have included in protocols themselves; and every IP implementation, end and intermediate system alike, was required to implement the protocol. To quote the IPv4 ICMP RFC, "[t]he purpose of these control messages is to provide feedback about problems in the communication environment, not to make IP reliable" [16].

We should stress that ICMP is both a protocol and a set of messages that *may* be sent by an IP system (end system or intermediate system) to indicate that certain faults have occurred in the processing of IP packets. Will an intermediate system that throws away a packet always send an ICMP message? No. Will an intermediate system that cannot forward a packet always send a "destination unreachable" ICMP message? Again, the answer is no.

Note that ICMP does not implement the fault detection itself but rather the various "fault detected" error messages shown in Figure 10.8 as well as certain configuration management requests and responses. In addition to faults detected by the IP protocol machine, those detected by the TCP and/or the UDP protocol machines are also reported using ICMP messages. Beyond this, ICMP for IPv6 has been extended in the area of configuration management (bandwidth monitoring) to include the functions of Address Resolution Protocol (see Part IV) for IPv4 networks and the IP Group Membership Protocol as well, although as with much of IPv6 the last is a subject of continuing definition. Table 10.1 lists the message types for ICMP for IPv4 and IPv6.

The IPv4 and IPv6 ICMP messages listed can be categorized in terms of bandwidth and workload/monitoring and control. We focus on some of the more important ones in the following subsections.

Destination Unreachable

When an IP packet is discarded because an intermediate system cannot forward it or because the destination end system does not implement the requested IP service, then this ICMP message is sent. This is bandwidth monitoring in that it gives the sending IP system feedback on the "topology" of the IP network—for example, that the destination end system is absent. Note that in

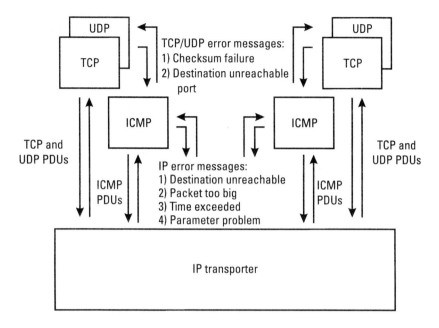

Figure 10.8 ICMP error messages: TCP, UDP, and IP.

IPv4 this message may also be sent by an intermediate system that cannot forward an IP packet with the *don't fragment* bit set but which is too large for any outbound network (transporter) to transport. ICMP for IPv6 defines a new *Packet Too Big* message for this.

Time Exceeded

This ICMP message may be sent by an intermediate system that discards a packet because the TTL/hop limit has been exceeded, which as we saw earlier is an indication of a likely routing loop. In addition, an IP system that discards packets awaiting reassembly will use this message to indicate the likely loss of one or more fragments.

Parameter Problem

An IP packet that violates the protocol or otherwise requests unsupported IP options will be discarded and the discarding system can use this message to report the fault to the source.

Source Quench

An IPv4 system that discards an IP packet because it lacks the resources (such as buffers) to process it may use this message to inform the source end system

Table 10.1
ICMP Messages: IPv4 Versus IPv6

IPv4	IPv6
Destination unreachable	Destination unreachable
—	Packet too big
Time exceeded	Time exceeded
Parameter problem	Parameter problem
Echo request	Echo request
Echo reply	Echo reply
Time stamp request	—
Time stamp reply	—
Information request	—
Information reply	—
Address mask request	—
Address mask reply	—
Source quench	—
Redirect	Redirect
—	Group membership query
—	Group membership report
—	Group membership termination
—	Router solicitation
—	Router advertisement
—	Neighbor solicitation
—	Neighbor advertisement

that it needs to reduce the rate at which it is sending. This is an attempt at workload (flow) control; if successful it actuates the arrival rate of traffic from the source.

Redirect

At its most elementary, an IP intermediate system that receives a packet will check if the network address is the same as that of the transporter from which it has been received. If it is then the intermediate system will use an ICMP Redirect to the end system to send packets directly to the destination end system.

Echo Request and Echo Reply

ICMP provides a loopback facility in the form of an Echo Request message, on receipt of which it is mandatory that an IP system (end or intermediate) send an Echo Reply message. This is the famous IP "ping" that many network administrators have used for years as a tool in troubleshooting network problems. As we saw in Chapter 4 and subsequently, a loopback, in this case via ICMP echo request/echo reply messages, is, in fact, an instance of bandwidth monitoring, specifically fault detection and/or fault isolation, in which the server in question is obviously the IP transporter.

The remaining IPv4 ICMP messages were not carried over to IPv6 ICMP because they were obsolete or otherwise superfluous; these include time stamp, information, and address mask requests and replies. For more details on the additional IPv6 ICMP messages consult Huitema [15], Loshin [17], or the applicable RFCs.

10.5.1 ICMP and MTU Discovery

We conclude our examination of ICMP by illustrating its use in MTU discovery, particularly important in IPv6 networks since intermediate system fragmentation is no longer supported. When an end system sends an IP packet, each intermediate system will attempt to forward it normally; however, if the packet is too large then an ICMP Packet Too Big (IPv6) or Destination Unreachable (IPv4) message is returned to the source, which must either resend the packet after fragmenting it itself or abandon the attempt. This process is illustrated in Figure 10.9.

10.6 TCP and UDP

The last protocols in the IP suite are TCP and UDP. UDP is a relatively simple protocol that, like IP, is connectionless and best effort in the service it offers its clients. Management in UDP for the most part is confined to multiplexing and demultiplexing traffic from multiple upper layer protocols (clients) which it does by means of ports, 16-bit numbers that uniquely identify UDP clients at source and destination IP systems (Figure 10.10); ports are similar to protocol type codes in IP and some of the data link protocols. Although UDP does include a checksum, and one that covers the data payload being transported (unlike IP's checksum, which is header only), use of the checksum is optional and is generally ignored. More effective for fault detection, the UDP header includes a field that indicates the length of the UDP PDU so that if any

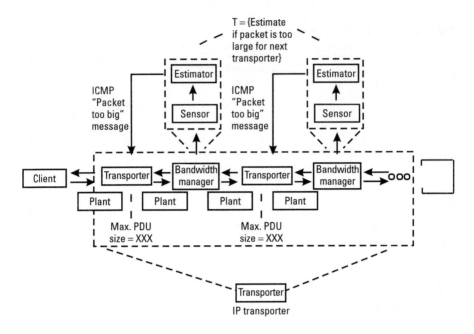

Figure 10.9 MTU discovery process.

0	4	8	16	31
UDP source port			UDP destination port	
UDP message length			UDP checksum	

Figure 10.10 UDP header.

fragments are lost the incomplete message is discarded. By definition, a UDP PDU is transported with no guarantees.

It may be surprising but, despite or perhaps because of its management-light nature, UDP is preferred by many application programmers who want to retain management of transport execution, for example, retransmission. Because UDP offloads so little from its clients (the upper layer protocols) it affords developers such latitude.

At the other end of the spectrum from the *laissez-faire* management of IP and UDP is TCP. All of the management that was kept out of IP was put into TCP. It is a connection-oriented, PAR protocol with flow control and other management designed to offer the most reliable transport service possible over all manner of underlying networks from the least to the most reliable. TCP will correct for out-of-order delivery and duplicate packets. In addition, TCP provides the same multiplexing and demultiplexing traffic from multiple upper layer protocols (clients) as UDP by means of the same mechanism of ports, that is, protocol type fields. Of course, TCP's management comes at the price of a PDU header much larger than UDP's (Figure 10.11).

The most important of these fields are the Sequence and Acknowledgment numbers, with which TCP protocol machines manage retransmission; the Window field, used in TCP flow control; the checksum, which is used to detect faults; and the six subfields within the Flags field (Figure 10.11(b)), which are used to indicate, for example, if the data are urgent. Thus, they provide an "out-of-band" signaling mechanism between the TCP protocol machines. We note that because TCP is full duplex, there is no provision for a workload scheduling mechanism to "turn around" the connection between the two protocol machines. That is, within the constraints imposed by flow control each protocol machine can send without requesting permission from or being polled by the other.

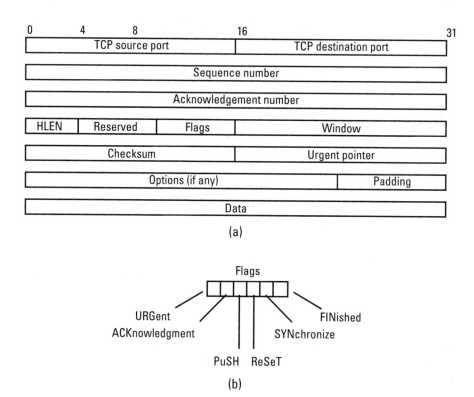

Figure 10.11 (a) TCP packet header and (b) flag subfields.

10.6.1 TCP Connection Management

We saw in Part II that connection-oriented data link protocols such as SDLC and HDLC used special frames (PDUs) to manage connection actuation—initiation, teardown, resetting, and so on. TCP does not use special PDUs for this purpose but rather exchanges equivalent management requests and data using several of the subfields of the Flag field of the TCP header. We should note that, like the ARPANET Host–Host protocol, TCP protocol machines are peers—either side can initiate connection setup and either can initiate connection teardown.

A TCP connection request is sent by a TCP protocol machine setting the SYN subfield in a TCP segment. This flag is so named because, as with any connection-oriented protocol, there is a need for the protocol machines to exchange state information. With TCP the state information includes type of payload data to be carried in the connection (via the source and destination port numbers); the initial sequence numbers that each side will use; and the maximum amount of data that each will accept. In addition, a TCP protocol machine can specify a maximum transmission unit, called the maximum segment size (MSS), that it is willing to accept. Because the TCP protocol specifies that there are three steps to this process (connection request, connection request response, and acknowledgment of the response), it is commonly known as a *three-way handshake*.

This brings us to another difference between TCP and the data link protocols, namely, the sequence numbering. Instead of each TCP PDU (called a *segment*) being assigned a sequence number that is one greater than the sequence number of the preceding PDU, sequence numbers are related to the size of the PDUs being sent. That is to say, the size of TCP's sliding window does not measure the number of segments outstanding but rather the number of bytes. Because the sequence numbers are 32 bits, this means that up to 4 GB can be outstanding before sequence numbers wrap. When TCP was designed in the late 1970s this seemed such a large number that few could foresee difficulties ever arising. However, in today's high-speed networks this is no longer the case and much research is going into solving the problem.

The teardown actuation of a TCP connection is effected by sending a segment with the FIN flag set. This indicates to the receiving protocol machine that no more data will be sent over the connection. However, since TCP is symmetric it is possible that the receiving protocol machine may still have data to send and can continue sending it.

10.6.2 Fault Recovery and Retransmission

Fault recovery in TCP relies on the same mechanisms as, for example, we saw with SDLC (Figure 10.12). In terms of the management we discussed in Chapter 1, these can be decomposed into closed-loop and open-loop maintenance:

1. *Closed-loop maintenance:* The sending protocol machine calculates a checksum for the payload it wishes to transport. At the destination the same checksum calculation is made, and a fault estimated if there is a discrepancy, which is then returned to the sender by the absence of a positive acknowledgment. As in SDLC, acknowledgments in TCP specify the *next* byte expected.

2. *Open-loop maintenance:* A timer is set by the sending protocol machine when it sends a PDU; should the receiving protocol machine not send an acknowledgment before the timer expires then retransmission is initiated.

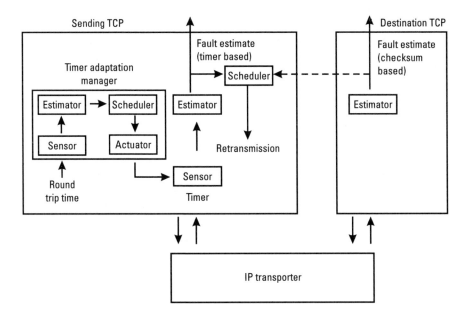

Figure 10.12 TCP fault detection.

The performance of such open-loop managers with their timer-based retransmission depends on how well the retransmission time-out (RTO) value matches the round-trip time (RTT)[3] between source and destination. If the RTO is much greater than the RTT, then retransmission will be initiated later than is optimal. On the other hand, if the RTO is much less than the RTT, then retransmission will be initiated too quickly, resulting in unnecessary traffic. In decision theory the former event is known as a rejection while the latter is called a false alarm.

We saw earlier that SDLC, HDLC, and LLC2 all provide extensive configuration parameters that can be actuated, albeit statically, to tune and optimize performance of the data link with respect to retransmission and fault recovery. The difficulty that confronted TCP's designers was that, unlike the data link protocols we considered in Part II, it was not possible to obtain good a priori estimates of the latency of the transporter, in TCP's case the underlying IP network(s) over which its traffic is carried since IP networks can range in size from a single high-speed LAN to a complex mesh of low-speed WAN circuits. To quote from TCP's RFC-793, "[b]ecause of the variability of the networks that compose the internetwork system and the wide range of uses of TCP connections, the retransmission timeout should be dynamically determined" [18]. Some authors refer to this as an *adaptive retransmission strategy*, the cornerstone of which is an estimation of round-trip times executed by the estimator within the timer adaptation manager in the TCP protocol machine (Figure 10.12).

Over the years a number of estimation mechanisms (algorithms) have been proposed, starting with RFC-793 itself, which gives the following formula for calculating an estimate of RTT that it calls the *smoothed* RTT (SRTT):

$$SRTT_{new} = (\alpha * SRTT_{old}) + [(1 - \alpha) * RTT]$$

This is recursive parameter estimation. And with each updated estimate for $SRTT_{new}$ the scheduler in the timer adaptation manager actuates the protocol machine's retransmission timer; the original specification called for the RTO to be twice the updated estimate for $SRTT_{new}$, but this was revised in 1989 to include take account of jitter (variance) in the RTT measurements. The new algorithm relies on the mean variance rather than standard deviation because the latter is more computationally intensive [19]. The new estimator is given by a pair of equations

$$SRTT_{new} = SRTT_{old} + g(RTT - SRTT_{old})$$

---

3. Also known as round-trip delay (RTD).

$$\text{Dev}_{new} = \text{Dev}_{old} + h\left(\left|\text{RTT} - \text{SRTT}_{old}\right| - \text{Dev}_{old}\right)$$

where Dev is the mean deviation and the coefficients g and h are "gain" constants, generally set to equal 0.125 and 0.25, respectively. The value of RTO is then calculated from the equation

$$\text{RTO} = \text{SRTT}_{new} + 4\,\text{Dev}_{new}$$

Of course, to derive any of these estimates, the TCP protocol machines must measure the RTT periodically. Conceptually, this is straightforward: Simply set a timer to measure the elapsed duration between when a segment is sent and when its acknowledgment is returned by the destination. However, over the years a number of complications have been discovered concerning so-called ambiguous acknowledgments, which occur when a segment has been retransmitted and then acknowledged, a circumstance that leaves the sending TCP protocol machine uncertain as to which copy of the segment is being acknowledged. The solution to this is known as Karn's Algorithm, named after its author, and has two parts. First, RTT measurements should be ignored for any segment that has been retransmitted; second, if an RTO timer expires and initiates retransmission, then the value of RTO is increased to "back off" the retransmission of further segments until the IP network stabilizes [20].

10.6.3 Flow Control

Recall that much of the ARPANET IMP–IMP protocol was concerned with scheduling the flow of traffic (workload management) from a source host to prevent overwhelming the limited buffers and processing capabilities of the destination IMP and host. And whereas the ICMP *Source Quench* message provides limited flow control, it is too coarse to constitute really effective actuation of the arrival of traffic from TCP/IP end systems. Therefore, much of TCP's initial design effort as well as subsequent modifications have been focused on flow control to maximize throughput while avoiding congestion.

We saw in Part II that, with data link protocols such as SDLC and its derivatives, a receiving protocol machine can actuate traffic by either explicitly indicating its (un)willingness to receive any PDUs using *Receiver (Not) Ready* frames or by withholding acknowledgments and thus exhausting available sequence numbers. This latter approach would not be viable with TCP because the sequence number space is so much larger. Instead, each time a TCP protocol machine sends a segment it includes in the window field the maximum number of bytes it is willing to specify. Because the field is 16 bits, this means that with standard TCP the largest window is 65,535 bytes. Allowing a receiving protocol machine to indicate its maximum size flow control is an actuation (of the size) of the sliding window.

Beyond this, however, TCP has been augmented to react to increasing response times to prevent congestion in end and intermediate systems alike. The two principal mechanisms are called *slow start* and *multiplicative decrease*. Both techniques rely on each TCP protocol machine maintaining a second flow control window, above and beyond the window advertised by its peer station with each segment. Called the *congestion limit window*, in normal times this equals the receiving protocol machine's advertised window. However, when congestion begins, as indicated by lost segments, the size of the congestion limit window is reduced by half and the RTO value is increased exponentially. This continues until either the congestion limit window equals the maximum segment size (MSS) fixed in the connection setup or until no further segments are lost.

Slow start, on the other hand, begins in the opposite direction. When a TCP connection is established, slow start dictates that the congestion window be actuated down to the MSS limit and increased gradually by one segment size per acknowledged segment sent. The same scheduling mechanism is used when recovering from congestion, when the congestion limit window has been reduced but is now to be increased. An additional nuance of flow control in TCP is that, when recovering from congestion-induced window constriction, a TCP protocol machine will slow the rate of increase in the congestion limit window size once the window has reached half its original size. At this point, the protocol machine enters a *congestion avoidance* phase, and the window size is incremented only when all outstanding segments have been acknowledged.

An unforeseen consequence of TCP's windowing management is that it is possible for two TCP protocol machines to get stuck allowing small window sizes: A receiving protocol machine advertises a small window size because of, for example, a temporary buffer shortage; the sending protocol machine sends a small segment, which is processed, and a new advertisement is sent for another small window, *ad infinitum*. This final complication is called the *silly window syndrome* (SWS). To prevent this, TCP's flow control management has been modified to prevent receiving protocol machines from advertising small windows and to prevent sending protocol machines from sending a segment unless either a full-sized segment can be sent; or a segment equal to 50% of the largest advertised window size can be sent; or all the data that need to be sent fit in a small segment and there is no unacknowledged segment(s).

10.7 Summary

In this chapter we have detailed the management mechanisms used in internet-working protocols, starting with the original ARPANET and up to TCP/IP. We discussed how BBN 1822 specified the physical interface and signaling used between hosts and IMPs as well as a set of messages to be used in host-to-IMP and IMP-to-host communication. We saw that the IMP-to-IMP communication had two parts: (1) a protocol used over the serial links connecting pairs of IMPs and (2) a protocol that was used to communicate between source and destination IMPs. The IMP-to-IMP protocol employed positive feedback from a receiving IMP to a sending IMP: a frame that arrived correctly (as indicated by checksums) was acknowledged. However, negative feedback was not sent: A corrupted frame was simply discarded by the receiving IMP, and the sending IMP initiated retransmission by using timers to detect lost or corrupted frames.

We then moved on to the internetworking model of data transport as first proposed by Cerf and Kahn and then modified by Postel to embrace a two-tiered modularization of management into a management-light Internetworking Protocol and a management-heavy Transmission Control Protocol. We saw that while IP was connectionless, offering only best effort delivery, TCP provided the most robust management possible. However, we stress that in this chapter we have seen nothing in TCP's management that we have not also encountered in the data link protocols we surveyed in Part II. This brings us back to one of the major themes of this book: All of the management mechanisms in computer networking protocols can be reduced to a relatively small set of primitives that recur again and again. TCP is merely the highest level of this recursion, encapsulating and managing end-to-end transport over IP.

References

[1]   RFC-907, Most Access Protocol Specification, July 1984, p. 1.

[2]   Kleinrock, L., W. E. Naylor, and H. Opderbeck, "A Study of Line Overhead in the ARPANET," *Commun. ACM,* Vol. 19, No. 1, January 1976, p. 5.

[3]   *Specification for the Interconnection of a Host and an IMP,* BBN 1822, p. 3-2.

[4]   McQuillan, J. M., and D. C. Walden, "The ARPA Network Design Decisions," *Computer Networks,* Vol. 1, 1977, p. 284.

[5]   Padlipsky M. A., *A Perspective on the ARPANET Reference Model,* RFC-871, 1982.

[6]   Cert, V., *The Internet Activities Board,* RFC-1120, 1989, p. 1.

[7] Cerf, V., and R. Kahn, "A Protocol for Packet Network Interconnection," *IEEE Trans. Commun.*, Vol. COM-22, No. 5, 1974, p. 648.

[8] Leiner, B., et al., "The DARPA Internet Protocol Suite," *IEEE Commun. Magazine*, Vol. 23, No. 3, Mar. 1985, pp. 29–34.

[9] *Comments on Internet Protocol and TCP,* Internet Engineering Note 2, August 1977, p. 1.

[10] Cerf, V., *Final Report of the Stanford University TCP Project,* Internet Engineering Note 151, p. 1.

[11] Postel, J. (Ed.), *DoD Standard Internet Protocol,* RFC-791, 1981, p. 3.

[12] *DoD Standard Internet Protocol,* Internet Engineering Note 123, Dec. 1979, p. 22; and *DoD Standard Internet Protocol,* Internet Engineering Note 128/RFC-760, January 1980, p. 22.

[13] RFC-791, pp. 2, 23.

[14] *IPv6 Specification,* ed. R. Hinden, RFC-1883, December 1995, p. 28.

[57] Huitema, C., *IPv6 The New Internet Protocol,* Upper Saddle River, NJ: Prentice Hall, 1996.

[16] *Internet Control Message Protocol,* RFC-792, 1981, p. 1.

[17] Loshin, P., *IPv6 Clearly Explained,* San Francisco, CA: Morgan Kaufmann, 1999

[18] *DoD Transmission Control Protocol,* RFC-793, 1983, p. 162.

[19] Stevens, W. R., *TCP/IP Illustrated,* Vol. 1, Reading, MA: Addison-Wesley, 1994, p. 300.

[20] Comer, D., *Internetworking with TCP/IP,* Vol. I, 3rd edition, Upper Saddle River, NJ: Prentice Hall,1998, p. 191.

11

End-to-End Management: SNA and APPN

11.1 Introduction

In this chapter we conclude our examination of management in end-to-end protocols by looking at IBM's Systems Network Architecture (SNA). We should study the management in SNA's protocols for at least two reasons. First, SNA illustrates a completely different type of network architecture than TCP/IP, namely, a terminal–host model of communication that relies on centralized (master/slave) management; second, SNA has a more extensive set of mechanisms for monitoring and controlling traffic than any other protocol architecture.

The chapter begins by discussing SNA's antecedents in the terminal–host networks that were deployed starting in the 1950s in such diverse areas as air defense (SAGE) and airline reservations (SABRE). Terminal–host networks differ from the computer–computer network model embraced by the ARPANET and its successors, most notably in the fact that support of any-to-any communication is not necessary because remote devices do not need to send data to each other. Even as the ARPANET researchers were advancing a model of peer communications and distributed management, terminal–host networking continued to be developed, culminating in the release of SNA in 1974.

We then introduce SNA's rather elaborate cast of characters and their management: network addressable units, paths and routes, path control network, subareas and domains, and so on. In sharp contrast to the almost *laissez-faire* management philosophy of the IP protocol suite, SNA relies on master/slave mechanisms and centralized management to realize a granularity of

349

control largely unobtainable in decentralized/distributed implementations. We detail these management mechanisms in the three principal end-to-end protocols of the SNA protocol stack, namely, the Path control, Transmission control, and Data Flow Control layers.

The chapter then moves on to discuss that management executed within an SNA network versus that management delegated to SNA end systems. Here, too, the contrast with TCP/IP's modularization of management could not be greater. Whereas the latter assumes that data links will be unreliable and that retransmission will only be attempted by end systems using TCP, SNA's management model is predicated on reliable data links, with fault detection in its end-to-end protocols limited to identifying lost packets. We will also look at one of SNA's greatest strengths, namely, its multilevel flow control.

Finally, we will briefly outline the management direction IBM has taken with its Advanced Peer to Peer Networking (APPN) and High Performance Routing (HPR) architectures. We will discuss the management changes that IBM has introduced in the new Path Control layer at the heart of APPN and HPR. These represent an effort to move SNA more toward the dynamic model of networking that has helped make TCP/IP so successful.

11.2 Terminal–Host Network Architectures

Like the ARPANET, the impetus behind the development of terminal–host networking was originally provided by the U.S. Department of Defense, which in the early 1950s implemented the SAGE (Semi-Automatic Ground Environment) network to collect and process radar information gathered from sites located across the United States. Soon afterward IBM and American Airlines developed the SABRE (Semi-Automatic Business-Related Environment) network to support thousands of airline reservation agents. (For more details on SAGE and SABRE, see [1, 2].)

Terminal–host networks were designed to meet a different set of requirements than today's Internet. Recall that this was the era of the mainframe computer. Before VLSI and the microprocessor, almost all processing and memory power was centralized in the data center. Terminals had little if any memory and no CPU, and even terminal servers were severely constrained in computing power. Given this relative abundance of computing power at the central site where the mainframe was located, it was natural that the data communications protocols that were developed were asymmetric in how the two ends, host and terminal, operated. The protocol mechanisms at the host side were responsible for scheduling access to the shared line, generally implemented in the form of polling, for error recovery and exception handling, and for managing the communications process generally. The resulting asymmetry of management, with the protocol station at the terminal side essentially a "slave" completely under the control of the master, namely, the host, was precisely what we saw with SDLC.

Many ramifications flow from such a design predicate but, beyond the locus of control, the most important concerns the resulting limitations on supported topologies and corresponding traffic flows. An immediate consequence for network design follows from such protocol design: two slaves, completely dependent on the master station at the host for managing communication, do not communicate directly in terminal–host networks (Figure 11.1). Again, consider the example of SDLC, where the task set of an SDLC data link consists solely of primary link station–secondary link station transport tasks, no secondary link station–secondary link station transport tasks.

This forces terminal–host networks into a hub-and-spoke topology, but this was no hardship in the data processing world up through the mid-1970s with its domination by mainframe computers. In addition, the on-line transaction processing applications that these networks address did not require terminal-to-terminal communications—two travel agents or bank tellers did not need to communicate with each other, only with the database that was located on the mainframe anyway.

At the same time as the ARPANET was beginning deployment in 1968, IBM introduced the Binary Synchronous (Bisync/BSC) protocol. This supported both the EBCDIC and the ASCII character sets and line control for

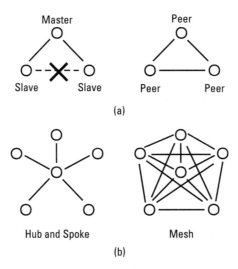

Figure 11.1 (a) Master/slave versus peer communications and (b) hub versus mesh topologies.

polling devices on multipoint links. Bisync was widely implemented by many vendors besides IBM, and continues to be a popular protocol for basic data communications applications such as automatic teller machines (ATMs) and point-of-sale (POS) devices. However, Bisync was largely superseded by what would prove to be the final word in terminal–host network protocols, IBM's Systems Network Architecture, which was introduced in 1974.

11.3   SNA: Concepts and Facilities

SNA is a study in contrasts to TCP/IP. Whereas the latter is highly dynamic and relies on peer management wherever possible, SNA is permeated by centralized management at all levels. At the heart of every SNA network, for example, is a small set of management applications that are centralized points of control for their respective domains, controlling and monitoring almost every detail of resource operation and traffic flow. As an architecture intended to service on-line transaction processing applications for largely commercial users, SNA's designers were less concerned about issues of survivability (recall, for example, that this was the focus of Baran's research on survivable networks without a single point of failure) than efficiency and stability. For corporations and other organizations unconcerned with hostile attacks on their data centers, SNA's emphasis on statically defined routes and centralized control of network resources was an acceptable trade-off in return for lower management information overhead.

In this section we lay out the management tasks of these control points, SNA's end-to-end protocols, and the various systems that make up an SNA network.

11.3.1   SNA Protocols and PDUs

SNA embraced the idea of a layered architecture introduced by the ARPANET designers, albeit with a different set of layers and correspondingly with a different modularization of management (Figure 11.2). As part of its layered architecture, SNA has three principal end-to-end protocol layers: the Data Flow Control, the Transmission Control, and the Path Control. A later revision of SNA split the Path Control layer into three sublayers.

In addition, above the Data Flow Control layer are two more layers of end-to-end protocols, Presentation Services and Transaction Services, but these are outside the scope of our survey. Finally, below the Path Control layer are the various data link protocols over which SNA can run and, in some cases,

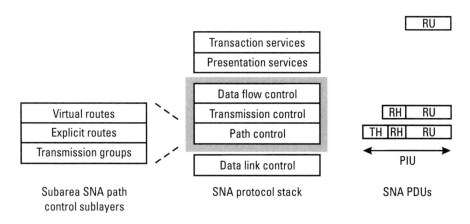

Figure 11.2 SNA protocol layers and PDUs.

other network protocols such as X.25 and Frame Relay over which SNA is tunneled (see Part IV).

11.3.1.1 The Path Control Layer

Like most L3 protocols, SNA's Path Control (PC) layer is responsible for concatenating data links and forwarding packets, which in SNA are called Path Information Units (PIUs), between end systems. The term *Path Control Network* (PCN) is used to refer to the collective set of PC protocol machines and data links, much like we used the term "IP network" in the last chapter. End-to-end transport of PIUs is the responsibility of the PCN, along with managing the global address space (more on this later).

The original SNA PC layer was monolithic and the end-to-end connection was called a *path*. In the early 1980s, however, IBM modified the PC layer by dividing it into three sublayers to support new functionality. These sublayers—Transmission Group Control (TGC), Explicit Route Control (ERC), and Virtual Route Control (VRC)—brought with them three new transport entities:

1. *Transmission groups* (TGs) are composite data links similar to the multilink PPP, in which two or more data links are bundled to increase the bandwidth and reliability. Multilink TGs were initially limited to similar types of data links but can now be heterogeneous, meaning, for example, that SDLC links can be grouped with FR circuits.

2. *Explicit Routes* (ERs) are one-directional point-to-point connections composed of one or more TGs concatenated together.

3. *Virtual Routes* (VRs) are composed of a pair of ERs that are comple-
    mentary, going in opposite directions relative to each other.

Among its advantages, this division of the PC layer allowed an SNA network to match its routing to asymmetric traffic flows or service requirements.
For example, if incoming messages to the mainframe(s) are relatively small in
size but result in much larger outgoing messages then the VR over which they
flow can be composed of two ERs with very different bandwidths, perhaps
composed of entirely different TGs. With the original SNA architecture, traffic
in either direction necessarily flowed over the same path.

SNA's address space is hierarchical: Global SNA addresses are divided
into two parts, called the subarea and element address fields, corresponding
approximately to IP's network/host division. SNA's addressing has been modified over the years from an original 16-bit global address to 48-bit addresses
most recently, with up to 15 bits allowed for elements. Also defined are smaller
address formats, called local addresses, that are either 6 or 8 bits long, and
which were included in SNA to reduce the storage requirements for smaller systems such as terminal controllers. Global addresses corresponding to these local
addresses are maintained by larger systems (called subarea nodes) to which the
terminal controllers are attached, and conversion between the two formats is
called the *boundary function.*

A major addressing difference between SNA and IP concerns how their
respective addresses are assigned to end and intermediate systems and why. We
saw in IP that the most important aspect of IP addressing was to alias the local
addresses of the underlying component networks, and that consequently each
interface of an IP system has its own IP address. In contrast, in SNA there
was no need, at least initially, to alias local network addresses because the two
data link technologies in the initial SNA were SDLC and the S/3X0 channel; as
we saw in Chapter 6, SNA does not have link station (i.e., local) addresses.
Instead, SNA assigns addresses on a system basis to so-called *network addressable units* (NAUs) within SNA systems. (We discuss NAUs in more detail
later.)

Given its evolving architecture and multiple types of addresses, it is perhaps not surprising that SNA defines no fewer than six different PIU formats,
which can be differentiated in terms of whether the addresses carried are global
or local; whether TGs, ERs, and VRs are supported; even whether the destination is an SNA or non-SNA system. A PIU's header, called the *transmission
header* (TH), contains a 4-bit field called the *format identifier* (FID) that carries
the packet type. When IBM divided the PC layer, for example, two new THs
were defined, called FID 4 and FID F, which included a number of new fields
to support the management of TGs, ERs, and VRs. FID 2 headers are used

when local addresses are employed; this means that the PIU in question involves a T2 SNA node (see later discussion).

Figure 11.3 shows the FID 4 and FID 2 transmission headers and their most significant fields, notably FID, sequence numbers, addresses, and packet size (reserved fields are shaded). Note the difference in addresses carried: The FID 4 header carries (48-bit) global addresses, whereas the FID 2 header carries (8-bit) local address. In addition, the FID 4 TH fields related to TG, ER, and VR management, particularly VR flow control, are identified although we will defer discussing their use until later.

Conspicuous by their absence from either FID 4 or FID 2 (or any other SNA) THs are any fields for carrying checksums. Recall from Chapter 9 that SNA's architects attempted to restrict checksum and timer-based fault detection to the data link layer, which is why SNA uses reliable data link protocols such as SDLC and LLC2; this is exactly the opposite of TCP/IP's modularization of management, in which only TCP maintained effective checksums and timers of any sort, and it was assumed that the data link protocols would be unreliable. In contrast, SNA's end-to-end protocols (PC, TC, and DFC) will only detect lost or missing PIUs. SNA's end-to-end protocols assume that the

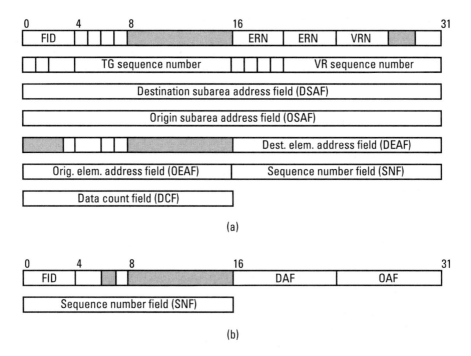

Figure 11.3 Transmission headers: (a) FID4 and (b) FID2.

data link protocols will detect and correct any latent faults, that is, packets that have been corrupted but *not* lost.

Finally, the PC layer may execute segmentation and/or blocking of PIUs into what SNA calls *Basic Transmission Units* (BTUs). Blocking combines multiple PIUs into one BTU. Segmenting, on the other hand, splits a PIU into multiple BTUs. This is done mainly for memory-constrained peripheral nodes. The BTU is the unit of data used that the PC layer requests the data link protocol transport.

11.3.1.2 The Half-Session Layers: Transmission Control and Data Flow Control

Above the PC layer are the Transmission Control and Data Flow Control layers, the combination of which is known as a *half-session.* In the grammar of SNA, two half-sessions plus a path or VR (i.e., an end-to-end connection through the PCN) constitute a session (more on sessions later). Half-sessions that manage the transportation of user data reside in SNA end systems, though we will see later that both end and intermediate system in SNA networks have half-sessions for other management reasons. An added reason for treating these two layers together is that in some respects they *are* melded, and that only one PDU header (called the Request/Response Header) is used by the two.

The actual process of transporting data begins when the DFC layer is handed a message to be transported by the Presentation Services layer above it. These messages, which may be client data (for example, an ATM debit request or an airline reservation response) or one of hundreds of management messages that SNA's architects have defined during the past two decades to manage connections, activate resources, and so on, are called *Request/Response Units* (RUs). An RU plus a Request/Response Header (RH) constitutes the uppermost protocol data unit, called the Basic Information Unit (BIU). RHs are 3 bytes long and their fields include data allowing fault recovery and flow control.

For each RU it receives from the PS layer the DFC protocol machine then generates a sequence number to uniquely identify the RU; this sequence number is used by both the PC layer (where it is carried in the SNF in the TH) and by a TC protocol machine to verify received sequence numbers. When the PC layer segments a PIU into two or more BTUs, each segment has the same TH, and the same sequence number; this enables the TC protocol machine at the receiving end system to reassemble the original RU.

Besides reassembly and sequence number verification, the TC layer also "paces data exchanges to match processing capacity [of the receiving end system] and enciphers data if security is needed" [3]. The TC layer is rather thin compared to other L4 protocols such as TCP and OSI's TP4. This is why some authors will, when comparing the protocol stacks, illustrate the relative management provided by showing the PC layer extending into L4 territory, with a corresponding reduction in the TC layer's thickness. This is understandable because the conventional role of the transport layer—to enhance the reliability of the lower layers—is not as important in SNA because it assumes that the underlying network is reliable. The specific options for an implementation of the TC layer are laid out in Transmission Services (TS) profiles.

The Data Flow Control (DFC) layer manages the end-to-end connection, that is, the session, scheduling it between the two clients (sessions are always point to point). Because a session may be half-duplex, and in fact most are, some management mechanism must be responsible for scheduling which end system is to send and which is to receive. This scheduling is executed by the DFC protocol machine. In addition, the DFC protocol machine also correlates request and response units. When a transaction request is sent in a request unit with a given sequence number the DFC protocol machine matches it to the subsequent response unit using the sequence number. Finally, the DFC protocol machine manages end-to-end retransmission; we discuss this in the next section. Just as the TC options are specified in TS profiles, options for the DFC layer are specified in Function Management (FM) profiles. Figure 11.4 shows the request and response headers and their respective fields.

11.3.2 Systems in an SNA Network

Whereas IP has two types of systems (end and intermediate), in SNA there are no fewer than five types of systems (or as they are sometimes called, nodes). The most important of these is the mainframe, which alone in SNA is referred to as a host (unlike IP, where any end system is a host); in SNA these are called Type 5 (T5) systems or also System/3X0 or simply S/3X0. Beyond running the very large database programs[1] that support on-line transaction processing, these T5 systems also play a crucial role in managing their SNA network(s) by means of the centralized management mechanisms in another host program called the virtual telecommunications access method (VTAM). In the early days of SNA there could be only one host in an SNA network; today, when multiple hosts can be in a network the region each manages is called a *domain*.

Next are SNA's intermediate systems, known variously as communications controllers or front-end processors (FEPs); these are called Type 4 (T4) systems. T4 systems are the layer 3 relays in SNA networks and come in two varieties: local, meaning they are directly attached to mainframes by means of a

---

1. These are referred to as *host application subsystems* (HAS) and include IBM's TSO, IMS, CICS, and DB2 programs.

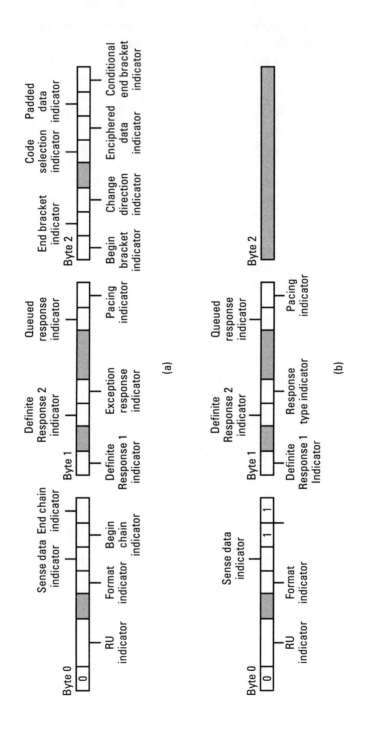

Figure 11.4 (a) Request and (b) response headers.

special high-speed data link called the S/3X0 channel; and remote, meaning they have no channel connection but rather are connected to other T4 systems via SDLC, Token Ring, or newer technologies such as Frame Relay. T4 systems from IBM run a special software package called the Network Control Program (NCP) that handles the concatenation of SNA data links. Although many in the IP community refer to SNA as a "nonroutable" protocol, this is emphatically untrue; the T4 systems are, in fact, the routers in SNA networks. T5 and T4 systems are both examples of subarea systems, so-called because they each constitute a subarea in SNA's addressing and because PIUs exchange global (subarea/element) SNA addresses.

Neither T5 nor T4 systems directly attach terminals, printers, and other devices such as ATMs and POS registers. Instead, these are attached to yet a third type of SNA system known as a cluster controller, which is a Type 2 (T2) system. As with T4 systems, T2 systems come in local and remote versions, the former being connected to mainframes (T5 systems) directly by means of an S/3X0 channel and the latter being connected to a T4 system from which its data are relayed to the mainframe. SNA also defines a Type 1 node, a small terminal server, but these are seldom encountered in SNA networks today. T2 and T1 are called peripheral systems. We should mention here that T2 systems have been subsequently enhanced into what are called T2.1 nodes; however, these are best discussed in the context of APPN, so we defer further explanation.

One convenient way to remember these system types is that as the type increases so does the management sophistication. The most capable, and indeed the central manager of an SNA network, is the T5 system running VTAM. A T4 system offloads routing from T5 systems, and has considerable autonomy in how it manages traffic flowing in the network. A T2 system, on the other hand, is essentially a slave to the T5 systems with which it is communicating; its primary responsibilities are to service terminals, printers, and other data entry/egress devices as a proxy for the mainframe, freeing up the latter from tasks such as character echoing and screen management.

Figure 11.5 illustrates the various nodes in an SNA network along with the span of routes (explicit and virtual) and route extension, which is the last leg of an SNA path that occurs when one of the end systems is a T2 node. Also shown are the subarea numbers of the T5 and T4 systems.

11.3.3 Network Addressable Units and Sessions

We mentioned in the earlier section on SNA addressing that, unlike IP, SNA addresses do not correspond to systems or their interfaces but rather to NAUs within SNA systems. But what is an NAU? It is, quite simply, a program that runs in an SNA system and is responsible for managing some aspect of the

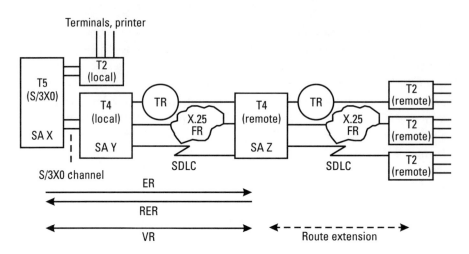

Figure 11.5 Traditional SNA systems and routes.

system's presence in the network; because of this, every system (node) in an SNA network must contain one or more NAUs. In addition, every NAU contains at least one half-session. In fact, we can put this more strongly: Half-sessions exist *only* in NAUs. That is to say, the TC/DFC protocol machines do not exist independently as stand-alone protocol stacks. SNA does not allow the implementation of "naked" half-sessions.

SNA defines three different types of NAUs, depending on the nature of the management tasks they execute. The three are system services control points (SSCPs), which are the centralized managers to which we have referred several times in this chapter; physical units (PUs), which are the agents used by the SSCP to manage an SNA system; and logical units (LUs), which are responsible for managing the end-to-end transport of client data. In the remainder of this section we look at these NAUs and the roles played by their respective management mechanisms in SNA networks.

11.3.3.1 Physical Units

Because it is arguably the simplest, we start with the physical unit. A PU is a program that is the agent for controlling the communications resources in an SNA system, including link stations and buffers. For example, the PU in an SNA node is the actuator for that node's link station(s), responsible for actuating such link station's status transition. When management information concerning a node's state is desired, it is the PU that provides it. These data are sent in management RUs over the session maintained between the PU in an active SNA system and the SSCP controlling its domain; likewise, management requests (RUs) for actuations and/or state information are sent by the SSCP to the PU over this session. This is how an SSCP would request a link station be turned on or off.

Just as there are five types of SNA systems, there are five types of PUs. Mainframes each contain a PU Type 5, communications controllers each contain a PU Type 4, cluster controllers each contain a PU Type 2.0, small terminal servers each contain a PU Type 1, and enhanced peripheral nodes each contain a PU Type 2.1. These PU types all have different management capabilities and responsibilities.

11.3.3.2 Logical Units

A logical unit is the service access point for SNA applications and end users. As with PUs, SNA has defined a number of types of LUs for different classes of end users, notably terminals (LU 1, LU 2, and LU 4), printers (LU 3), and programs (LUs 6.0, 6.1, and 6.2). The management provided by LUs is strongly master/slave; with the exception of LU 6.2, LUs are always either primary (PLUs) or secondary (SLUs). PLUs are always located in the T5 systems, whereas SLUs are located in T2 systems. Likewise, with the exception of LU 6.2, LUs cannot activate sessions on their own initiative but rather must request permission from the SSCP managing their domain. Such LUs are said to be dependent. This is done via the LU–SSCP session.

Beyond the restriction that a PLU can only have a session with an SLU (and vice versa), sessions between LUs are limited to their same kind: An LU 2 can only have a session with an LU 2, an LU 3 with an LU 3, and so on. The exception to this is that every active LU must have a current session with the SSCP controlling its domain.

11.3.3.3 System Services Control Point

If SNA can be characterized as a network architecture built around centralized management (monitoring and control) then the system services control point is the realization of that management. Located within the VTAM program we just mentioned, an SSCP is involved in almost every aspect of managing an SNA network except for the routine transfer of end user data over LU–LU sessions, which is managed autonomously by the LUs involved, and the actual routing of PIUs, which is the task of T4 nodes.

It may be easiest to understand the role of system services control points in an SNA network by way of analogy with the telephone company. An SSCP is the equivalent of the phone network's 411 (directory assistance), 611 (fault reporting), and 911 (emergency services) rolled into one. However, the role of an SSCP goes beyond even this: As we will see below, its configuration at sysgen provides the principal topology and directory information used in an SNA network's operations.

Within its domain the SSCP is supreme and is responsible for managing an SNA network's resources, for which it relies on the various PUs and LUs. If two or more SSCPs are active in an SNA network then the network is partitioned into disjoint domains; an SSCP cannot share control of its domain with another SSCP. As we just said, an SSCP that is the controlling SSCP in a domain (as opposed to a backup SSCP, which later versions of SNA allow for resilience) will maintain SSCP–PU sessions to every SNA node in the domain and SSCP–LU sessions to every end system. Because sessions between dependent LUs require the intervention of an SSCP to set up, in this way the SSCP can control the total number of users in the network. If, for example, many of the PCN intermediate systems (i.e., T4 nodes) are heavily utilized, and starting a new session would require actuating an idle VR, then the SSCP many deny the requesting LU the session.

We should note that several other types of control points are defined in SNA but only the SSCP is an NAU. For example, PU Types 2 and 4 contain what is called the physical unit control point (PUCP). A PUCP executes a part of the SSCP's management tasks so that it can actuate and monitor a node's resources (Figure 11.6).

A third type of management session occurs between SSCPs. Such SSCP–SSCP sessions are used to coordinate cross-domain management. For example, if two LUs seek to exchange data but one is in the domain of SSCP A and the other is in the domain of SSCP B then the two SSCPs must coordinate their actions by means of exchanging special SNA management RUs.

Finally we should note that end user (LU–LU) sessions are parameterized by performance characteristics that are lumped into a single parameter called Class Of Service (COS). Within VTAM a COS table is maintained that maps

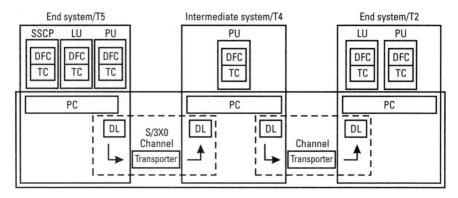

Figure 11.6 SNA systems with NAUs.

sessions to underlying VRs able to provide the desired bandwidth, throughput, security, and so on. The order in a COS table reflects the priority parameterization of VRs. VTAM provides software-controlled parameter actuators for modifying the ordering of VR/priority entries by writing an access method exit routine (a type of VTAM procedure). To be invoked each time an LU–LU session is activated, this allows LU–LU traffic to be channeled across different VRs to control LU–LU session traffic within the same COS.

11.4   Management in the Path Control Network

Now that we have defined SNA's components, we can look at its management proper. As we said earlier, the main management task of the Path Control Network, apart of course from concatenation of the individual S/3X0 channels, SDLC data links, and Token-Ring LANs, is flow control or, as it is called in SNA, *pacing*. We discuss pacing within the PCN in this section as well as other management complications that arose when IBM enhanced the PC layer by introducing transmission groups, explicit routes, and virtual routes instead of monolithic paths.

11.4.1 Transmission Group Management

The first area of PCN management we want to look at is that associated with transmission groups. TGs, bundling multiple data links so they appear to be a single data link, are an instance of bandwidth management. A multilink TG appears to its upper layer protocol (Explicit Route Control) client(s) to be a data link with greater bandwidth and reliability than otherwise would be the case were the client simply using one of the TG's component data links.

But the introduction of multilink TGs also brought a potential for new faults. In SNA as originally defined it was impossible for PIUs sent by a half-session to arrive out of order, for the simple reason that all the PIUs flowed along a fixed path composed of single data links. But this changed with multilink TGs: Even if the links are identical in speed, it is possible that PIUs can arrive out of sequence due to scheduling issues, faults, retransmissions, and so on. For example, two identical SDLC links that are used round-robin to transmit PIUs can produce out-of-order delivery if one link suffers a transient fault that occasions retransmission; in such a case, the packets will no longer be in order.

To prevent PIUs from being delivered out of order, the TGC protocol machines (which are within the PC layer) use TG sequence numbers (TGSNs) carried in the FID 4 headers to resequence them. Each subarea node that supports multilink TGs must reorder them before sending the PIUs on to the next stage (data link). We quote from IBM's SNA overview:

> Path control assigns transmission group sequence numbers to path infor-
> mation units (PIUs) before transmitting them across a transmission group.
> The assignment of sequence numbers to PIUs is called *PIU sequencing*.
> Because data link control can route related path information units (PIUs)
> over different links in a transmission group, path control in another node
> might not receive the PIUs in the order in which they were sent. With the
> PIUs sequenced, path control on the other side of the transmission group
> can use the sequence numbers to reorder any out-of-sequence PIUs before
> continuing to route the data through the network. This ensures that the
> arrival order at the destination session endpoint matches the sending order
> of the origin endpoint. [4]

11.4.2 Actuation of Virtual and Explicit Routes

The next area of management we want to discuss is the actuation of ERs and VRs. When SNA's architects defined these they made an important addition to the FID 4 TH, namely, the inclusion of three fields called the initial ER number (IERN), the ER number (ERN), and the VR number (VRN). Note that the IERN and ERN to date have been treated identically (i.e., always have the same value). These fields are 4 bits each, meaning that between any pair of subarea nodes in an SNA network up to 16 ERs can be defined. In addition, because VRs are identified by their VR number and their transmission priority and because SNA allows three transmission priorities (0, 1, and 2) this means that up to 48 VRs can be defined between any pair of subarea nodes.

Why is this so important? Because it means that SNA, a virtual circuit architecture with statically defined routing, could at last offer alternative rout-ing around failures much like IP. In other words, allowing up to 16 different routes between source and destination meant that an SNA network could theoretically suffer 16 fatal faults of TGs without disconnecting an origin subarea/destination subarea pair. Consider that every VR is composed of a pair of ERs (called the ER and the reverse ER or RER). The traffic of these ERs in turn is carried by a series of TGs. If one of these TGs fails because its last (or only) data link suffers a fatal fault, then the ER is lost and so is the VR. How-ever, the session may not be lost, or at least it can be reestablished, if a second VR was defined at sysgen that flows over an ER that does not use the failed TG. The result is a degree of resilience that might surprise even the most ardent advocate of datagram protocols.

Actuations of ERs and VRs are initiated by the origin subarea node, which sends the appropriate SNA management RUs to each subarea node through which the ER, RER, and VR flow. For example, to activate an ER, the origin subarea node sends an RU called NC_ER_ACT to the path control logic in each of the subarea nodes through which that ER threads its way to the destination end system. This actuation of VRs and ERs is automatic, meaning no operator involvement is needed (or allowed). In addition, the SSCPs have automatic schedulers that make the decision based on session traffic flow and/or messages indicating that a TG and consequently an ER have been lost. Such messages are generated by the PUs in T4 (and, where applicable, T5) systems and sent to the controlling SSCP in the domain. The SSCP likewise automatically deactivates explicit routes when a PU on a component link or node is deactivated or becomes inoperative due to fatal faults.

11.4.3 RPacing

The last area of management in the PCN that we discuss is that of pacing, that is, flow control. The reason for flow control in the PCN is that SNA's architects were very worried about managing the buffers of the communications controllers (T4 systems) so as to prevent reassembly lockup, similar to what happened to the early ARPANET. In a virtual circuit architecture like SNA, however, the consequences would be even greater since a locked-up subarea node would effectively disable all the virtual circuits flowing through it, wreaking havoc in a network and defying easy attempts to isolate the cause.

The flow control solution they devised is called *RPacing* and operates on a VR basis between the two terminating (subarea) nodes of a Virtual Route as well as the subarea nodes through which the VR's ERs flow. RPacing is automatic—no operator intervention is required. The actual flow control mechanism used is called a pacing window, which limits the number of PIUs that can be sent by one of the VR endpoints before permission is granted for further sending. This permission is called a *pacing response*. When one end subarea node of a VR has sent the maximum number of PIUs allowed by its pacing window it must block itself until it receives a pacing response from the destination subarea system. Note that pacing is independent in each direction of the VR, meaning there are actually two pacing windows involved per VR.

A second level of flow control is effected by actuating the size of the pacing window itself. When a VR is activated its pacing windows are specified in terms of the minimum and the maximum number of PIUs that the two subarea terminating systems can send. These can either be specified at sysgen or default values are calculated automatically by the software. The default for the window sizes are minimum size = the number of TGs of the underlying ER (i.e., its hop count); and maximum size = 3 times the minimum size. The size of the pacing window is actuated in response to congestion encountered in the subarea nodes through which the VR flows. If a T4 system determines it is becoming congested then it will actuate a reduction in the pacing window size to reduce the flow of PIUs into the network.

RPacing and window management are effected using certain fields included in the FID 4 header. Recall from our earlier discussion that the FID 4 TH contains a number of fields related to managing the RPacing window for scheduling PIUs on each VR. These include:

- *VR pacing request* (VRPRQ), set by the sending subarea node to notify the destination subarea node that this is the first PIU in a pacing window;

- The *VR pacing count indicator* (VRPCI), set by the sending subarea node to notify the destination subarea node that this is the last PIU to be sent, that is, the pacing window is now closed;

- The *VR pacing response* (VRPR), set by the receiving subarea node to give permission to reopen the pacing window;

- The *VR change window indicator* (VRCWI), set by any subarea node through which the VR flows to actuate the pacing window size (either incrementing or decrementing it by 1 PIU);

- The *VR change window reply indicator* (VRCWRI), set by the sending subarea node to confirm it has actuated the size of the pacing window; and

- The *VR reset window indicator* (VRRWI), set by any subarea node through which the VR flows to immediately close the pacing window size to its minimum.

Use of the last three fields is determined by the congestion level of the various subarea nodes through which the VR flows. Each subarea node contains a scheduler that regulates the flow into the PCN based on feedback from the congestion estimator. Depending on the estimate provided by the congestion estimator, the congestion scheduler will take appropriate action:

1. No congestion → do nothing;
2. Moderate congestion → reduce pacing window size; or

3. Severe congestion → reset pacing window size to minimum (slam window shut).

So how is the congestion estimate determined? There are two answers for this, depending on whether boundary function conversion is involved. For VRs that do not involve route extension (no boundary function) the NCP in a T4 node simply measures the queue sizes for each TG queue (there are three queues per TG, one for each transmission priority level) as well as the total size for all three queues. Using either congestion threshold's defaults or values defined at sysgen, the estimator decides that moderate congestion has occurred if any one of the three queues for a TG hits its respective congestion threshold. If the total count hits its threshold then the estimator decides that severe congestion has occurred. Any VR with ERs that use that TG is then estimated to be congested.

For subarea nodes that are executing the boundary function, a second estimator is used as well. This is because such nodes must maintain a separate storage area called the *boundary pool* (BPOOL) to hold PIUs destined for T2 systems. When the BPOOL exceeds its congestion threshold then any VRs sending PIUs to these systems are deemed congested.

We should mention a local form of congestion control that complements RPacing's end-to-end workload management. Every T4 system has an NCP configuration parameter called *slowdown* that is specified in terms of the percentage of its buffers that are free. When a T4 system hits its slowdown threshold it will go into a mode called *slow-polling,* where it will issue Receiver Not Ready messages to all the secondary link stations attempting to send it data. This is to allow the T4 to clear some of its buffers before accepting more data.

11.5 NAU Management

11.5.1 Session Actuation

As we said earlier, when an SNA network is started up the SSCPs will seek to establish sessions to all the PUs and LUs that are to be to activated initially; after this, other NAUs may be activated or deactivated by network operators or by automation features built into management applications such as IBM's NetView.

As for actuation of LU–LU sessions, the mechanisms involved are determined by whether the LUs are dependent or independent. Dependent LUs require the intervention of the SSCP to actuate their sessions. An SLU will send a management RU called *Init-Self* to the SSCP controlling its domain over the

SSCP–LU session. The SSCP will consult with its directories to determine the network address of the requested PLU. If it is in its domain, the SSCP will send over the SSCP–LU session an RU inquiring if the PLU will accept a session with the requesting SLU. If it accepts then the SSCP will use its COS table to determine the optimal (VR, priority) pair and set up the session. (If the PLU is in another domain the SSCP uses its SSCP–SSCP session to the controlling SSCP for that domain to ask it to query the requested PLU; if the response is acceptable then the two SSCPs cooperate to actuate the session.)

We discuss actuation of sessions between independent LUs in the later section on APPN.

11.5.2 Fault Detection and Recovery

As we have stressed throughout this chapter, SNA expects that the primary execution of fault detection and recovery will reside with the various data link protocols in the network. However, there is provision for end-to-end fault management in SNA in case PIUs are lost completely. This functionality resides in the DFC layer, which manages recovery by means of what are called *chains* and *brackets*. Both chains and brackets are delimited using fields in the RH. Another task of the DFC protocol machine is to specify the nature of the response expected to an RU; this also is indicated with fields in the RU header.

A chain is a collection of RUs that a sending half-session groups together for atomic delivery: Either all the RUs in chain must be delivered to an upper layer destination or none should be. A chain, therefore, is the unit of recovery and retransmission between end systems.

Carrying this one step further, brackets are collections of chains exchanged between the two end systems which are "related" in the sense of constituting a transaction that is to be executed atomically; if any chain in a bracket fails to be transported successfully then the effects of the previous chains must be rolled back. This is the responsibility of another component within an LU, the *resource manager*, which maintains a journal of transactions; using this journal, it is possible to roll back transactions that are incomplete.

11.5.3 Session-Level Pacing

SNA also uses a destination-to-client flow control called *session-level pacing* to coordinate clients (half-sessions) from overwhelming the destinations, especially the storage. Furthermore, SNA distinguishes pacing according to its direction, whether it is inbound or outbound pacing and whether it is one- or two-stage pacing. Like RPacing, separate windows are maintained for the two directions. Two-stage pacing is used if one of the end systems is a T2 node; the second stage of pacing is that employed on the route extension.

11.6 APPN and High-Performance Routing

In the early 1980s the commercial success of IBM's minicomputers led to demand for a version of SNA that would meet the needs of small systems networking, most notably to enable SNA networks to operate without mainframes and the all-important SSCP. This was realized by enhancing the Control Point in T2.1 systems, itself an enhancement of the Physical Unit Control Point within the PU 2.0. The result was first known as Low Entry Networking (LEN), later renamed Advanced Peer to Peer Networking (APPN). APPN itself was subsequently modified extensively, with a new set of PC protocols designed to accommodate the new high-speed, low-BER (bit-error rate) data links such as ATM and FR. This redesigned APPN was first called APPN+ but later renamed High-Performance Routing.

11.6.1 The T2.1 Architecture

We start by briefly outlining why the T2.1 management mechanisms were devised. As we saw earlier, the PU hierarchy of management sophistication (PU 5 down to PU 2) was in large part driven by the costs of computer processing and memory when SNA was originally designed in the early 1970s. However, as the cost of hardware dropped and small minicomputer systems proliferated, the original set of four system types was augmented in the early 1980s by a fifth, a Type 2.1 (T2.1) system which could manage more of its own communications tasks independently. For example, two T2.1 systems that are directly attached by a data link such as a Token-Ring LAN or an SLDC link can set up LU–LU sessions (if the LUs are independent, i.e., LU 6.2) without the management services of a T5 system; this is impossible for two T2 systems.

The heart of T2.1 architecture is the ability to process a Bind RU sent by a LU 6.2 NAU. To do this IBM enhanced the Physical Unit Control Point in from a T2 system, renaming it the Peripheral Node Control Point (PNCP). This was later shortened to simply the control point (CP).

LU 6.2 was developed in conjunction with the T2.1 architecture to be able to function independently of the SSCP. LU 6.2, also known as Advanced Program-to-Program Communication (APPC), supports distributed transaction programs that could dynamically schedule workload among components on various computers. This clearly would be incompatible with having to function as a dependent LU. Further evidence of the tight coupling of LU 6.2 and T2.1 is that the CPs are, in fact, written as LU 6.2 programs.

11.6.2 Advanced Peer to Peer Networking

With the T2.1 architecture in place, IBM had an extensible architecture that it could use to develop new SNA protocols, which amounted to little more than writing additional LU 6.2 transaction program modules. Building on the CP foundation, additional management capabilities were implemented that resulted in a new architecture called Advanced Peer to Peer Networking. Foremost among APPN's achievements was that it allowed APPC sessions to be set up between *nonadjacent* systems, eliminating T2.1's requirement that the two systems be directly attached. In fact, APPN replaced the SNA hierarchy of systems with a two-tiered hierarchy similar to that of TCP/IP. An APPN network consists of end nodes (ENs) connected to network nodes (NNs) which are responsible for relaying PIUs.

In addition to this feature, called Intermediate Session Routing (ISR), APPN's architects implemented a dynamic path control layer that eliminated the need for the extensive static configurations that characterize traditional SNA. To do this, APPN defined a set of APPC distributed transaction programs for topology management, coordinated by means of topology database updates (TDUs) sent over the CP–CP sessions. An APPN network is actually a collection of LU 6.2 sessions between APPN end and intermediate systems.

This new dynamism came at a price: APPN had to abandon the PC protocols of subarea SNA—there are no ERs or VRs, nor any of the associated management mechanisms such as RPacing. In addition, while APPN retained SNA's virtual circuit orientation there is a major difference in how end-to-end paths are set up and how addresses are used. When an EN requests a session to another node the NN to which it is attached will consult its topology database and determine the path the session will follow through the APPN network. The NN then will send an SNA bind RU that carries a *Route Selection Control Vector* (RSCV), a specification of the source-routed end-to-end path through the APPN network. Finally we note that the APPN architects modified the FID 2 TH to use a new addressing format: The OAF and DAF fields plus an additional bit are put together to define a 17-bit circuit identifier.

An additional complication was that, because APPN was really designed with APPC in mind, there was no simple way for subarea SNA traffic to be directly transported across an APPN network. IBM was forced to devise an inelegant tunneling mechanism called dependent LU Requester/Server

(DLUR/S) to transport subarea Bind RUs in LU 6.2 tunnels across an APPN network to an SSCP for session establishment.

11.6.3 High-Performance Routing

Although APPN succeeded in demonstrating that SNA could incorporate ideas such as dynamic routing, its management mechanisms were found to have too high an overhead to scale adequately for large networks. IBM's architects sought to introduce a new architecture more conducive to high-speed forwarding and to streamline the management. What they came up with are three new protocols called Automatic Network Routing, Adaptive Rate-Based flow/congestion control, and Rapid Transport Protocol.

11.6.3.1 Automatic Network Routing

Like APPN's route selection mechanism, Automatic Network Routing (ANR) uses source routing. However, ANR is a connectionless protocol that uses what has come to be known as tag or label forwarding. The chief advantage of this is that intermediate HPR nodes can forward an ANR PIU merely by consulting the ANR labels (ALs) it carries; there is no need to maintain routing tables or consult them in the forwarding. We discuss this in more detail in the chapters of Part IV.

11.6.3.2 Adaptive Rate-Based Flow/Congestion Control

HPR's second change to APPN's PC layer was to alter the session-level pacing from stage by stage to end to end. The mechanism, called Adaptive Rate-Based (ARB) flow/congestion control, relies on estimators in the sending HPR end system to decide when the HPR network is getting congested, which it does by monitoring round-trip delay (similar to the TCP mechanisms we discussed in Chapter 10). In addition, the receiving HPR end system sends feedback concerning the state of its processing and buffers. This information is also used by the sending HPR end system to schedule actuations of the rate of traffic it sends into the HPR network.

11.6.3.3 Rapid Transport Protocol

The last part of HPR is the Rapid Transport Protocol (RTP), a connection-oriented reliable transport protocol that executes segmentation, retransmission, and reordering of HPR PIUs. RTP was derived from so-called lightweight transport protocols such as XTP and, unlike traditional SNA, was designed to accommodate unreliable data links. Like ARB, RTP only involves end systems:

no intermediate system fault detection, reassembly, and so on are performed in an HPR network except at the end systems.

11.7 Summary

In this chapter we examined the management mechanisms in SNA's principal end-to-end protocols, namely, the Path Control, Transmission Control, and Data Flow Control layers. We traced many of SNA's characteristics to its inheritance from the broad class of terminal–host network architectures starting with the SAGE and SABRE networks. The most important of these was that there was neither need nor benefit realized from supporting any-to-any communication among end systems. We contrasted this with TCP/IP and its antecedent, the original ARPANET.

We also explored the consequences of SNA's other principal management decisions, namely, locating most fault management in the data link protocol rather than higher up the protocol stack, and using a centralized management architecture without concern for single points of failure. We documented the evolution of SNA's architecture and even its PIUs, the two most important being the FID 4, used in subarea SNA, and FID 2, used to and from peripheral SNA systems. Finally we briefly reviewed the latest changes to SNA, APPN, and HPR.

References

[1]   Moreau, R., *The Computer Comes of Age,* Cambridge, MA: The MIT Press, 1984.

[2]   Bashe, C., et al., *IBM's Early Computers,* Cambridge, MA: The MIT Press, 1986.

[3]   *SNA Technical Overview,* 5th ed., IBM Corporation, 1994, pp. 4, 132.

Part IV:
Concatenation Management

12

Concatenation Management: Basics

12.1 Introduction

Having considered management in the physical, data link, and end-to-end protocol layers we now come to concatenation, the most complex area of management in computer networking. The goal of this chapter is to use the MESA model as a framework with which to unify various concatenation techniques, including L3 concatenation (routing) and L2 concatenation (bridging). Note that we defer discussion of specific L3 and L2 concatenation protocols to Chapters 13 and 14, respectively.

As we will see, an enormous amount of ingenuity has been exercised by protocol architects to devise concatenation mechanisms that can adapt to changing network and/or traffic conditions while minimizing the attendant overhead. Such concatenation mechanisms have ranged from the open-loop flooding to closed-loop routing protocols, although few implementations are purely of one type or another. In the last chapter, for example, we saw that with SNA feedback is combined with static routing to produce a hybrid concatenation management.

The chapter begins with an overview of routing and bridging, the two principal means of concatenation used in networks today. We then discuss tunneling, a third type of concatenation that is being used increasingly to define virtual private networks as well as to transport SNA PDUs over IP backbones. We also discuss the faults to which the intermediate systems that execute these concatenation tasks are liable; the consequences of such faults include end-to-end

PDUs being misforwarded, proliferating without limit, or simply disappearing into "black holes."

We then move to construct a taxonomy of the workload management that is at the heart of all concatenation. As we have said in earlier chapters, concatenation is workload actuation of kind, scheduling the realization of global or end-to-end transport tasks by mapping these to the transport tasks of component transporters (such as data links). Our management taxonomy for classifying concatenation mechanisms examines whether or not the concatenation is executed with respect to a model of the network and/or traffic; whether the concatenation uses open-loop or closed-loop scheduling; whether the concatenation scheduling is datagram or virtual circuit in nature; and finally whether the intermediate systems executing the scheduling decide the entire end-to-end schedule or merely the next stage transport task.

Finally, we briefly discuss the complement of the concatenation problem, namely, the network design problem. Whereas concatenation actuates traffic, scheduling its execution by available servers (transporters), network design actuates the bandwidth of transporters in response to the demand of traffic.

12.2   What Is Concatenation?

Recall that we briefly discussed in Chapter 9 the basic concepts of concatenation and the workload management tasks that an intermediate system (also known as a relay) must execute to "glue" together two or more transporters (generally speaking, data links). In this chapter we resume that discussion and examine the management of concatenation—the mechanisms by which two or more transporters are amalgamated into a composite whole. As we will see, common to all forms of concatenation is scheduling: realizing the end-to-end transport tasks out of the transport task sets of component transporters.

The scheduling decisions are complicated by the fact that, except for the simplest network topologies (see Section 12.4), there will generally be two or more possible schedules that can realize a given end-to-end transport task. In these instances, the workload managers within concatenating intermediate systems must choose among several alternatives. The optimality criteria may range from schedule length (hop count) to effective bandwidth of alternatives (favoring longer schedules composed of high-bandwidth transporters over shorter schedules composed of low-bandwidth transporters), to response time, security, cost, or other so-called *policy* routing parameters. The cost functional based on these parameters is referred to as the *metric*.

12.2.1 Routing and Bridging

Put succinctly, routing is concatenation at layer 3 and bridging is concatenation at layer 2. Prior to the development of LANs and their global L2 address spaces, a network was composed of two or more data links joined at layer 3. This layer 3 concatenation was called routing and a layer 3 intermediate system was called a router. The task set of this layer 3 transporter (network) was defined by the L3 addresses, but the end-to-end transport tasks were realized in terms of scheduling of one or more transport tasks drawn from the task sets of the data links (L2 transporters) that were components of the network (see Section 9.2).

However, with the development of LANs came a new type of intermediate system, called a bridge, that operated at layer 2 and enabled a station (end system) on one LAN to send data to a station on another LAN without routing (L3 concatenation) coming into the picture. Consider, for example, the simplified two-LAN scenario shown in Figure 12.1, where each LAN has only two stations, namely, an end system and an interface of the L2 intermediate system (i.e., bridge). The task set of the data link on the left is {0000.1034.223A ↔ 0000.4F11.99EB}, and the task set of the data link on the right is {AA00.0400.297E ↔ 0000.87BF.E111}. But because the L2 intermediate system forwards PDUs between the two LANs, the task set of the composite transporter includes the task {0000.1034.223A ↔ 0000.87BF.E111}, yielding the desired end-to-end transport.

Various bridging mechanisms have been devised over the years. Early bridges simply forwarded arriving PDUs out every interface, other than the one on which they arrived, what is called flooding; such operation was said to be *promiscuous*. Soon, however, manufacturers added processing capabilities to bridges, allowing them to monitor traffic on the LANs to which they were attached and thus learn the location of various end systems from the source

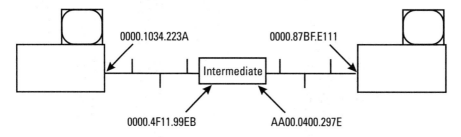

Figure 12.1 Two end systems on two LANs.

MAC addresses of the frames they received. These so-called *learning* bridges would construct simple forwarding tables with which they would decide if a frame was to be ignored (if the destination MAC address was known to be local to the LAN on which it arrived) or forwarded.

The taxonomic challenge presented by bridging is that this is little different from conventional L3 concatenation. Recall that one form of workload management, which we referred to as workload actuation of kind, changes an RFS into one or more new types of RFS; in contrast, workload actuation of degree merely actuates the arrival rate of RFSs and/or time slices an RFS into two or more RFSs of the same type. Workload actuation of kind is precisely what occurs in both routing and bridging.

Figure 12.2 shows this concept at its most abstract. We have a composite transporter T_3 composed of two-component transporters T_1 and T_2 and a workload manager (within the intermediate system) that maps the tasks of the composite to the tasks of the components. Note that the task set of the composite T_3 is the Cartesian product of the task sets of the two components T_1 and T_2, and it can be broken down into three disjoint subsets:

1. The tasks that map exclusively to the task set of the component T_1;
2. The tasks that map exclusively to the task set of the component T_2; and
3. The tasks that map to tasks in both the task sets of the components T_1 and T_2.

What is the difference between L2 and L3 concatenation? From Figure 12.2 it is impossible to tell the protocol layer at which the workload

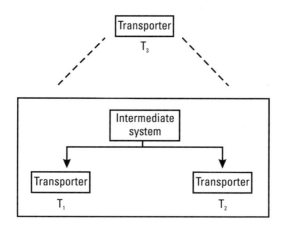

Figure 12.2 Composite transporter with components including workload scheduler.

manager is executing its scheduling (concatenation) decisions. And this is precisely the point: Because this scheduling can be effected at any layer where global addressing is employed, we need a term that is neutral with respect to the layer where the concatenation is executed. The term we have chosen is *concatenation*, which is generic enough to encompass routing, bridging, and so on. In addition, "concatenation" harkens back to the early days of internetworking, when Cerf adopted the term *catenet* to refer to networks composed of networks internetworked together [1].

Last, we should briefly discuss the world of IP switching, including the new standard called MultiProtocol Label Switching (MPLS). Many vendors are promoting products they claim to be layer 3 switches. These intermediate systems range from layer 2 switches (Chapter 14) with routing supervisors that supplement the forwarding decisions to hybrid IP/ATM products that seek to marry the two technologies more efficiently than does simple IP over ATM (defined in RFC-1577 [2]). Various proprietary techniques have been defined by vendors such as IBM, Toshiba, Cisco, and Ipsilon but these have been superseded by the IETF-sponsored MPLS. MPLS abandons ATM's enormously complex system of signaling (the Network–Network Interface protocol) and uses instead the routing protocol mechanisms developed for IP networks. (We consider these in Chapter 13.) For a detailed exploration of MPLS and IP switching, please consult the book by Davie, Doolan, and Rekhter [3].

12.2.2 Tunneling

Beyond bridging and routing, there is a very different type of concatenation that can be used, namely, tunneling. When two transporters are to be concatenated via a third transporter that is to remain "invisible," traffic between the two may be tunneled across the third. As we saw in Chapter 9, "normal" encapsulation implies that a layer n PDU will be placed within one or more layer $n-1$ PDUs. With tunneling, however, the encapsulation may be much more varied:

- An L2 PDU may be encapsulated in an L2 PDU to create a *virtual private network* (see later discussion) using tunneling protocols such as Microsoft's Point-to-Point Tunneling Protocol, Cisco's Layer Two Forwarding (L2F) protocol, or the new IETF standard called the Layer Two Tunneling Protocol (L2TP).

- An L2 PDU such as SDLC or LLC2 frame can be encapsulated in an upper layer PDU such as a TCP segment. This is how Data Link Switching (DLSw) enables SNA and NetBIOS traffic to transit IP backbones.

- An L3 PDU such as an X.25 packet can be encapsulated in a higher level PDU such as an SNA RU. This is how an IBM communications product called XI lets SNA networks carry X.25 traffic in places like Europe where many X.25 devices are deployed.
- Another early application was remote bridging (see Chapter 14), where two Ethernet bridges were connected by an HDLC data link, with the Ethernet frames encapsulated in HDLC frames; PPP's Bridging Control Protocol (BCP) works the same way.

With tunneling the encapsulated PDU is called the payload protocol and the encapsulating protocol is called the carrier. Figure 12.3 shows the choices of carrier protocol level for encapsulating an L2 payload protocol; L2 PDUs (frames) are received by the two tunneling intermediate systems from their respective end systems and placed in carrier PDUs at a layer in a carrier protocol stack such as layer 4 (DLSw) or layer 2 (L2TP). The *boundary* of the tunnel is defined by the "edge" of the carrier protocol stack. Note that within this boundary the carrier network and its protocols are invisible to the end systems and their payload PDUs. In fact, one of the challenges of tunneling is correlating events between the tunneled and tunneling networks for fault determination, for example.

Today one of the most important applications for tunneling is to implement virtual private networks (VPNs), which tunnel IP traffic using IP as the carrier protocol as well; VPNs enable private IP networks, using unauthorized

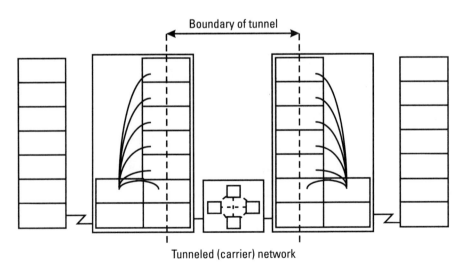

Boundary of tunnel

Tunneled (carrier) network

Figure 12.3 L2 tunneling.

IP addresses, to be connected over the global Internet, something that would otherwise require either readdressing or the use of IP network address translation (NAT). Because the tunneling network is invisible to the IP networks that are outside, and likewise the tunneled traffic is merely payload data carried in the carrier protocol's PDUs, there is no difficulty with address spaces.

12.2.3 Concatenation Faults

Like any server, an intermediate system concatenating two or more transporters will have finite bandwidth (forwarding rate) and finite reliability. Whatever concatenation mechanism(s) are used, we must consider the possibility of concatenation faults, which may include forwarding loops, "black holes" in which PDUs simply disappear without a trace, and explosive proliferation of PDUs due to forwarding replication. We can define a concatenation fault quite simply as any scheduling decision that fails to correctly forward a PDU toward its destination or, if it is a multicast PDU, destinations. However, we must distinguish between fatal and latent concatenation faults. An example of a fatal fault is an intermediate system discarding a PDU due to buffer constraints, whereas an example of a latent concatenation fault is when two intermediate systems keep sending PDUs to each other due to topology or other state information (utilization, delay, and so on) inconsistencies.

As we saw in Chapter 9, detecting and recovering from fatal concatenation faults is the province of the end-to-end protocols employed in the network; mechanisms such as timer-based retransmission coupled with end-to-end CRCs can detect and recover from end-to-end PDUs lost due to fatal concatenation faults. Latent faults, however, present additional complications that defy such straightforward recovery mechanisms as retransmission. While the end-to-end protocol mechanisms can detect latent faults, generally speaking end systems are powerless to recover from these because they cannot actuate the concatenation decisions of the intermediate systems. An exception to this is source routing as employed in 802.5 Source-Route Bridging (SRB), where a lost LLC2 connection can result in a new topology discovery (see Chapter 14 for more details).

12.3 Taxonomy of Concatenation Mechanisms

We want to unify the various approaches to concatenation such as routing, bridging, and tunneling within the framework of the MESA model and workload and bandwidth management. Toward this end we will use three principal taxonomic criteria:

1. Does the scheduling mechanism use a model of the global discrete event system or not? If it does, is the model's state information updated (closed loop) or not (open loop)? Is the model explicit or not? If it is, is the feedback (measurements and/or estimates) strictly local or is it global? Is scheduling centralized or is schedule creation distributed among intermediate systems?

2. Does a schedule created for an initial PDU persist for all subsequent PDUs (virtual circuit forwarding) or is a schedule created for each PDU (datagram forwarding)?

3. Does an intermediate system attempt to create an end-to-end schedule or merely schedule the next-stage transporter (data link)? In other words, what is the scope of the scheduling decisions made in intermediate systems?

Figure 12.4 shows the concatenation taxonomy tree that results from these criteria.

Beyond this, concatenation management must be evaluated according to several different and often conflicting criteria. Obviously, this should include how vulnerable the concatenation mechanisms are to concatenation faults. Then there is efficiency and performance. As we have stressed throughout this book, management always comes with a price in terms of overhead, and much of the skill in designing concatenation mechanisms and even network topology is to maximize performance while minimizing management overhead.

We must also keep in mind the computational and storage requirements attendant to various concatenation techniques. This, in fact, was one of the

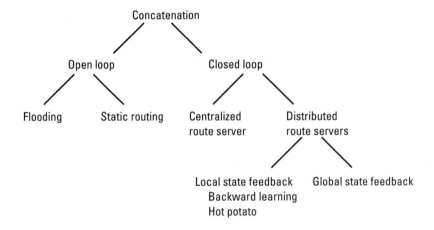

Figure 12.4 Concatenation taxonomy tree.

principal arguments between advocates of transparent and source route bridging, the two different variants used to concatenate LANs at layer 2. The former requires more computation and storage than the latter, and source route bridging partisans pointed to cheaper bridges as one of its benefits (see Chapter 14 for further discussion of this).

For now, though, let's turn to our taxonomic analysis.

12.3.1 Workload Management: Models Versus Feedback

On initial consideration it may seem that the first taxonomic discriminant should be simply whether the scheduler is closed or open loop—that is, is feedback employed? However, a simple binary discriminant is not enough; both flooding and static routing are open loop but nonetheless constitute very different levels of sophistication in their respective concatenation scheduling. Therefore, we want to use a compound discriminant that has two components: (1) the use (or not) in concatenation of a model of the discrete event system—the network (transporter) and its traffic (workload); and (2) the use (or not) of feedback to update the state of the network and/or the state of the traffic arrivals. Taking these two together results in four classes of scheduling:

1. *Class 0 (stateless without feedback):* open-loop scheduling without any model;

2. *Class 1 (stateless with feedback):* closed-loop scheduling without any model;

3. *Class 2 (stateful without feedback):* open-loop scheduling with static model; and

4. *Class 3 (stateful with feedback):* closed-loop scheduling with dynamic model.

We should note, however, that Class 1 concatenation is effectively meaningless. Although a model can be useful without feedback (Class 2), feedback without a model cannot; closed-loop monitoring implicitly assumes or requires that there is a model of the server and/or workload to be updated. Recall that in Chapter 1 when we discussed control systems we touched on the part played by the model of the plant being managed and on the intimate relationship between the model and feedback. Put succinctly, if no model is used by the scheduler then there is no state information to update, hence there is no reason to monitor the plant (the network and/or its traffic). Since with stateless (model free) concatenation no feedback is necessary or even possible, this leaves us with three effective classes of concatenation to consider.

Of course, because there can be two parts to any model of a discrete event system, there are really *nine* possible levels of scheduling corresponding to the combination of the network and traffic models; we abbreviate these with the notation (x,y), where x is the class of the scheduling with respect to the network and y is the class of the scheduling with respect to the traffic. For instance, many concatenation management implementations use scheduling that is stateful with feedback with respect to the network model but which is stateless (indeed, completely ignorant) with respect to the traffic; this is $(3,0)$ scheduling. Examples of $(3,0)$ scheduling include the principal routing protocols used in IP networks, namely, RIP and OSPF (see the next chapter for more on these protocols). We assume in what follows that no traffic state information is being employed (unless otherwise specified).

Which scheduling offers optimal concatenation? The answer depends on a number of factors, including the network topology, the reliability of the network's components, and the volatility of the traffic. The last two determine the "load" that the concatenation management mechanisms must handle, which in conventional control systems is referred to as the bandwidth; the more frequently faults occur and/or traffic patterns change, the higher the required *management* bandwidth. And, of course, we cannot ignore the costs of the management overhead itself: The higher the bandwidth of the control system, the more resources it will consume, especially if global rather than merely local feedback is employed and intermediate systems must exchange significant amounts of state information (see later discussion).

With Class 0 (stateless) concatenation the workload scheduler makes its decisions without regard to the past or current state of the plant (network and traffic). Class 0 concatenation techniques include flooding as well as randomized forwarding (the intermediate system randomly chooses one of its k outbound links down to forward a PDU). The great virtue of stateless concatenation is that it reduces the computational and storage loads placed on intermediate systems; but its potential proliferation of PDUs makes it not only very expensive in terms of transporter bandwidth but rules out its use for most topologies without some mechanism like hop counts being included in the end-to-end PDU header to constrain replication.

Flooding should not be confused with another early concatenation technique, Baran's *hot potato* algorithm (see, for example, [4]), in which an intermediate system schedules a PDU for transport out of the transporter with the shortest queue. Queue lengths, as we briefly discussed in Chapter 1, are determined by the traffic and the bandwidth of the transporter involved; monitoring queue lengths, therefore, is a proxy for monitoring both the network and its traffic. Hot potato forwarding, therefore, is $(3,3)$ concatenation, albeit of a very simple sort.

With Class 2 concatenation each intermediate system relies on a static model of the network and/or its traffic on which to base its scheduling decisions. Widely employed even today in such protocols as subarea SNA, with Class 2 concatenation the forwarding tables are calculated at the beginning of network operation and not changed while the network is operational. The downside is that static topology models are vulnerable to changes in the states of the component transporters and intermediate systems: If a component of the network suffers a persistent fault (fatal or latent), the forwarding (scheduling) decisions of the intermediate systems will not adapt to this, even if alternative end-to-end paths are available.

We should note that pure open-loop concatenation is seldom encountered today because all but the simplest systems (end and intermediate) are instrumented to monitor directly attached communications channels, data links, and so on, meaning that at least *local* feedback is available. The result is that most Class 2 intermediate systems effect a hybrid concatenation, using static routing but with some feedback influencing the forwarding choices.

Another shortcoming of Class 2 (static) concatenation concerns growth. The addition of data links, intermediate systems, end systems, and so on requires taking the network down and modifying the static route definitions in the network's intermediate systems. This has been a deficiency, for example, in SNA networks, where "adds, moves, and deletes" are generally limited to weekly or even monthly sysgens. However, Class 2 concatenation is more likely to include traffic models in the determination of the concatenation schedules (such as forwarding tables) precisely since the generation of the intermediate systems forwarding schedules is infrequent (see later discussion).

Finally there is the closed-loop scheduling of Class 3, in which the models of the network and/or traffic are updated using feedback that is local and/or from other remote systems. Figure 12.5 shows an intermediate system instrumented with both bandwidth (network) and traffic (workload) monitors and receiving state feedback from other intermediate systems.

Why employ feedback? For the same reason we have seen management required in the previous chapters of this book: Because servers, both transporters and intermediate systems, have finite bandwidth and finite reliability. Closed loop means that the scheduling (forwarding/concatenation) decisions are made based on the state of the composite transporter and/or its traffic. In most instances, a closed-loop workload manager bases its scheduling on topology reconstructions created by the bandwidth monitor. Without any feedback about such faults, the intermediate system may just attempt to forward PDUs toward a next hop that is no longer operative.

The adaptivity of a concatenation technique determines how well the forwarding mechanisms respond to events in the network and/or traffic that alters

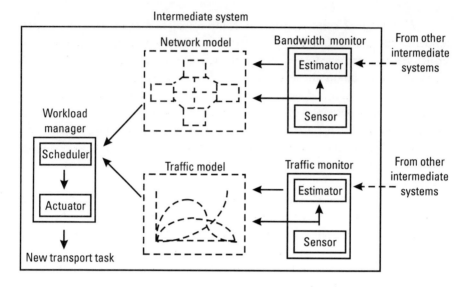

Figure 12.5 Network and traffic models within an intermediate system.

the dynamics and the discrete event (transport) system. We should stress that not all adaptation requires feedback (Class 3) concatenation. In fact, Class 0 concatenation can be very adaptive: Flooding will discover all possible end-to-end paths (schedules) to the destination end system, but at the cost of potentially explosive replication. On the other hand, pure Class 2 (no feedback, not even local state information) is not adaptive. Here we face a seeming paradox: Open-loop concatenation that is model based is not adaptive, but open-loop concatenation that is model free *is* adaptive. How can this be so? The answer rests in the fact that the former by definition relies on a priori state information (static topology and/or traffic definitions), whereas the latter is deliberately engineered to operate without any reference to the state of the discrete event system.

We say network is self-healing if, when a component (link or intermediate system) fails, the traffic is rerouted, assuming of course that an alternative route exists. If we recall the definition of fault management (detection, isolation, and recovery) then clearly this alternative routing recovers functionality of the whole (task set), if not, it recovers the condition of the failed component. Obviously, this is not true fault management in the sense that the failed component is not repaired or replaced. Rather, it is using the network's redundancy to effect a work-around that bypasses the failed component. This is not maintainability but rather reliability. The former is concerned with (reducing) the time to repair, the latter with increasing the time between failures in the first place. By introducing redundancy in the form of a multiply connected topology, component failures can be prevented from producing system failures. To summarize, redundancy is fault management via workload management, whereas repair is fault management via bandwidth management.

12.3.1.1 Types of Feedback

Beyond differentiating feedback about discrete event systems into information about the network and/or the traffic, we can further distinguish the feedback in terms of its scope and its nature. With respect to scope, feedback can be divided into four categories depending on how the state information is disseminated:

1. It is kept local; no global reconstruction of topology.

2. It is sent to a centralized estimator for reconstruction of global topology (state).

3. It is sent to those intermediate systems that are "nearby" or *adjacent*, that is, a subset of the set of all intermediate systems, for reconstruction of the regional topology.

4. It is sent to all intermediate systems for a reconstruction of the global topology.

The first option is the minimal level of closed-loop scheduling, in which the forwarding decisions are made in light of local conditions (data links up or down, queue sizes, and so on), but without any other information concerning the current state of the global discrete event system. This is also referred to as *isolated* concatenation (routing) because there is no coordination between intermediate systems. As we said earlier, the days of intermediate systems that are not instrumented to even this minimal extent have long since passed; and even otherwise open-loop concatenation uses local state information in deciding the next forwarding stage.

Beyond local state feedback, there is the global state reconstruction of the network or traffic state(s). Whereas local state information can be obtained from an intermediate system's own sensors or estimators, reconstructing (estimating) the global state information (topology and/or traffic) is much more complicated. Recall that in Chapter 1 we illustrated this with the parable of the blind men and the elephant, each of whom could feel just one feature (the trunk, the tail, the ears) and made correspondingly wild estimates about the beast (thinking it a tree, a snake, and so on); only when they exchanged their impressions (i.e., local state information) was the correct reconstruction achieved. These mechanisms for exchanging local topology (and, if applicable, traffic) data are generally known as *routing protocols.*

Consider the two intermediate systems exchanging their local state (topology) information in Figure 12.6. An intermediate system can measure or estimate its local transporters but not those to which it is not connected. For example, the Token Ring connected to the left-hand intermediate system is invisible to the right-hand intermediate system, which knows about its local topology (the Ethernet and the two serial links) because it is attached to these and it can monitor its own configuration.

Similarly, the left-hand intermediate system can monitor its local configuration and therefore it knows about its logical topology (the Token Ring and the two serial links). When the two intermediate systems send their respective measurements to each other, the topology estimators in each can use these nonlocal measurements to estimate/reconstruct the global topology. In each case the challenge is to exchange sufficient topology information as to allow the workload managers in each intermediate system to schedule the next component transport task necessary to realizing the network (end-to-end) transport.

Although many mechanisms for estimating global state information have been devised, they all amount to the blind men and the elephant, with intermediate systems exchanging the local state information. The question, however, is the scope of the information exchange, which can vary radically from, at one extreme, a single, centralized global state estimator that receives local state information from every intermediate system in the network to, at the other extreme, decentralized estimators located in every intermediate system that exchange their respective state information with every other intermediate system in the network.

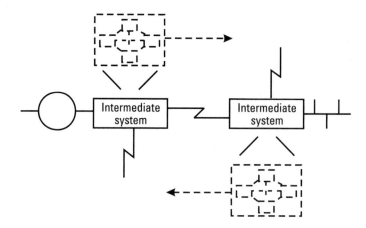

Figure 12.6 Network with two intermediate systems.

Between these two extremes are many shades of "compromise," in which the set of all intermediate systems is broken down into subsets, within which there is full exchange of local state information but between which the exchange is limited, either in frequency or in granularity (see later discussion). As we will discuss in the next chapter, this modularization is at the heart of the scalability of modern networks, especially IP and its hierarchy of areas and autonomous systems, along with the division between interior and exterior routing protocols.

Why not share state (topology and/or traffic) information as widely as possible? Quite simply, the cost due to management overhead grows prohibitive. In a network of n intermediate systems, if each intermediate system exchanges its local topology information with every other intermediate system then $n(n - 1)$ updates will be sent. Another factor that helps determine the overhead of the topology exchange is its frequency. Routing protocols can also be differentiated into opportunistic versus periodic in their updates. With the former an intermediate system sends a state update when something changes; with the latter the exchanges occur regularly, irrespective of whether there is new state information. There are advantages to each approach, although it should be noted that with opportunistic updating phenomena such as "route flapping" (when a link frequently goes up and down) can cause enormous amounts of routing protocol traffic.

In addition, there is the granularity of state information. Transparent bridges, for example, reconstruct the global topology of the transparent bridges but not anything about end systems or downstream transporters (networks). As we will see in the next chapter, routing protocols come in two varieties, distance vector and link state, which use different techniques to represent the topologies. At one extreme we have link state routing protocols such as OSPF, which exchange explicit topology models listing IP networks (data links) and intermediate systems. From these topology models the workload manager calculates the shortest schedule (sequence of transporters) to the destination network and end system. Distance vector routing protocols, on the other hand, do not exchange explicit topology data but rather their forwarding tables, which contain the best estimates of shortest schedules to various destinations.

Finally, we have already noted the intersection of concatenation (rerouting) with fault management. Likewise, we have the intersection of topology reconstruction with configuration management, on the one hand, and global traffic reconstruction with performance management, on the other. Configuration management is concerned with collecting data about the network's resources; clearly, the topology of the network falls within this category. Likewise, performance management, collecting information on the utilization of the network, overlaps with any collection of traffic statistics by the routing protocol.

12.3.1.2 Centralized Versus Distributed Scheduling

We now come to the *raison d'etre* for concatenation, namely, scheduling the next tasks that are sufficient to realize the end-to-end transport task. Every intermediate system in a composite transporter (network) contains a workload manager that creates the schedule for the transport tasks in either the next stage (task) to be executed or, if source routing is employed, the entire schedule at once. In most instances except with promiscuous forwarding (Class 0 concatenation), the workload manager relies on a forwarding/routing table to make its scheduling decisions. The question is one of where and when the routing or forwarding table is created.

To answer this, we must first note that there are in fact two echelons of scheduling involved in concatenation. The first echelon is the actual concatenation executed by intermediate systems; we refer to these first level workload managers as concatenation managers. The second echelon, however, executes the creation of the routing or forwarding tables/rules used by the schedulers in the first echelon; we refer to these second-level workload managers as "route" calculators or "route" servers, where the quotation marks are intended to indicate that route should not be taken to exclude other forwarding mechanisms such as bridging, tag switching, and tunneling. To the extent that the intermediate systems are adaptive, it is this second level of scheduling that actuates changes to the concatenation managers' schedules in the intermediate systems. Figure 12.7 shows this hierarchy.

Note that the two-tiered scheduling hierarchy holds for both open-loop and closed-loop concatenation. With Class 2 concatenation (stateful without

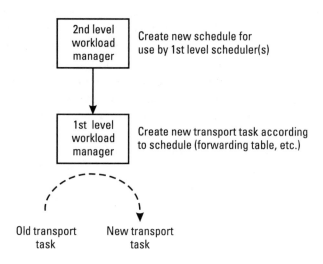

Figure 12.7 First- and second-level workload managers.

feedback), for example, the routing tables would be calculated a priori by a route server and then disseminated to the individual intermediate systems. However, the picture is even more complicated if feedback is employed (Class 3 concatenation), since it may be used in scheduling at the first echelon, the second echelon, both, or neither.

Just as global state reconstruction can be realized with a centralized estimator or with estimators in each intermediate system in the network, the same is true with respect to creating the scheduling information (routing tables) used by the intermediate systems in the network to concatenate the component transporters. Indeed, since the global reconstruction is independent of the routing table calculation, it is conceivable (although unlikely) that one could be centralized and the other distributed. For example, a centralized estimator could reconstruct the global topology based on local state information sent to it but then send this global topology to route servers (second-order concatenation managers) located in each of the intermediate systems.

The question then becomes this: How is this second-order scheduler implemented? Many early networks did rely on a centralized route server located in a routing center to calculate routing/forwarding tables and disseminate these to the "front-line" intermediate systems. Indeed, subarea SNA still works this way: The NCP load modules, including the path definitions that specify the VR/ER forwarding, are generated in the sysgen process at an S/3X0 mainframe and then downloaded to the FEPs. Until the next sysgen occurs, these forwarding tables are fixed.

However, the entire trend of concatenation in the internetworking era has been toward distributed route servers, with each intermediate system calculating its own routing tables just as it estimates for itself the global (or, in the case of hierarchical/modularized routing protocols, regional) state information. As we saw in Chapter 10 part of the rationale for the ARPANET was to research distributed mechanisms for calculating forwarding/routing tables: Each intermediate system is its own route server. This goes back to Baran's RAND research on highly survivable communication networks for military applications.

Figure 12.8 shows centralized and distributed route calculation. Note that although in each case we indicate with dashed lines the possible exchange of local state information, it is feasible to implement either centralized or distributed route calculation without any such feedback.

12.3.2 Workload Management: Schedule Persistence

The next major taxonomic criterion is what we call the persistence of a concatenation manager's scheduling decisions. By this we refer to whether each

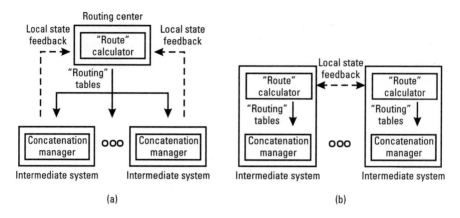

Figure 12.8 Routing table calculation: (a) centralized versus (b) distributed.

end-to-end PDU is scheduled individually or is a scheduling decision made for one PDU applied to subsequent PDUs between the same end systems? As we briefly discussed in Chapter 9, this difference in persistence is at the heart of one of the enduring debates in networking, namely, the relative merits of virtual circuit and datagram forwarding mechanisms.

With datagram protocols each end-to-end PDU is forwarded separately, meaning any two end-to-end PDUs may take very different sequences of transporters; that is why TCP/IP's architects had to include resequencing in TCP, to recover from out-of-order delivery. Virtual circuits, on the other hand, are predetermined or rely on establishing a "pattern" of forwarding with the initial PDUs that is then followed by intermediate systems as they forward subsequent PDUs. Virtual circuit forwarding can use various mechanisms, including circuit identifiers, labels, or flow numbers to speed the scheduling of the next stage by the concatenation managers in the intermediate systems. This is why we argued in Chapter 10 that IPv6 flows, notwithstanding the rhetoric of the IPv6 architects, are very much in the virtual circuit mode of forwarding. Virtual circuit concepts likewise underpin many of today's most sophisticated forwarding techniques, known variously as tag or label switching (including MPLS).

We need to correct another misconception about virtual circuits and datagrams, beyond those we discussed earlier. Virtual circuit protocols tend to be associated with static routing, on the one hand, whereas datagram protocols are assumed to use dynamic routing. This is incorrect. It is entirely possible, for example, to lay out an IP network entirely with static routes, and it is quite common to use static routes in part of the network where, for example, only a single data link connects a remote LAN to the rest of the network. A datagram network may be completely nonadaptive, if static route definitions (schedules) are used in the intermediate systems; and a virtual circuit network may be very adaptive if a routing protocol is used to update the forwarding mechanisms of the intermediate systems.

Virtual circuits and datagrams can be "unified" as the extremes of a spectrum of end-to-end schedules and their persistence. Running from most to least persistent, we have:

- *Fixed virtual circuit:* The scheduling of end-to-end transporters is totally nonadaptive. They are composed of the same components for the lifetime of the network.
- *Dynamic virtual circuit 1:* The transporters are adaptable between execution but fixed while running. Workload managers can reconfigure end-to-end schedules not currently in use.
- *Dynamic virtual circuit 2:* The transporters can be reconfigured while executing. The adaptation may be scheduled to be open loop or closed loop (opportunistic or event driven, respectively).
- *Datagram:* The transporters are reconfigured for each PDU.

Subarea SNA is an example of fixed virtual circuit concatenation management; it allows end-to-end PDUs to be forwarded on up to 16 different end-to-end virtual circuits (Explicit Routes) between pairs of subareas nodes but these ERs are fixed at sysgen. An example of a Dynamic Virtual Circuit 1 is APPN, which, as we saw in Chapter 11, establishes an end-to-end path dynamically when a session is requested by an end node. IPv6 flows are an example of Dynamic Virtual Circuit 2 concatenation management, where end-to-end scheduling corresponding to a given flow label can change as the topology of the network changes. IP with dynamic routing is the paradigmatic example of datagram concatenation management. This continuum is shown in Figure 12.9.

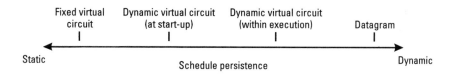

Figure 12.9 Continuum of schedule persistence.

12.3.3 Workload Management: Next-Stage Versus End-to-End Scheduling

Finally, we must look at the scope of the schedule that an intermediate system's concatenation (workload) manager creates. At a minimum, every concatenation manager needs to determine the next stages for realizing the end-to-end transport, and we have just discussed a wide spectrum of techniques used. However, when source routing is desired then the initial intermediate system or the sending end system must calculate the *entire* schedule. In addition, those concatenation mechanisms that construct spanning trees to every destination network, such as link state routing protocols, effectively create end-to-end schedules although, generally speaking, they only use the next-stage component of the schedules.

We should note as well that source routing with its end-to-end schedules can be used with both datagram concatenation as well as with (dynamic) virtual circuits. We saw, for example, with APPN that the intermediate system (network node) to which an end system (end node) is attached calculates the schedule of transport tasks that realizes the desired end-to-end task (APPN path); the intermediate system then sets up the path by sending an SNA bind RU carrying a Route Selection Control Vector (RSCV) to the intermediate systems along the path. Similar source routing is available as an option with IP.

12.3.4 Hierarchical Versus Flat Concatenation

Before finishing with concatenation we should briefly discuss hierarchical versus flat concatenation and the relationship between concatenation and the address space architecture. A hierarchical address space is naturally conducive to abstraction and "divide and conquer" modularization of the network and by extension of both the reconstruction of global state information and the calculation of concatenation schedules. And this is the key: To reduce the nonlinear growth of overhead as the size of the network grows, hierarchical addressing offers natural segmentation points at each level of a hierarchy, dividing the respective estimation and scheduling tasks into decoupled subtasks with considerably less total overhead. How much less overhead? If we divide a network of n intermediate systems into two equal component networks of $n/2$ intermediate systems each then we see that the overhead for the two new networks of a routing protocol that scales according to $O(n^2)$ is

$$2\left(\left(\frac{n}{2}\right)\left(\frac{n}{2}-1\right)\right)=\left(\frac{n^2}{2}-n\right)$$

As long as the routing protocol overhead between the two networks is less than $(n^2/2)$ then the transport load will be correspondingly reduced. Note that the advantages of such "divide and conquer" reductionism are even greater for routing protocols that scale according to $O(n^3)$ or higher powers.

This is in part why hierarchical addressing is so common. Postal addresses, phone numbers, even the use of first and family names, all of these are examples of hierarchical addressing. Likewise, hierarchical concatenation is used constantly in everyday life. When a letter is mailed out of state, the local post office does not attempt to determine the end-to-end schedule down to which letter carrier will handle final delivery; rather, it sends the letter to the main post office for that state, from which it is forwarded to the next concatenating office, and so on. Both SNA and IP use hierarchical addressing: SNA with its subarea/element addresses and IP with its network/host addresses. Forwarding decisions need only be made on the basis of the subarea or IP network number, reducing the size of the routing/forwarding tables markedly.

Flat address spaces, on the other hand, are unwieldy and limit attempts at more efficient concatenation. Take, for example, the 802 MAC address space; excepting locally assigned addresses, the 802 address space is globally significant but it is *not* hierarchical. Two end systems on the same LAN are likely to have MAC addresses that are completely unrelated. Layer 2 intermediate systems (MAC bridges) could build topology models, and indeed, as we said earlier, this is how learning bridges work; but no abstraction or summarization of the model would be possible because of the flat address space. This alone precludes the use of a full routing protocol among L2 intermediate systems, because the size of the topology models exchanged would be enormous. Instead, as we will see in Chapter 14, transparent bridges make a simple estimate using their local topology data about whether to forward a PDU, with the spanning tree algorithm used to prevent loops among bridges.

The modularization that is possible with hierarchical addressing can be likened to software module design (abstract data types). The low-level topology information is hidden within a "module." The high-level topology is sufficient to get the traffic to the module interface. At the module interface the topology information "explodes" and finishes the routing. The "what" is defined by the set of addresses known to be within the module—that is, the transport tasks corresponding to these addresses. One or more gateway routers would have the topology knowledge/information to supply the "how," the location of and path(s) to the desired destination.

With such modular routing, the gateway router does the "information hiding"—it abstracts the implementation details (the topology of transporters and their interconnection) and merely "advertises" the tasks it can execute (destinations it can reach). In this it functions like the entry point to a software module (abstract data type). Put another way, information on the global topology is partitioned in such a way that it exploits some aspect of its structure to reduce the amount of information exchanged. Consider if, for example, all the network addresses in a module are subnets of a Class B network. Assuming that no other subnets of the network are outside this cloud then the topology information that need be distributed/transported to other routers outside the cloud can be abbreviated to identifying the Class B element itself. This is information hiding by modularization.

By using abstraction to hide implementation (topology) information in a modularization of the overall network, the scheduling task is simplified as well as the exchange of routing information and its storage. An intermediate system will calculate the optimal schedule not to the ultimate destination (server), but to the module in which the destination is known to be located. Once the RFS (PDU) arrives at an entry point to the module, the rest of the routing decision making (scheduling) can be executed.

Hierarchical concatenation works particularly well with IP's next-stage scheduling because this does not require the source to specify the route completely, only to the next hop. With both distance vector and link state routing protocols, however, each router will calculate the complete route even though it only requires the next hop. What hierarchical routing does is hide part of the topology and hence reduce the routing calculations. Topology information hiding may have a negative impact on scheduling only if there are multiple gateways/schedulers to the module in question *and* if one of these is better located with respect to the destination than the others. In this situation, the scheduling/realization of the geodesic (shortest path) by the source will be defeated by the information hidden within the module.

We discuss IP hierarchical routing mechanisms in the next chapter.

12.4   Bandwidth Management: Network Redesign

The last topic of this chapter is the relationship of the concatenation problem to its complement, the network design problem. With concatenation we are given a composite transporter composed of a set of component transporters (data links and so on) and the management task is to map the end-to-end transport tasks (traffic) to the transport tasks of one or more of the component transporters. With the latter, on the other hand, we are given a set of clients and the management task is to determine the optimal bandwidth (capacity) of the transporter(s): How much effective bandwidth should be implemented to meet the traffic offered by the client? What trade-off should be made between nominal bandwidth, reliability, and maintainability? In other words, what is the transporter's optimal realization—its composition, topology, and so on?

Such bandwidth actuation may occur throughout the life cycle of the network. Of course, the process begins with the initial design of a network and its topology; in this case the bandwidth actuation starts with nothing—a null server—and creates the network and its components. However, just as dynamic concatenation implies that the workload manager monitors the state of the transporter and its components, and changes the end-to-end scheduling according to changes in the transporter(s), an ideal bandwidth manager should monitor changes in the workload arrivals (traffic) and actuate the bandwidth of the transporter and its components to reflect these. This process is generally known as tuning, redesign, or adaptation.

12.4.1 Bandwidth Modularization

Another parallel between the workload management of concatenation and the bandwidth management of network redesign is modularization. Bandwidth modularization is another way of saying that we build networks out of multiple components. But why are transporters realized as such composites? Because various factors preclude monolithic implementations for all but the smallest transporters. This, indeed, is the original meaning of the term *network*: a composite of two or more data links joined together. This is why the OSI model designated as the *network* layer that level of the protocol stack responsible for concatenation. Some of the factors that cause us to realize transporters as composite (i.e., with two or more components) include the following:

- *Distance:* We saw in Chapter 2 that as channels increase in length their reliability falls. It is often necessary to use two or more transporters, perhaps concatenated at the physical layer via some type of repeater, simply to span the distance between end systems.

- *Traffic intensity:* As traffic loads grow, it is necessary to add bandwidth to avoid performance degradation. However, it is not possible to increase bandwidth arbitrarily. Generally speaking, bandwidth can be incremented only in discrete "chunks," and then only to some finite maximum.

- *Reliability:* Even if an infinite bandwidth channel could be implemented, it would still constitute a single point of failure. By implementing a composite transporter with alternative routing between two end systems we increase the reliability of the whole.

- *Management overhead:* With multiaccess data links, for example, the overhead from managing the contending clients grows with the number of clients, whether the data link is scheduled with polling or by random access. By implementing additional data links and dividing the clients among these, this overhead can be scaled back (reduced).

- *Localization:* Most traffic patterns exhibit localization. For example, if 80% traffic generated by users in the accounting department is localized, then by allotting them their own LAN it is possible to reduce the workload of an intermediate system to forwarding "through" traffic destined for end systems outside the accounting department.

To discuss composite transporters we need to employ some of the concepts and terminology of graph theory. A graph is composed of two types of entities: nodes (or vertices) and edges (or arcs). If a direction is associated with an edge, then the graph is said to be directed; otherwise the graph is undirected. End systems and intermediate systems are obviously nodes; transporters are edges. Because we will assume that all component transporters (data links) are duplex if the transporter is point to point or any to any if it is, multipoint, it follows that the graph model is undirected.

The network design problem then consists of finding a set of edges that links all the nodes in such a way as to optimize cost relative to performance (or vice versa). Given *k* locations that must be serviced by a network, there are many possible topologies of the (composite) transporter. The two edge minimal topologies are linear and hub-and-spoke topologies. Figure 12.10 shows the two minimum topologies. Conversely, there is a unique maximal topology, namely, the completely connected graph with $n(n - 1)/2$ edges; this is also known as a full mesh. The obvious goal is to find the optimum topology between these extremes.

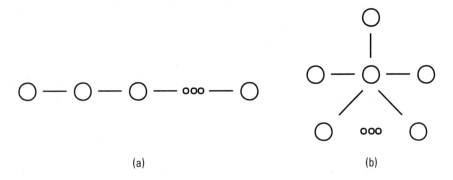

(a)   (b)

Figure 12.10 Two minimum topologies: (a) linear and (b) hub-and-spoke.

The mathematics of network design are well developed but unfortunately constraints of space preclude us from exploring them too extensively. Very briefly, a principal parameter that estimates reliability is the connectivity of the composite as measured by its cut-sets. This is the number of edges and/or nodes that must be removed from the server's corresponding graph to disconnect the composite. Notice that the occurrence of such a set of faults does not disable the execution of all external tasks. Strictly speaking, a cut-set must be defined relative to an external task. A network is said to be *n-connected* if the removal of *n* edges and/or nodes will split the network into two parts. Such a combination is called a *cut* and the collection of all these is called the *cut-set*.

Consider, for example, the simple network in Figure 12.11: The tasks 1 → 4, 4 → 1 have the cut-set {[(1,2),(1,3)],[(1,2),(3,4)], [(2,4),(3,4)], [(1,3), (2,4)]}. On the other hand the tasks 2 → 3, and 3 → 2 are unaffected by the first and third sets. Cut-sets also determine the bandwidth of the network as demonstrated by one of the central results of graph theory, the Fulkerson-Ford theorem, in which it is proved that the maximum flow in a network (its maximum bandwidth) is determined by the minimum cut in its cut-set. For more details on this and related topics, please consult the reference section for books on network design such as [5–8].

12.5 Summary

In this chapter we moved into the area of internetworking proper and discussed the various mechanisms that have been developed to concatenate transporters and forward traffic between them. We stressed that concatenation is workload actuation of kind. The end-to-end transport task is realized in two or more stages using the transport tasks of component transporters (generally, data links). As we saw, these include routing, bridging, and tunneling.

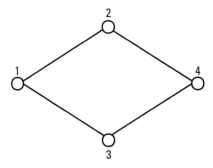

Figure 12.11 Network with four point-to-point links.

We saw that in classifying a concatenation mechanism we must ask two related questions: What state information does the concatenation workload manager know and when does it know it? That is to say, what model (if any) of the discrete event system does the intermediate system use, and how (if at all) is this model kept current with the changing state of the network and/or traffic? We discussed the three basic classes of concatenation that are defined by these two discriminants; and, with closed-loop concatenation (Class 3), we saw that reconstruction of global state information can be realized with a central estimator or with decentralized estimators in every intermediate system.

We saw that the same choice applies with the creation of the forwarding tables used in these intermediate systems: Are these created at a central site and disseminated or is this second-order scheduling distributed? The persistence of a scheduling decision was then examined, and we identified four distinct levels of adaptivity ranging from pure virtual circuits to pure datagrams, with dynamic virtual circuit architectures such as APPN in between the two. Our examination of scheduling then concluded by looking at its close dependency on the address space architecture (flat versus hierarchical) and the scaling advantages of hierarchical concatenation. Finally, we briefly touched on the network design problem and its complementary role to concatenation.

References

[1] Cerf, V., *The Catenet Model for Internetworking,* Internet Engineering Note 48, July 1978.

[2] Laubach, M., *Classical IP and ARP over ATM,* RFC-1577, Jan. 1994.

[3] Davie, B., P. Doolan, and Y. Rekhter, *Switching in IP Networks,* San Francisco, CA: Morgan Kaufmann, 1998.

[4] Tanenbaum, A. S., *Computer Networks,* 2nd ed., Reading, MA: Addison-Wesley, p. 296.

[5] Stoer, M., *Design of Survivable Networks,* New York: Springer-Verlag, 1992.

[6] Cahn, R., *Wide Area Network Design,* SanFrancisco, CA; Morgan Kaufmann, 1998.

[7] Shier, D., *Network Reliability and Algbraic Structures,* Oxford University Press, 1991.

[8] Seidler, J., *Principles of Computer Communications Network Design,* Chichester: Ellis Horwood, 1993.

13

Layer 3 Concatenation: Routing and L3 Switching

13.1 Introduction

With our MESA model of concatenation management now in hand, this chapter will focus on layer 3 concatenation, that is, routing and layer 3 switching, and specifically dynamic routing realized with routing protocols. As we saw in Chapter 12, such closed-loop concatenation is intimately coupled to the nature of the model of the discrete event system (network and traffic) being managed. Here we compare the different modeling and scheduling approaches used by distance vector and link state routing protocols and discuss their respective advantages with respect to properties such as convergence and stability. We also consider the two by looking at IP's Routing Information Protocol (RIP) and Open Shortest Path First (OSPF) protocol.

The second theme that dominates any discussion of closed-loop concatenation and routing protocols is modularization and its role in constraining management overhead as the network size increases. The modularization of concatenation is effected in the first place with the division of a network into autonomous systems and the corresponding division of concatenation tasks between the interior and exterior routing protocols. Toward this end, exterior routing protocols exchange state information of a reduced granularity, sometimes referred to as reachability rather than routing information. Complementing our discussion of RIP and OSPF, which are interior routing protocols, we will look at the Border Gateway Protocol, which is an example of a (distance vector) exterior routing protocol.

As we will see, both interior and exterior routing protocols achieve routing efficiencies by exploiting IP's hierarchical address space to "divide and conquer" the tasks of reconstructing global state information and updating schedules in the routing tables of intermediate systems. We will look at the evolution of IP addressing from "classful" to "classless" and the accompanying development of what is called Classless Interdomain Routing (CIDR). CIDR allows the aggregation of multiple IP networks into *supernets,* abstractions that reduce the overhead of reconstruction and scheduling.

13.2   Routing Modularization

13.2.1 Autonomous Systems and the Topology of the Internet

The Internet as we know it today is highly modularized in the way traffic is forwarded and end-to-end transport tasks are realized. In this section we want to discuss the topology of the Internet and why its modularization proved so important in its growth. In contrast, the Internet's predecessor, the ARPANET, was monolithic from the point of view of routing. There was no partitioning of the topology nor was any hierarchical routing employed. Every interface message processor (IMP) had in its routing table an entry for every other IMP along with the next hop to get there. The ARPANET routing protocol used was the Gateway to Gateway Protocol (GGP), an early distance vector protocol that we will discuss in more detail later.

With the introduction TCP/IP and the internetworking model of Cerf and Kahn, however, the incorporation of disparate networks created strains. First of all, treating these networks as part of a single routing domain was problematic because it meant that they all had to implement and use the ARPANET's GGP to exchange global state information. Given that many of these networks served as experimental testbeds for new protocols and architectures (like THE ALOHA SYSTEM we discussed in Chapter 8), this was regarded as an unacceptable constraint. Equally, however, there was the scaling problem: The overhead of the routing was growing too quickly, with too much network bandwidth being consumed by routing protocol exchanges. (As we will see later, GGP's poor scalability was one of several reasons that led to the development in 1979 of the first link state routing protocol for deployment in the ARPANET.)

A more fundamental change, however, came in 1982. This was the hierarchical decomposition of the Internet into distinct routing domains called *autonomous systems* (ASs), each of which was assigned a globally unique 16-bit identifier (the AS number). Originally this was a two-tiered hierarchy with a core AS, built around the ARPANET, to which were connected all the remaining ASs, called stub ASs. The core AS was the only transit AS, that is, the only AS that forwarded traffic between other ASs (Figure 13.1). Later this hierarchy was extended to encompass multiple levels but a strict tree topology was always required—no mesh connections between ASs could be implemented. This restriction was only lifted with the next generation of the Internet in 1989 (discussed more later).

Hand in hand with this decomposition of the Internet into ASs came the division of concatenation between interior routing protocols such as OSPF or RIP and exterior routing protocols such as the Exterior Gateway Protocol (EGP) or its successor, the Border Gateway Protocol. An interior routing protocol (also known as an interior gateway protocol) is used within an autonomous system. Traffic between autonomous systems, on the other hand, is forwarded using an exterior routing protocol (also known as an exterior gateway protocol) that exchanges less state information (see later discussion). In part, the assumption behind this division was that most traffic originating within an AS was likely to be destined for end systems within the AS as well, and that the overhead of full topology exchange would be justified, whereas between autonomous systems the lower traffic volume allowed a reduced granularity of state information.

Such modularization is similar to the decomposition of a monolithic application into multiple subroutines (or abstract data types) with corresponding information hiding. The philosophy was that within an autonomous system

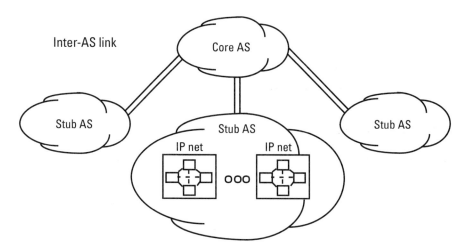

Figure 13.1 AS hierarchy: core and stub ASs.

[t]he protocols, and in particular the routing algorithm which these gateways [routers] use among themselves, will be a private matter, and need never be implemented in gateways outside the particular domain or system. [1]

We should note that over the years the criterion for creating an autonomous system has changed. Originally an autonomous system was an administrative unit like a corporation or government agency that was logically integrated; however, now it is tied to the policy routing and route aggregation (more on this later).

13.2.2 Reachability Versus Routing Information

The reduced granularity of state information that is associated with exterior routing protocols is known as *reachability*. Reachability refers to state information that specifies *what* networks are within an AS without specifying any details about *how* to get to them, that is, topology models or routing tables. Reachability information is exchanged by exterior gateways (routers), that is, L3 border intermediate systems contained within their respective autonomous systems:

[r]outing information (routes) exchanged via EGP consist of *network-layer reachability information* (NLRI, expressed as a list of IP network number), the network layer address of the IS that should be used as the next hop when forwarding to the destinations depicted by the NLRI, and a metric. [2]

Thus, rather than knowing the topology of the entire network, intermediate systems in a given AS know how to forward PDUs within their own AS; and they know the other IP networks that are reachable in the other ASs and which exterior gateways/border intermediate systems are the portals to these.

13.2.3 Classless Interdomain Routing

The abstraction that came with the introduction of autonomous systems and reachability information proved quite successful in reducing the overall concatenation management overhead. Nonetheless, by the early 1990s a new problem was looming: The growth of the Internet was rapidly exhausting the Class B address space, and the prospect of assigning multiple Class C network addresses to meet the demand for new networks threatened to overwhelm the core Internet routers, the size of whose routing tables had grown to the point where performance was being affected by long look-up times.

This proliferation of network numbers was a direct consequence of IP's division of the 32-bit IP address space into Class A, B, and C networks. Recall that, in part, this addressing scheme was designed to expedite forwarding, with the number of bits used for network addresses encoded in the most significant bits of the most significant octet of the address. When a classful IP address is "decoded," the most significant bit of the most significant octet indicates if the network address uses the next 7 bits (Class A) or more than the next 7 bits. The next most significant bit tells if the network address uses the next 14 bits (Class B) or more than the next 14 bits, and so on. The highest 3 bits, therefore, tell the routing algorithm how many of the 32 bits it should treat as specifying the network address, and this was intended to improve performance of IP intermediate systems.

The difficulty with this classful addressing was that Class A networks, which allow up to 16 million hosts, were too big while Class C networks, which allow up to 254 hosts, were too small for most corporations and organizations. Class B networks, which allow up to 65,000 hosts, were also too large for all but the largest corporations but the "fit" was much better than Class A or Class C; and as the availability of Class A networks was almost nil, the rapid exhaustion of Class B addresses was inevitable; hence, the recourse to Class C addresses and the pending routing table explosion. What was needed was a technique to constrain this growth while still enabling new networks to be deployed.

The solution was suggested by the way subnetting had been deployed in IP networks. Recall subnetting itself was a work-around introduced in response to the coarse granularity of IP's classful addressing: Few if any networks supported 65,000 hosts but many corporations had multiple physical networks (data links) that might have anywhere from a few to a few hundred users each, and subnetting allowed these networks to share a single Class B IP network number. At the same time, carefully laying out the addressing used by these subnets allowed the details of their topology to be abstracted. Intermediate systems outside the IP network in question would route traffic to it based on its IP network number, completely oblivious to the subnet addressing employed within. Figure 13.2 shows an example of subnetting and route summarization where variable length subnet masking (VLSM) is employed.

This technique of route summarization is also known as route aggregation. Route summarization/aggregation is fundamentally about reducing the overhead of management information exchanges by using variable granularity of resolution. This translates into setting up a hierarchy of schedulers, with routing (topology state) information available on a need-to-know basis. This "aliasing" may reduce the size of the routing table drastically depending on the extent of the respective aliases ("fan-out"). This can be seen as just continuing

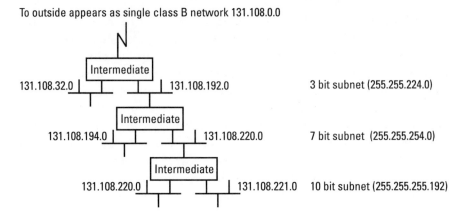

Figure 13.2 Route summarization with VLSM.

the abbreviation of routing tables that is enabled by hierarchical addressing where, rather than having a routing entry for every host, the table has an entry for each network address (Figure 13.3).

The full potential of route aggregation to reduce the size of IP routing tables was, however, limited by the coarse granularity of summarization that a three-tiered address hierarchy permitted. The solution to both this and the exhaustion of the IP address space entailed abandoning IP's address classes (A, B, and C) and instead moving to so-called *classless* addressing. The 32 bits were divided between a *prefix*, which subsumed the concepts of both network and subnet numbers, and a host number. Thus, for example, the network addresses in Figure 13.2 would in classless notation be written as 131.108.0.0/16, 131.108.192.0/19, 131.108.220.0/23, 131.108.221.0/26, and so on.

Classless addressing was formally introduced in 1993 as part of CIDR, which as we will see later is not a routing protocol but rather a modification of existing routing protocols to support classless addressing and aggressive route aggregation. For example, RIP was modified to RIPv2, OSPF to OSPFv2, and BGP3 to BGP4 to support CIDR addressing and aggregation.

Destination network	Next network/forwarding interface
18.0.0.0	192.14.7.0/E0
197.13.148.0	19.0.0.0/T1
52.0.0.0	128.19.0.0/E1
⋮	⋮

Figure 13.3 Forwarding table for IP routing.

13.2.4 Moy's Taxonomy

Before we discuss actual routing protocols, we want to consider a taxonomy that John Moy, one of the principal developers of the OSPF routing protocol, has proposed for comparing routing protocols [3]. The discriminants include these:

- *Type:* Interior or exterior? Distance vector or link state?
- *Encapsulation:* IP, TCP, UDP, or data link?
- *Path characteristics:* What transporter parameters are used in scheduling the next hop?
- *Neighbor discovery and maintenance:* How do intermediate systems initially find and then keep track of each other?
- *Routing data distribution:* What state information is exchanged and when?
- *Route deletion:* How does the routing protocol adapt to a lost destination prefix (network)?
- *Routing table calculation:* How does the second-order scheduler create the schedules in the routing table?
- *Robustness/reliability:* How does the routing protocol avoid concatenation faults due to hardware and/or software errors?
- *Aggregation:* Will the routing protocol automatically aggregate multiple prefixes where possible?
- *Policy controls:* Can the second-order scheduling incorporate policy considerations such as transit forwarding restrictions when calculating routes (schedules)?
- *Security:* How does the routing protocol prevent deliberate tampering (faults)?

Moy's first criterion is whether the routing protocol is used interior or exterior to an autonomous system, and whether it employs distance vector or link state mechanisms to create the (first-order) schedules in the routing tables of the intermediate systems. As we will see below, all four combinations have been tried with various routing protocols, although some argue that distance vector mechanisms are more suited to the reduced granularity of exterior routing protocols and link state mechanisms to interior routing protocols where extensive topology reconstruction is advantageous if not mandatory (more later).

We can map Moy's remaining discriminants onto the MESA workload/bandwidth model quite well. Path characteristics, Neighbor discovery and maintenance, Routing data distribution, and Route deletion all pertain to the bandwidth management employed in the intermediate systems, specifically the state variables (path characteristics) and mechanisms (all the rest) of the global state reconstruction collectively executed by the bandwidth estimators. Routing table calculation, Robustness/reliability, Aggregation, and Policy controls are all concerned with the workload management, specifically how the scheduling is collectively executed by the workload schedulers in the intermediate systems that are running the routing protocol.

13.3   Interior Routing Protocols

Let's now consider interior routing protocols, which before the Internet was partitioned into autonomous systems were the *only* routing protocols. Recall from Chapter 12 that two component tasks are executed by a routing protocol: estimation of global state information from the local state information (measurements and/or estimates) generated by bandwidth and/or workload monitoring at various intermediate systems; and creating the schedules for realizing all (source routing) or part (next-hop routing) of the end-to-end transport tasks. In graph theory the scheduling task is called the *shortest path problem.*

Not all routing protocols exchange explicit topology information; instead, some exchange their local routing tables, from which globally consistent routing tables are created. The former is the approach behind link state routing protocols, which rely on explicit topology models of the network, whereas the latter is the approach used in distance vector routing protocols, where no explicit model is used. For this reason, these are sometimes referred to as the distributed database and the distributed computation methods, respectively.

The algorithms that have been devised for finding the shortest paths between nodes of a graph were not originally intended for distributed execution; rather, it was assumed that they would be executed "monolithically." Of course, we saw in the last chapter that networks with centralized estimation and the scheduling mechanisms are possible and had even been implemented (e.g., Tymnet), but with IP we are only interested in distributed estimation and scheduling. Fortunately, most shortest path algorithms can be adapted for distributed execution, and these are at the heart of the various routing protocols in use. The obvious decomposition is to have each intermediate system find the shortest paths from it to the other intermediate systems in the network (or AS). Distance vector routing protocols use a distributed version of the Bellman-Ford algorithm, whereas link state routing protocols generally use the Dijkstra algorithm.

We should note that there are two models of distributed computation, namely, synchronous and asynchronous. Synchronous distributed computation requires that each intermediate system coordinate its execution with the others, often entailing complex signaling mechanisms such as semaphores to keep things moving in lock-step fashion. For example, with a synchronous routing algorithm all the intermediate systems must start their computations at the same time, clearly not feasible in a large network. Asynchronous distributed computation, on the other hand, allows each intermediate system to execute its calculations as it sees fit, incorporating updated state information as it arrives. All the routing protocols that we consider here rely on asynchronous execution.

Routing protocols can be compared in terms of their *convergence* properties, the process of synchronizing of the routing schedules and/or concatenation state information (e.g., topology models) of the intermediate systems. Related to convergence are sensitivity and stability: When a link's cost changes (for example, if the link fails) how quickly does the routing protocol reflect this change? Does the routing protocol converge accurately or do oscillations due to overshoot plague the network?

In the discussion that follows we assume the network is represented with a undirected graph. Directed graphs (digraphs) are required if any of the following hold:

- Simplex data links (transporters) are used in the network;
- Full- or half-duplex data links with asymmetric bandwidth are used in the network;
- The state information includes traffic (workload).

With the first two conditions, the bandwidth between two nodes can differ in the opposing directions. The last condition, on the other hand, reflects the fact that traffic flows are almost always asymmetrical. In addition, there is significant volatility, which is a reason why few routing protocols include traffic state information in their scheduling.

For purposes of illustrating the distance vector and link state approaches, we use the example network shown in Figure 13.4 consisting of eight intermediate systems in a partial mesh; for simplicity we assume unit cost for each link.

13.3.1 Distance Vector Routing Protocols

As we just said, the basic idea behind distance vector routing protocols is that the intermediate systems in an autonomous system will exchange with each other their respective routing tables. In the network of Figure 13.4,

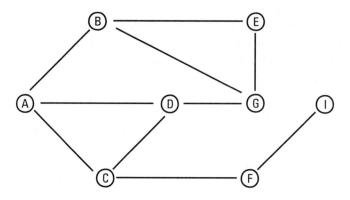

Figure 13.4 Example network.

intermediate system A will exchange routing tables with intermediate systems B, C, and D; intermediate system C will exchange routing tables with intermediate systems D and F in addition to intermediate system A; and so on. Why is this called *distance vector*? The origin of the term is somewhat clouded and there are at least two explanations. First is that the combination of direction and a "cost" such as distance or delay constitutes a vector as physicists use the term; for example, speed is a scalar but velocity, which includes direction as well as speed, is a vector. The second explanation is that the exchanged routing tables constitute vectors in the mathematical meaning of the term; that is, they are multidimensional state variables.

Whatever the origin of the term, *distance* can be any generalized cost that is used in the routing (scheduling) metric; it can even be a negative value, although this is not the case in transportation problems such as those using computer networks. Different routing protocols have parameterized links in terms of reliability, delay, bandwidth, load, and other variables; we discuss some of these later. If every link is assigned a cost (or weight) of 1 then the routing metric will minimize the hop count, that is, the total number of links in an end-to-end path (schedule). On the other hand, if each link is assigned a cost inversely proportional to its bandwidth then the end-to-end schedules created will, traffic loading aside, be the fastest possible.

An intermediate system that is executing a distance vector routing protocol maintains a routing table that has entries for destination networks in which is recorded the best current estimate of the distance to that network and the next intermediate system to which PDUs should be forwarded (i.e., the next hop). Some routing protocols and implementations will also store one or more of the next best routes for either load balancing or quick cut-over in case of a disruption in the end-to-end path. However, the performance advantages must be weighed against the increased table size.

When a distance vector intermediate system (DVIS) is started, it knows of only those IP networks (data links) to which it is directly attached; these are given the distance of 0. As it receives routing updates (or routing advertisements) from other DVISs containing their respective routing tables, the IP networks listed will either be already known or unknown to it. If an IP network is unknown, then the receiving DVIS will create a new entry in its routing table and, for the distance/metric estimate, it will put the updated estimate (that carried in the received table augmented by the estimate of the distance between it and the sending DVIS). On the other hand, if the IP network is known, then the receiving DVIS will compare its estimate to the updated estimate: If the updated estimate is greater than that already known then it will be ignored, unless the protocol has provision for storing multiple paths; on the other hand, if the cost is less than that known the table will be modified accordingly to reflect the new information.

Distance vector routing protocols illustrate that it is not the topology of the composite that we are ultimately interested in but rather the optimal routes between clients and destinations. Explicit topology reconstruction such as that used by link state routing protocols is only a means to an end and is unnecessary with distance vector routing protocols.

13.3.1.1 Finding Shortest Paths (Optimal Routes)

The heart of the path estimation (route calculation) in a distance vector routing protocol is Bellman's equation from dynamic programming.[1] This equation, a cornerstone of control theory and operations research, is used to solve multistage decision and optimization problems. (This is why distance vector protocols are sometimes called "Bellman-Ford.") In its basic form, Bellman's equation is given by

$$D_i - \min_j \left[D_j + d_{ji} \right], \quad \text{for } i \neq 1$$

---

1. The term *dynamic programming,* like *linear programming,* does not refer to software but rather to "a statement of the actions to be performed, their timing, and their quantity (called a 'program' or 'schedule'), which will permit the system to move from a given status toward the defined objective" [4]. Interestingly, like much of the original investigations behind the ARPANET, the research behind linear and dynamic programming was conducted at the RAND Corporation and sponsored by the U.S. Department of Defense.

where D_j is the estimated distance (cost) to the ith node (network) and d_{ji} is the distance (cost) of the link from the jth node to the ith node. If there is no link between the ith and jth nodes, then the cost is set to infinity; this is also the case if a link goes down. The initial condition is

$$D_1 = 0$$

which merely indicates that the root node is at a distance of 0 from itself.

Dynamic programming rests on what is often called the *principle of optimality:* that any subpath of an optimal path is itself optimal. For example, in Figure 13.4 if the optimal path from intermediate system A to intermediate system E is A–B–G–E then the optimal path from B to E must be B–G–E. Otherwise, if another path, say, B–E, were better then the optimal path from A to E would be A–B–E. This is a commonsense observation that nonetheless has deep ramifications, one of which is that it allows us to "prune" the set of possible paths as we iterate the Bellman-Ford algorithm.

The Bellman-Ford algorithm begins by calculating the optimal path (schedule) to get to nodes (intermediate systems) that are one link away. This, of course, is trivial: Those nodes directly connected are one link away; those nodes that are not directly connected to the root (the local intermediate system) are initially unknown. The process next finds optimal paths (schedules) that consist of two edges (stages). If a two-stage path costs less than a one-stage path then the latter is eliminated from the list of optimal paths. This pruning is an application of the principle of optimality.

In a network with N links the process of estimating path length continues for up to a maximum of $N-1$ iterations (since no path in the network can have more than $N-1$ links in it). When it terminates, the Bellman-Ford algorithm yields the *shortest path spanning tree* for each node in the graph.

The Bellman-Ford algorithm that solves the Bellman equation is typically realized with a centralized implementation, which assumes a complete knowledge of the graph's topology. Obviously, with computer networks each node has ready access to its local topology information, but global topology must be explicitly reconstructed (as it is with Dijkstra's algorithm; see later discussion). With distance vector routing protocols, on the other hand, no explicit topology model or topology information is exchanged, merely routing tables among neighboring intermediate systems.[2] How can this work?

---

2. Intermediate systems (nodes) that are one link (edge) away from each other, that is, nodes that are directly connected, are called *neighbors.*

It turns out that there is a distributed, asynchronous form of the Bellman-Ford algorithm in which every intermediate system calculates its own shortest path spanning tree independently of each other, and which only relies on periodic routing table exchange. Under very limited assumptions (routing table exchange, aging out of old path length estimates, and nonnegative link costs) it can be shown that the distributed, asynchronous form of the Bellman-Ford algorithm will converge to the correct solution (see, for example, [5, 6]).

The performance of the Bellman-Ford algorithm is highly dependent on the topology of the network. Worst case, distance vector routing protocols must iterate $N-1$ times to discover all of the optimal paths in an N-node network, where each iteration entails $N-1$ comparisons for each of $N-1$ nodes. This means that the algorithm scales according to $O(N^3)$. However, it can be shown that the scaling is often better than this: If A is the total number of edges (links) in a network of N nodes (intermediate systems), then the Bellman-Ford algorithm scales according to $O(mA)$, where $m < N$ is the length of the longest shortest path in the network. Since

$$A \leq N(N-1) \text{ for directed graphs}$$

or

$$A \leq \frac{N(N-1)}{2} \text{ for undirected graphs}$$

it is clear that the less fully meshed the network, the lower the overhead of executing Bellman-Ford.

13.3.1.2 Concatenation Faults in Distance Vector Routing Protocols

Because distance vector routing protocols typically require that intermediate systems age out their routing tables and that they recalculate routes as updates come in from adjacent intermediate systems, inconsistencies such as routing loops generally do not last long. However, distance vector routing protocols can suffer from faults, with certain topologies causing particular difficulties. One of the most nettlesome distance vector faults is called "counting to infinity." Consider the simple network shown in Figure 13.5, where we have three

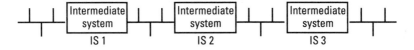

Figure 13.5 Example of "counting to infinity."

intermediate systems connected linearly (for example, nodes C, F, and I in Figure 13.4).

Assume for simplicity that the metric is simple hop count and that each intermediate system starts with accurate routing tables. If the link between IS 2 and IS 3 fails, then IS 2 will change its routing table to reflect the fact by setting the distance to IS 3 as infinity. However, IS 1 may still advertise that it can reach IS 3 in two hops (via IS 2, of course). IS 2, receiving this update, will adjust its routing table again, advertising that it can reach IS 3 in three hops via IS 2. As routing updates bounce between IS 1 and IS 2, the metric will increase each time until each exceeds the maximum allowed by the routing protocol, hence the term "counting to infinity."

Two techniques used to prevent such faults are called *split horizon* and *poisoned reverse*. With split horizon, an intermediate system will modify its routing update to another intermediate system by excluding any estimates on the cost to reach a destination if the second intermediate system is the next hop. Poisoned reverse likewise modifies the routing table but by advertising the destination as unreachable (i.e., setting its distance to infinity). A third approach to the problem is path holddown: If a link is estimated to be lost then the routing protocol will for a time "hold down" the link's state and ignore any routing updates that indicate the link is usable. After the path hold down interval expires, the routing protocol will accept new estimates of the link's condition.

13.3.1.3 The ARPANET Gateway to Gateway Protocol

As we indicated earlier, the original routing protocol employed in the ARPANET was a distance vector protocol called the Gateway to Gateway Protocol (GGP). Routing tables were exchanged among intermediate systems and the distributed, asynchronous form of the Bellman-Ford algorithm was run in each IMP to calculate the short paths to every other IMP.

The GGP did have one major departure from what we have considered until now, namely, the metric was estimated delay as measured by the queue length on outgoing links. From elementary queuing theory we know that queue length is determined by the traffic intensity, itself determined by the rates of arrival and service. It was found, however, that the inclusion of such traffic state information in the scheduling caused major difficulties with instabilities due to rapidly changing state information. The result was large oscillations. Although the successor (link state) routing protocol also used traffic state information, the new protocol reduced the "gain" on the feedback considerably to damp the oscillations (see later discussion). Because of this phenomenon, subsequent routing protocols have avoided using traffic information in their scheduling.

13.3.1.4 Routing Information Protocol

The most common distance vector routing protocol in use today is the Routing Information Protocol (RIP), which was developed by Xerox PARC as part of its Xerox Network Systems (XNS) architecture. A version of RIP modified to distribute IP routing tables was developed at the University of Illinois and included in BSD Unix in the early 1980s, propelling the protocol to widespread popularity and greatly stimulating the growth of the Internet. In fact, RIP was widely deployed long before it was formally documented with RFC-1058 in 1988. RIP-2, which was defined with RFC-1388 in 1993, added support for CIDR addressing and aggregation, security, and other improvements in efficiency.

Unlike the GGP, RIP takes no account of traffic in its path length estimation. Instead, RIP uses a simple hop count in its path selection. Each link is given a cost of 1, irrespective of its bandwidth. In addition, RIP limits the maximum path to 15 hops; 16 is the value used for unreachable (i.e., 16 is "infinity" to RIP implementations). This has proven to result in a constraint on the size of RIP networks, limiting their diameter to at most 15 networks; in fact, since the best route may involve more links than the "direct" path, the size limit may be considerably more binding.

RIP sends its routing tables encapsulated in UDP, using port 520. These tables are limited to holding the best route for each destination IP network; multiple routes are not allowed. Entries in RIP routing tables are time-stamped for aging. If the entries are not refreshed within 180 seconds then they will be removed; this removal will then trigger the intermediate system to send its own routing update. RIP routing updates are limited to 512 bytes, which given the format of the RIP messages means that an update can only include 20 entries. Larger tables require multiple update messages.

In addition to triggered updates, RIP requires periodic exchanges of routing tables. Every 30 seconds a RIP intermediate system broadcasts its routing table to its neighbors. In addition, RIP updates may be triggered, if there are any changes to the local state information (either a local interface comes up or goes down) or if a route in the routing table ages out or if a routing table is received from a neighboring RIP intermediate system that causes the local routing table to change one or more of its entries. Finally, RIP includes a message that can be sent by a newly started (or restarted) router to any neighbors requesting they send their routing tables immediately rather than waiting for either a triggered or a periodic update.

RIP relies on split horizons to eliminate counting to infinity. We should note that RIP's periodic updates have been found to be a source of instability in the Internet [7]. The problem is that RIP intermediate systems in a network have been found to self-synchronize to each other, resulting in all systems sending their periodic updates at approximately the same time every 30 seconds. This surge of traffic was found to adversely affect the performance of time-critical Internet applications such as voice and video. Various measures have been proposed for reducing or eliminating this behavior.

We should note that one of the most ambitious interior routing protocols, the Interior Gateway Routing Protocol (IGRP) developed by Cisco Systems, is a variant of RIP. IGRP augments the state information to include delay, bandwidth, reliability, and load for every IP network. Shortest paths can be selected according to a metric that can be tuned by a network administrator to give various weights to these factors. In practice, however, several of these parameters are defaulted to zero, simplifying the calculations and limiting the potential for traffic feedback instabilities such as were seen with the GGP. Finally, we should note that IGRP has been superseded by Enhanced IGRP (EIGRP), to provide CIDR support, among other modifications. EIGRP incorporates a new routing technology referred to as *diffusive routing* and a routing algorithm called the Distributed Update Algorithm (DUAL) [8, 9].

13.3.2 Link State Routing Protocols

Given that distance vector routing protocols scale in the worst case according to $O(N^3)$ it is not surprising that many researchers have sought a more efficient way to find the shortest paths for a graph and hence construct routing tables. As it happens, a more efficient algorithm was devised by Dijkstra in 1959 to find all the shortest paths originating at a given node; this algorithm scales according to $O(N^2)$ or better. Like the Bellman-Ford algorithm, Dijkstra's algorithm is properly considered an instance of dynamic programming since it is structured as a sequential (i.e., multistage) decision process.

However, there is a major restriction on use of the Dijkstra algorithm: The global topology must be known before the algorithm can be executed [10]. In contrast, the Bellman-Ford algorithm only requires local state information plus neighbors' periodic estimates of their shortest paths to other intermediate systems. For this reason, it is necessary to combine Dijkstra's algorithm with an explicit topology model of the network, reconstructed by intermediate systems exchanging their respective local state information describing their local topology. This is different than the state information exchanged in Bellman-Ford algorithms, which are estimates of the shortest paths (schedules) rather than explicit topology.

In terms of the MESA tasks, link state routing protocols execute bandwidth estimation (reconstruction) of the global topology of the network and workload scheduling to find the shortest paths to the other intermediate systems. We now look at these two parts.

13.3.2.1 Reconstructing Global Topology

The global topology model that every link state intermediate system maintains is variously known as the topology database or the link state database. Estimation of the global topology is done by each intermediate system sending out routing updates that consist of its local state information plus any estimates of the global topology that it has reconstructed. Like the topology models, these routing updates are known by various names. In OSPF they are called link state advertisements (LSAs), in OSI's IS-IS protocol they are called link state packets (LSPs), in APPN they are called topology database updates (TDUs), and so on. In most of these routing protocols, these routing updates consist of the entire routing table, so care must be taken not to send them too often or the management overhead will crowd out user traffic. In addition, to reduce the size of topology updates several of the most important of these routing protocols employ hierarchical decomposition within an autonomous system, dividing it into areas and limiting full topology exchange within these (see later section for a discussion of OSPF's area hierarchy).

The global topology models in a link state routing protocol can therefore be thought of as replicated information in a distributed database system; and a principal task of link state routing protocols is to synchronize the topology databases of the intermediate systems. Figure 13.6 shows the topology model maintained by one intermediate system, obviously reconstructed using local

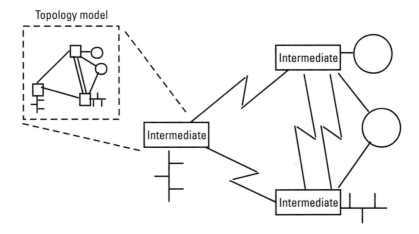

Figure 13.6 Topology model for link state routing.

topology state information from the other two intermediate systems. Assuming that the leftmost intermediate system has also sent its routing updates, these will likewise have identical models of the network; if not, then when it does send its routing update the remaining two intermediate systems will synchronize their models.

When an intermediate system first comes up, it can do two things with respect to exchanging topology information. First, it can send out a topology update that typically includes the intermediate system's identifier; the links to which it is attached and their respective state information (cost, operational status, bandwidth, and so on); and the network addresses of the links. Second, it may broadcast a request to other intermediate systems to send their respective topology updates immediately, rather than wait for periodic updates, so that it can synchronize its model as quickly as possible. And as a link state intermediate system receives topology updates, it adds the information to its topology model (database) and this, too, is sent out with future topology updates. To summarize, the bandwidth monitor (sensor and estimator) in each link state intermediate system executes three tasks:

1. Measuring local topology information (typically obtained from the intermediate system's configuration files listing the IP addresses in the networks to which it is attached);

2. Reconstructing the global topology from local measurements and the topology estimates received from other link state intermediate systems; and

3. Sending out the global topology estimates thus created to the other intermediate systems.

The dissemination of global topology estimates from the individual intermediate systems relies on flooding. A link state intermediate system that receives a routing update in one interface will modify its copy of the topology model if necessary and then flood the update out of its remaining interfaces to downstream intermediate systems. The importance of synchronizing the topology models as quickly as possible is that it minimizes the probability of routing faults due to inconsistent routing tables.

Much of the complexity in a link state routing protocol is to be found in the mechanisms implemented to ensure that this flooding occurs reliably while at the same time not slowing the process down too much. A principal concern, particularly as the network grows in size, is that old updates do not get confused with newer updates as they are flooded throughout a large network with a complicated topology. If this happens, it can be extremely difficult to straighten matters out. Indeed, when this occurred with the ARPANET's link state routing protocol the only remedy that worked was to power cycle all the IMPs in the network and thus clear out the buffers of all copies of the updates.

At least two provisions are commonly made to prevent this. First is the aging of routing updates, much like IP's TTL field: An old routing update will stop being forwarded when its age expires. Second is the use of a carefully managed sequence number space. This allows a link state intermediate system that receives a routing update to compare the sequence number the update carries with the sequence number of the last update that resulted in changes to the topology model. Both age and sequence number fields are carried in the routing update message headers.

13.3.2.2 Finding Shortest Paths (Optimal Routes)

The second part of a link state routing protocol, namely, finding the shortest paths (i.e., schedules) and, at a minimum, the next hops for forwarding end-to-end PDUs, is executed by the workload schedulers in the link state intermediate systems. We should note that though most link state routing protocols use the Dijkstra algorithm for calculating the shortest paths, Moy points out that other algorithms can be used as well [11]. However, just as with Dijkstra, these other algorithms require that the global topology be known or estimated beforehand. Their execution is predicated on the explicit topology models created and continually maintained (updated) by the bandwidth monitor. For this reason we concentrate here on Dijkstra's algorithm.

Dijkstra's algorithm begins its execution much like the Bellman-Ford algorithm: by finding the shortest paths to nodes that are one edge away. For simplicity let's assume that the local node (from which the shortest paths are to be found) is node 1. The initial conditions for the algorithm are that the distance of node 1 to itself is 0 and that the only node in set P is node 1. As it proceeds sweeping outward, every node is labeled with an estimate of the shortest path to it. The principle of optimality is used to prune paths just as it is with Bellman-Ford.

When the Dijkstra algorithm decides that the minimum distance estimate from node 1 to a given node has been found, the node is said to be permanently labeled, and added to a set P of permanently labeled nodes. For all nodes j that are not permanently labeled, the algorithm then finds the next closest node (call it node i); node i is then added to set P. The algorithm's iteration terminates when P contains all the nodes in the network.

If nodes remain that are not in P then the distance estimates are updates according to the formula

$$D_j = \min\left[D_j, D_i + d_{ij}\right]$$

where d_{ij} is the distance of node i to node j and $D_j = d_{ij}$. The algorithm then iterates until all the nodes have been permanently labeled and placed in P.

The number of minimizations is proportional to N, the number of nodes in the network; and there are $N - 1$ iterations. Hence the Dijkstra algorithm scales according to $O(N^2)$. Under certain assumptions concerning the topology of the network and the implementation of the data structures used to store the nodes that are in set P and those not yet in set P, the scaling of Dijkstra's algorithm can be limited to $O(n \log(n))$ [3, 11, 12].

13.3.2.3   Concatenation Faults in Link State Routing Protocols

Because they rely on explicit topology models rather than on estimates of shortest paths as relied on by distance vector routing protocols, link state routing protocols are more robust. For example, they are not vulnerable to such conditions as counting to infinity. However, inconsistencies in the topology models in the various intermediate systems can cause routing loops until they are repaired by the exchange of topology data. The robustness of a link state routing protocol is dependent on how well it maintains the topology models, including such measures as checksums to detect faults in the updates themselves.

13.3.2.4   Open Shortest Path First

We now examine the OSPF protocol, the most important link state routing protocol in use today. Although the first link state routing protocol was deployed in the ARPANET in 1976 to replace the GGP we discussed earlier, it was not until OSPF was developed in the late 1980s that a link state routing protocol could seriously challenge distance vector routing protocols such as IGRP and RIP. The protocol was first defined in RFC-1131.

As with many other important Internet protocols, there is a wealth of background material on OSPF, including its design objectives and what criteria determined the shape of the final protocol. The overwhelming need was for a new routing protocol that could scale with the exponential growth the Internet was already experiencing (nearly 10 years before the explosion of the World Wide Web). This meant, among other things, the following:

1. *Efficiency:* A more efficient interior routing protocol than RIP was sought because the overhead of RIP's routing table exchanges was growing too great, using too much bandwidth and displacing too much data (i.e., nonmanagement) traffic.

2. *Convergence:* The convergence problem for distance vector routing protocols was felt to be intractable without recourse to clumsy mechanisms such as poisoned reverse, split horizons, and so on.

3. *Metrics:* Cisco's IGRP had demonstrated the advantages of routing metrics that went beyond simple hop count. In addition, the limit of 15 hops to which RIP limited network reachability was considered unacceptable.

4. *Modularization:* Even within an AS it was felt there had to be some protocol support for hierarchical routing and route aggregation.

In addition to these objectives, the new protocol was to implement some form of security to prevent unauthorized routers from introducing spurious routing or topology information. Support for type of service (TOS) routing was also a goal, although as we discussed in Chapter 10 this feature has never been widely deployed and in fact was dropped from IPv6.

The efficiency and convergence issues led to the decision to employ link state routing in the new protocol, which was dubbed the Open Shortest Path First because it was nonproprietary (i.e., open) and because it would employ Dijkstra's shortest path first algorithm. After 5 years of design and development, OSPF was selected in 1992 as the IETF's recommended interior routing protocol.

Unlike RIP, which used UDP encapsulation of its routing traffic, the OSPF architects opted to encapsulate directly in IP, using IP protocol number 89. The other options, namely, direct data link encapsulation or transport layer encapsulation in UDP or TCP, were rejected for being too lean and too rich, respectively. Direct data link encapsulation would have entailed the routing protocol handling its own fragmentation of routing updates that were larger than the maximum L2 PDU size for a given data link protocol. In addition, the plethora of data link protocols would have made the task of interfacing to each too onerous. Transport layer encapsulation, on the other hand, was rejected as coming with too much overhead and offering services, such as reliable transport, that the routing protocol was going to implement on its own (see discussion of reliable flooding later). Thus the decision to go with IP encapsulation.

With respect to link costs and metrics, OSPF chose to allow network administrators maximum flexibility. Every link can have a cost from 1 to 65,535, with lower cost links being preferable to higher cost ones. If all links are given the same cost then OSPF will minimize hop count in its routing calculations. One formula for calculating link cost is

$$\frac{10^x}{\text{link bandwidth}}$$

where x is chosen relative to the highest speed link in the network. For example, if 100 Mbps is the fastest link, then setting x equal to 8 results in a range of link costs from 1 (for a 100-Mbps link like FDDI) to 1785 (for a 56-Kbps WAN link). We should note that OSPF does not take into account traffic. Although some texts claim OSPF's metric can be set to minimize delay, this is questionable since delay is jointly determined by the bandwidth and traffic; and without monitoring traffic (workload) OSPF cannot minimize delay.

OSPF continues the modularization begun earlier with the division of the Internet into autonomous systems and the use of exterior routing protocols. OSPF divides an AS into a two-tiered hierarchy consisting of *areas*, with area 0 functioning as the backbone (or transit area) connecting the remaining areas in a hub-and-spoke topology. In some cases the connectivity to the backbone may be through what OSPF calls *virtual links*, which allow areas not contiguous to the backbone area to nonetheless be part of an OSPF network.

This modularization results in several classes of OSPF routers: internal routers, which are entirely within a given area; and area border routers, which are part of two areas (one of them area 0). A great deal of effort went into defining the inter-router coordination necessary to maintain synchronized topology models (link state databases). OSPF limits full topology exchange to routers in the same area. Routing information between areas is summarized to simple reachability information, which effectively is a distance vector protocol. This provides additional reduction in the routing overhead, and since OSPF arranges its areas to hub-and-spoke topologies, the issue of loop prevention does not arise [3].

The three different parts of the OSPF protocol are (1) the Hello protocol, (2) the reliable flooding protocol, and (3) the exchange protocol. The OSPF Hello protocol is used by routers to discover other routers in the same area. Every 10 seconds a router sends Hello packets out on all of its interfaces, which also function as keep-alives. When an OSPF router receives a Hello packet, it sends a Hello response and the routers proceed to use the exchange protocol to synchronize their link state databases. Routers that have synchronized their respective link state databases are said to be fully adjacent. The reliable flooding protocol is used among adjacent routers and is responsible for forwarding link state advertisements; it includes such mechanisms as aging as well as acknowledgments to ensure receipt of synchronization messages.

The contents of these messages are link state advertisements (LSAs). OSPF defines five different types of LSAs: router, network, summary for IP network, summary for border router, and external. For example, interarea exchanges use the summary for border router LSA to summarize routing information.

13.4   Exterior Routing Protocols

In our earlier discussion of exterior routing protocols, it was pointed out that the first of these, the Exterior Gateway Protocol (EGP) had limitations that constrained the Internet's AS hierarchy to a tree topology. Because of these limitations, at the same time as work began on OSPF to succeed RIP as the Internet's interior routing protocol research was also begun on a new exterior routing protocol called the Border Gateway Protocol (BGP). BGP was first defined in RFC-1105 (1989) while the current version, BGP-4, was defined in RFC-1771 (1995). As we will see in this section, BGP's greater sophistication removed many EGP-imposed limits on the Internet and the structure of the hierarchy of its autonomous systems.

13.4.1   Distance Vector Versus Link State Exterior Routing Protocols

It may seem surprising after our discussion of the scaling advantages of link state routing protocols, but the architects of BGP opted to use distance vector routing. However, certain characteristics of an exterior routing protocol tend to favor distance vector routing. For example, as we just remarked concerning OSPF and the exchange of interarea routing information, summarizing routing information naturally leads to a distance vector protocol. In fact, the increased granularity of state information that link state routing protocols exchange is a double-edged sword: while it makes it more suited for use as an interior routing protocol, it makes it less suited for use as an exterior routing protocol.

13.4.2   Border Gateway Protocol

The development of BGP was prompted by the limitations of EGP, most notably the limitation EGP placed on the Internet's hierarchy of autonomous systems into a strict hub-and-spoke topology. EGP also lacked security features to prevent routers from injecting spurious routing information into the Internet, and had no support for policy routing.

Once the decision was made that BGP would use distance vector routing, the principal issue confronting BGP's developers was loop prevention. Rather than incorporating the work-arounds such as split horizon and poisoned reverse, BGP chose a very different approach. BGP routers (called *peers*) exchange complete path information rather than simple next-hop routing tables. These *path vectors* list the sequence (schedule) of ASs that provide transit services in forwarding IP PDUs to a given destination network (in CIDR, prefix). What this means is that BGP is not a next-hop routing protocol. The path vector is, in effect, a form of source routing. As we will see in a moment, loops are prevented by BGP peers examining the path vector advertising an AS path to see if their own AS is listed.

As with other distance vector routing protocols, the actual creation of path vectors is done by executing the Bellman-Ford algorithm. Recall that although distance vector routing does not have access to a global topology model, the Bellman-Ford algorithm nonetheless creates a shortest path spanning tree that gives the entire end-to-end path (schedule) necessary to reach any node in the graph, where here the nodes are autonomous systems. However, because BGP does not list just basic network layer reachability information (prefix and distance), calculation of the shortest paths is somewhat more elaborate.

Several important aspects of BGP's design were driven by the fact that exterior routing tables, particularly as one got close to the core of the Internet, were rapidly growing to enormous sizes, today often listing 40,000 or more prefixes. This meant that the periodic exchange of routing tables as done in traditional distance vector routing protocols was to be avoided. Obviously, when two BGP routers first established contact (see later discussion) a full exchange was required, but afterwards it was decided that BGP would only exchange incremental updates.

The absence of the periodic full exchanges meant that BGP routers would not have any fallback to automatically correct for lost or corrupted data in their respective routing tables. For this reason, reliable transport was felt to be mandatory and BGP's designers therefore chose to use TCP encapsulation. TCP would handle any retransmissions of BGP routing updates and ensure that BGP routers received these uncorrupted.

A second consequence of the decision not to periodically broadcast full routing tables in BGP was the necessity of storing backup or alternatives routes (path vectors). Therefore, each BGP router maintains not just a routing table listing network prefixes (and path vectors to these) but also a complete list of alternative routes in what is called its *Routing Information Base–Inbound* (RIB-In). When a BGP router receives routing tables and/or updates from its BGP peers, all the routes (path vectors) for a given prefix are entered into the RIB-In unless a loop is detected—that is, a path vector includes the router's own AS number.

Obviously, the RIB-In is a superset of the actual BGP routing table, and in fact the latter is generated from the former. When a route in a BGP router's routing table becomes inoperative due to some link or router failure, the

RIB-In is consulted to find the next best route. The mechanism in a BGP router that selects routes (path vectors) from the RIB-In for inclusion in the routing table is called the *Decision Process*. The BGP Decision Process executes another level of workload scheduling, above and beyond that executed by the Bellman-Ford algorithm used to generate the AS path vectors. In addition, the Decision Process plays a crucial role in implementing routing policies with BGP.

13.4.3 Policy Routing in BGP

The demise of the monolithic Internet and its division into a hierarchy of autonomous systems, many of them Internet service providers offering transit services, created a whole new set of complications. Network administrators were confronted by restrictions on usage that were often not amenable to the simple graph-theoretic mechanisms used in standard routing protocols. For example, the early Internet was restricted from carrying commercial traffic, which entailed special routing considerations separate from noncommercial traffic.

Policy routing in the Internet refers to the inclusion of criteria other than topology or traffic in the route selection process. This can include various administrative preferences or prohibitions. BGP supports policy routing by allowing network administrators great flexibility in specifying the factors the decision process uses to generate the BGP routing table. A number of RFCs have been written exploring the implementation of routing policies in the Internet.

13.5 Summary

We saw in this chapter that any discussion of closed-loop concatenation and routing protocols is dominated by two main themes: models and modularization. To cope with the explosive growth of the Internet, we saw how techniques such as hierarchical routing and route aggregation have been employed to limit the size of routing tables, culminating in the development of Classless Interdomain Routing.

Following this we discussed the distinction IP makes between what are called interior routing protocols, which provide great detail on topology and connectedness, and exterior routing protocols, which limit their information to reachability, that is, what IP networks can be reached in that autonomous system without specifying any aspect of how to get there. Such details about *how* are left to the interior routing protocols.

We then moved on to examine how distance vector and link state routing protocols operate, along with their respective strengths and weaknesses. We discussed the Bellman equation from dynamic programming and the central role it plays in the Bellman-Ford algorithm used in distance vector routing. To illustrate distance vector routing, we discussed the Routing Information Protocol as well as the Border Gateway Protocol.

Link state routing, on the other hand, though likewise grounded in dynamic programming and sequential decision processes, is associated with the Dijkstra equation for discovering shortest paths. The Dijkstra algorithm generally scales better but is predicated on knowing the global topology of the network, whereas the Bellman-Ford algorithm merely requires exchanging local topology information along distance estimates. This is why the two are often referred to as the distributed computation and the distributed database approaches, respectively. To illustrate link state routing, we examined the Open Shortest Path First routing protocol.

References

[1] Rosen, E. C., *Exterior Gateway Protocol,* RFC-827, Oct. 1982, p. 2.

[2] Rekhter, Y., "Inter-Domain Routing: EGP, BGP, and IDRP," in *Routing in Communications Networks,* M. Steenstrup, Ed., Upper Saddle River, NJ: Prentice Hall, 1995, p. 106.

[3] Moy, J. T., *OSPF: Anatomy of an Internet Routing Protocol,* Reading, MA: Addison Wesley, 1998, p. 276.

[4] Dantzig, G., *Linear Programming and Extensions,* Princeton, NJ: Princeton University Press, 1974.

[5] Bertsekas, D., and J. Tsitsiklis, *Parallel and Distributed Computation,* Upper Saddle River, NJ: Prentice Hall, 1989, pp. 293–302.

[6] Bertsekas, D., and R. Gallager, *Data Communications Networks,* 2nd ed., Upper Saddle River, NJ: Prentice-Hall, 1987, pp. 325–333.

[7] Floyd, S., and V. Jacobsen, "The Synchronization of Periodic Routing Mechanisms," *Proc. ACM Sigcomm '93,* San Francisco, September 1993.

[8] Garcia-Luna-Aceves, J. J., "A Minimum-Hop Routing Algorithm Based on Distributed Information," *Computer Networks ISDN Syst.,* Vol. 16, No. 5, May 1989.

[9] Cheng, C., et al., "A Loop-Free Extended Bellman-Ford Routing Protocol without Bouncing Effect," *ACM Computer Commun. Rev.,* Vol. 19, No. 4, Sep. 1989.

[10] Bertsekas and Tsitsiklis, *Parallel and Distributed Computation,* p. 303.

[11] Moy, J. T., "Link-State Routing," in *Routing in Communications Networks,* M. Steenstrup, Ed., Upper Saddle River, NJ: Prentice Hall, 1995.

[12] Huitema, C., *Routing in the Internet,* Upper Saddle River, NJ: Prentice Hall, 1995, p. 106.

14

Layer 2 Concatenation: Local and Remote Bridging and L2 Switching

14.1 Introduction

In this chapter we look at L2 concatenation, specifically local and remote bridging of 802 LANs. The chapter begins by first looking at the 802 MAC addresses, and the consequences that flow from the fact that these constitute a flat address space. We then consider bridges and switches, bridges and routers, and the question of bridge complexity in terms of processing and storage requirements. As we will see, there are two very different approaches to concatenation used at layer 2: transparent bridging, which relies on intelligence in the intermediate systems (bridges) to hide the details of concatenation from end systems, and source route bridging, which uses simpler bridges but at the price of placing more of the concatenation burden on end systems.

From there we move on to discuss remote bridging, which is an L2 concatenation technique used to concatenate two or more LANs by tunneling their traffic across another data link, typically a WAN link. We will focus in particular on PPP's support for remote bridging and will discuss the Bridging Control Protocol (BCP). This affords us the opportunity to briefly discuss tunneling and its management implications such as visibility of the tunneling network to the client network(s) and their traffic that it tunnels.

14.2   Bridging Foundations

14.2.1 IEEE 802 MAC Address Space

Recall from Chapter 5 that with the first data link protocols the addresses only had local significance, but that this changed when Xerox decided that the higher bandwidth of Ethernet allowed the luxury of 48-bit addressing, an address space sufficiently large that it allowed globally significant addresses. This decision, later embraced by the IEEE 802 committee when it defined MAC layer addressing, was to have a deep and lasting impact on the nature of networking and connectivity. Not least, it enabled layer 2 concatenation: Without globally significant addressing, there was no way to determine if an L2 PDU (frame) was destined for an end system on its local LAN or if the target was an end system on a distant LAN. In short, bridging and its related technologies like L2 (frame) switching were made possible by the global significance of 48-bit addressing.

A complication, however, arose almost immediately. Because LANs were intended to be "plug and play," it was deemed imperative to keep configuration requirements to a minimum or none at all. This meant that each LAN adapter card had to be assigned a *universal* (i.e., globally unique) 48-bit address at the factory; and although the standard does allow such universal MAC addresses to be overridden by locally assigned addressing (LAA), outside of Token-Ring LANs used in SNA networks this is seldom if ever employed. An immediate consequence of this design decision is that the 802 address space defined was flat. Two end systems on the same LAN will have MAC addresses that, unless overridden by assigning locally defined addresses, are completely unrelated to each other.

Why does this have such ramifications? Because a flat address space limits the modularization that, as we saw in the last chapter, has been integral to the scalability of IP routing protocols. In a hierarchical addressing schema, the network (or, in SNA, subarea) part of the address can be thought of as aggregating or aliasing the addresses of the end systems that are on that network. For example, a set of one or more IP host addresses is effectively aliased by a network address, which is all that need be stored in an IP routing table. And, of course, there is a corresponding reduction in the granularity of information needed to describe the topology of the network.

This flat address space is responsible for one challenge to concatenation at layer 2, namely, the difficulty in constructing a global topology model such as we saw in link state routing or even an implicit model such as we saw with distance vector routing. This presents a number of challenges including the suboptimal utilization of transporter bandwidth and, most importantly, the looping of frames as one bridge/switch forwards it to another. This brings us to the second major challenge to layer 2 concatenation: Because no major data link (L2) protocol includes any header field for recording time-to-live or hop-count information, the danger of loops, black holes, and other concatenation faults looms large.

14.2.2 Bridges: Simple Versus Complex

As experience was gained with primitive Ethernet bridges in the late 1970s and early 1980s, the networking community became entangled in yet another inter-necine conflict, this one over the amount of "intelligence," that is, processing power and storage, that should be required of bridges. On one side were the advocates of the learning bridges then being deployed in Ethernet networks. Such bridging, which was called *transparent* because the L2 concatenation exe-cuted was invisible to end systems, was enhanced with the introduction of the spanning tree protocol developed by Digital Equipment Corporation to allow complex topologies with redundant paths.

On the other side were advocates of simpler bridges requiring less com-putational power, who pushed for the approach called source route bridging. Source route bridging requires that end systems participate in the topology discovery process by sending out *explorer* frames to find the end-to-end route to the destination MAC address. These explorer frames were forwarded by the bridges using flooding constrained by a path vector known as the routing infor-mation field (RIF).

The different levels of sophistication translated into very different costs for the two types of bridges. Transparent bridges have to have more processing power to execute the learning and spanning tree algorithms than source route bridges, which simply need to forward frames according to their RIF fields.

14.2.3 Bridges Versus Switches

An L2 switch is a bridge that has had many of its tasks implemented in silicon, with substantial improvement in bandwidth but otherwise no real functional enhancements. Both bridges and an L2 switch contain managers executing workload actuation of kind, transforming the requested transport task into two or more intermediate transport tasks. A switch is basically a faster bridge.

14.2.4 Bridges Versus Routers

We should say a word about the issue of concatenation at layer 2 versus con-catenation at layer 3. This is a complicated trade-off that has changed over time as ASICs and other hardware assists have been incorporated. The traditional advantage of bridges has been their relative simplicity and lower cost per port. L2 concatenation is also protocol independent: Because no L3 or higher layer addressing is referenced in the bridging scheduling, the concatenation is completely independent of upper layer protocol addressing. With the development of high-speed L2 switches, performance has also been maximized, frequently hitting "wire speed," that is, the maximum forwarding rate a given data link will allow.

On the other hand, the greater management sophistication that comes with L3 concatenation offers its own benefits in terms of optimally utilizing network resources. The greater topological sophistication of L3 concatenation led in the early 1990s to the maxim "route where you can, bridge where you must." However, when high-performance L2 switches were developed this was changed to "switch where you can, route where you must." Now, with L3 switches routing IP at wire speed the comparison with dumb (i.e., L2) switches is again one of economics and price per port.

14.3 Transparent Bridging: From Promiscuous to Learning Bridges

Before there were transparent bridges or even learning bridges, there were bridges that promiscuously forwarded frames. Recall from Chapter 12 that our concatenation taxonomy started by asking whether the workload scheduling that realized the end-to-end transport tasks relied on a model of the global discrete event system and if so whether feedback was employed to update the model's state information. Promiscuous bridging is an example of model-free or stateless concatenation: Although a promiscuous bridge may monitor the state of the data links to which it is attached, it is stateless with respect to the global topology and instead relies on flooding to ensure that L2 PDUs are forwarded to their respective destinations.

Such early bridges simply forwarded frames between its interfaces without any discrimination; for example, if a frame arrived at a bridge that had four interfaces then three copies of a frame would be sent out to the interfaces it had not arrived at. (This bears some similarity to split horizons in routing protocols.) From promiscuous bridges it was a short step to bridges that could "learn" the topology of the LANs at least to the extent of recording the source addresses of the frames that appear on their interfaces. For example, Figure 14.1 shows a four-port bridge along with the address caches containing source addresses from "learned" from frames seen on the respective Ethernet interfaces.

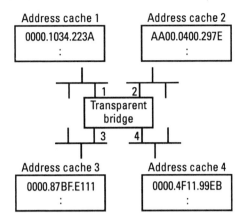

Figure 14.1 Four-port bridge with address caches.

With this information in their caches, learning bridges could forward traffic more efficiently than their promiscuous cousins. For example, if a frame arrives from Ethernet 1 with the destination address AA00.0400.297E, then the transparent bridge (or switch) would forward it directly to Ethernet 2, since that address had already been learned and was in Address Cache 2. Of course, if a frame arrives with a destination MAC address that is not in any address cache (either because it has never been learned or because it was aged out) then the bridge/switch will revert to promiscuous concatenation, flooding the frame out its other interfaces where it will presumably eventually find its destination.

We should note that an area where difference can arise between bridges and L2 switches is in the size of the address caches: So-called "single-station" switches allow only one end system per switch port, whereas no bridge ever built limited LANs to a single station. However, single-station switches actually map quite well onto the topology of 10BaseT hubs, which are already limited by the physical layer protocol to a single station per port. In fact, vendors exploiting this symmetry helped catapult Ethernet and Ethernet switching to an insurmountable lead over Token-Ring LANs in the mid-1990s.

What is going on with learning bridges and switches (L2 intermediate systems)? A nonpromiscuous bridge (or switch) uses information accumulated in the course of execution to reconstruct topology information. If the bridge is isolated (that is, there are no other bridges to which it is connected), then the topology information reconstructed will be strictly local. However, if two or more bridges are operating in tandem to concatenate multiple LANs, then the picture is more complicated. A MAC frame appearing at an L2 intermediate system's interface may have originated with an end system on that LAN *or* it may have been forwarded to that LAN by an upstream L2 intermediate system. To the downstream intermediate system, however, there is no way to distinguish one type of MAC frame from another, and the source MAC address simply gets entered in the corresponding address cache.

Thus, the topology model maintained by the estimators within the L2 intermediate system is actually a global model. Whereas L2 concatenation at its simplest (no learning) is a way of dynamically creating composite transporters without any knowledge of global topology existing anywhere in the network, transparent bridges do estimate global topology in a very primitive, solipsistic way: Stations on the network are aggregated into regions corresponding to the bridge's interfaces. In the preceding example, the transparent bridge with four interfaces, for example, has constructed a global topology model which has the network broken down into four "regions." The transparent bridge, as it monitors the MAC source addresses of frames arriving at its interfaces, records/learns the interface on which these stations are to be found; in this way, a topology map, albeit a crude one, is constructed.

What L2 intermediate systems do not do is exchange their topology models with each other. As we said earlier, the difficulty of even attempting to do this is that, because of the flat nature of the 802 address space, there is no way to summarize or aggregate the contents of the address caches. Any topology exchange would require exchanging every MAC address in each cache, something clearly not feasible in a substantial network with a large number of end systems.

If network designers had been content with bridged LAN networks that were limited to simple tree topologies, then the development of transparent bridging could have stopped with these basic learning mechanisms. However, such network designs are vulnerable to faults since there are by definition no redundant paths—that is the whole point of the topology restriction, after all. But even our admittedly brief exploration of network design at the end of Chapter 12 should have been adequate to illustrate the importance of redundancy in the design of network topologies. Without redundant paths, the failure of any single data link or intermediate system will result in a disconnected network.

The problem that arises when L2 concatenation is being used to forward traffic is that redundant paths can introduce loops, explosive replication of PDUs due to flooding, and other anomalies. If, for example, in Figure 14.1 there was a downstream bridge/switch attached to Ethernet 4 that was also connected to Ethernet 2 then a MAC frame could circulate on the loop defined by the two Ethernets and the two intermediate systems. Also, because there is no time-to-live or hop-count field in an Ethernet header (or any MAC header), nothing would cause the looping to ever stop. When this is combined with PDU replication from flooding, the potential for network meltdown is clear.

Any solution to this requires some form of global synchronization of the intermediate systems, specifically of their respective models/estimates of the global topology. Because we just said this cannot involve their respective models of the end systems (address caches), we appear to have a contradiction. However, transparent bridges can exchange explicit global topology information about the network, if the location of bridges relative to each other is limited. And that is what transparent bridges do: Using a special type of routing protocol, each L2 intermediate system contains an estimator that reconstructs a global topology of where all the bridges are in the network; and, using a shortest path first algorithm to superimpose a spanning tree on these, they cooperate to eliminate forwarding loops.

Let's illustrate this using the complex mesh of Ethernet LANs and transparent bridges shown in Figure 14.2. Without any mechanism to prevent loops, frames would bounce around endlessly, multiply, and eventually cause a meltdown. Instead, the bridges have discovered each other via this routing protocol and have constructed a spanning tree on top of the network. This begins with one bridge being "elected" to be the root of the spanning tree, after which the remaining bridges are arranged on the tree. (We should stress that this is only one possible spanning tree—other trees will result from different roots being elected, different bandwidth LANs, and so on.)

Based on the spanning tree thus determined, a transparent bridge will then determine out of which interfaces or ports it is allowed to forward frames. To prevent loops, the algorithm selectively disables or *blocks* interfaces. For example, in Figure 14.2 the blocking of just two ports is sufficient to prevent any loops in the network.

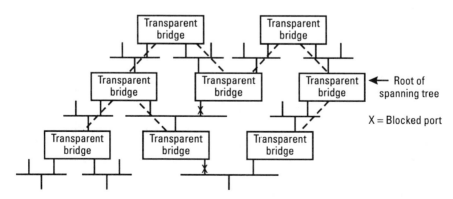

Figure 14.2 Transparent bridges and spanning tree.

As with the routing protocols we discussed in Chapter 13, the algorithm used in transparent bridging's spanning tree is completely distributed. The bridges exchange their topology information in bridge protocol data units, which are encapsulated in using the SNAP header 01000010 (= 0'x'82). Finally, note that there are two spanning tree protocols, the original one defined by DEC and the one defined by IEEE 802.1d bridging standard.

14.4   Source Route Bridging

Recall in our concatenation taxonomy that we said concatenation mechanisms could be divided into those that resulted in stage-by-stage (next-hop) forwarding and those that created the entire schedule of transport tasks sufficient to realize a given end-to-end transport task. The generic term for the latter is source routing; we discussed an example of source routing in Chapter 11, IBM's APPN protocol. The APPN protocol uses the Dijkstra shortest path first algorithm and a link state routing protocol to create source routed virtual circuits. The same techniques can be employed at layer 2 with suitable modifications, and indeed are at the heart of the source route bridging protocol used to concatenate Token-Ring LANs.

The Token-Ring community rejected transparent bridging for many reasons. Some of these were technical but the political factors cannot be discounted. Just as we saw in Chapter 8 that IBM would not adopt Ethernet and instead chose Token Ring, so too it would not accede to transparent bridging, which like Ethernet was widely identified in the computer industry with IBM's rivals, most notably DEC. And so, after source route bridging was rejected by the IEEE 802.1 committee as the official 802 MAC bridge standard, IBM took it to the 802.5 Token Ring committee, which ratified its use with Token-Ring LANs. This is why one *never* encounters source route bridging outside Token-Ring LANs.

The difference between source route concatenation at layer 2 and source route concatenation at layer 3 is directly attributable to the limitations of the 802 addressing schema. Just as we saw that the "flatness" of the 802 address space precluded any effective exchange of address caches among transparent bridges to construct a global topology model specifying end system locations, so too it prevents source route bridges and/or switches from reconstructing any global topology model to construct the end-to-end route. Note that with source route bridging the schedule consists of L2 intermediate systems (source route bridges or switches) and Token-Ring LANs, whereas with L3 source routing the route (schedule) created consists of L3 intermediate systems and networks.

Even if the address space were not an impediment, using any sort of routing protocol in source route bridging would have been difficult because of its central design predicate, namely, simple bridges without substantial processing resources. Otherwise the argument would have gone to transparent bridging. The whole point of source route bridging is to keep bridges simple and inexpensive. This is likewise reflected in the fact that, whereas transparent bridges connect an arbitrary number of Ethernet LANs, the canonical token-ring bridge is strictly a two-port device connecting two and only two Token-Ring LANs. Of course, multiport Token-Ring bridges and switches are common today but these rely on the fiction of creating a virtual ring within the intermediate system to which the *k* Token-Ring LANs are bridged by *k* virtual two-port source route bridges. This is illustrated in Figure 14.3 where the four-port source route bridge of the left is logically composed of four two-port source route bridges connecting the actual rings 1, 2, 3, and 4 to a virtual ring 5.

So what did the architects of source route bridging devise to support arbitrary (i.e., meshed) topologies and allow end-to-end schedules to be constructed? Rather than incorporate bandwidth (topology) estimators and workload schedulers within the L2 intermediate system a la transparent bridging, it was proposed to let the network "discover" itself with the source route bridges forwarding flooding special discovery PDUs that recorded their respective paths as they made their way to the destination end system. As these PDUs, known as explorer frames, fan out throughout the network, the result is an exhaustive breadth-first search of the composite network that discovers all the concatenation schedules (i.e., the end-to-end path made up of Token-Ring LANs and source route bridges) that will reach the destination end system.

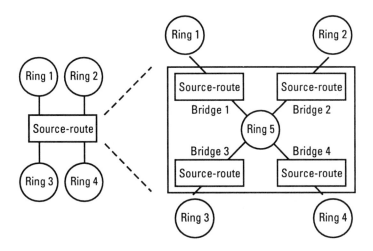

Figure 14.3 Multiport source route bridge as *k* virtual source route bridges.

For source route bridging to work correctly, a certain amount of manual configuration is necessary. After all, for explorer frames to record the paths they take there must be some way of differentiating one Token-Ring LAN from another and, because multiple bridges may connect a pair of rings, of identifying which source route bridge was transited. Specifically, the Token-Ring LANs must be assigned 12-bit-long ring numbers and the source route bridges assigned 4-bit-long bridge numbers (see Figure 14.3 for an example of this). The standard requires that ring numbers be unique but bridge numbers are unique only within the context defined by the two rings they connect. A little arithmetic indicates that a bridged Token-Ring LAN network can have up to 4096 rings, and that between any pair of rings there may be up to 16 source route bridges.

When an explorer frame is issued by a source route bridging end system, it contains a special header, inserted after the MAC header, that is called the routing information field (RIF). A RIF records the sequence of ring and bridge numbers (Figure 14.4). The task of source route bridges in this discovery is minimal, consisting mainly of replicating explorer frames and, before forwarding the explorer frames downstream, inserting their respective bridge and ring numbers (the globally unique identifiers for these transporters) with which they have been (manually) configured earlier.

Once these explorer frames reach the destination end system, a special *direction* bit is set in the RIF header by the destination end system indicating that the destination has been attained; and then the destination end system sends these explorer frames back to the source end system, where the RIF specifies the *exact* route the returning explorers follows. In all this, the source route L2 intermediate systems do not have to make any scheduling decision on the creation of end-to-end concatenations. These intermediate systems merely read the RIF in each source routed frame and forward the frame on to the next ring (data link) indicated. Another function of the RIF's contents is to prevent loops, much like the BGP path vectors we discussed in Chapter 13.

Back at the source end system, there are several choices as it receives these returning explorer frames carrying the routes they have discovered in their respective RIFs. An obvious choice is simply to take the first RIF that is returned, on the assumption that its explorer frame found the quickest path in

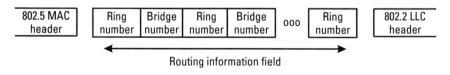

Figure 14.4 Routing information field.

the network. A second choice is to store two or more RIFs and either keep the excess as backups in case connectivity to the destination end system is lost or to divide traffic between the multiple routes, effectively exploiting the concurrency potential offered by redundant topology of the Token-Ring network. However, the latter strategy can backfire if used naively. For example, if LLC2 encapsulation is employed, this protocol does not tolerate well the out-of-order arrival of frames. In any case, it is important that an end system age its cache of RIFs since this will "turn over" any bad paths that may have been discovered, for example, when the network was particularly heavily loaded.

Finally we should note that an immediate consequence of this distributed, "breadth-first" topology discovery is that the overhead on the network can reach major proportions, particularly if a large number of source route bridging end systems seek to discover routes to their respective destinations simultaneously. In fact, this phenomenon was noticed early on with NetBIOS (which must be bridged and is generally run over Token-Ring LANs) clients of Microsoft's LAN Manager server application. Several times a day (9:00 A.M., after lunch, and so on) severe broadcast storms would overwhelm networks.

To reduce the overhead associated with the route discovery, the 802.5 standard defined two different types of explorer frames. These are known as all route explorers (AREs) and specifically routed explorers (SREs). What we have described up to now is how AREs function. SREs, on the other hand, represent a hybrid between source route bridging and transparent bridging in which the source route bridges run a spanning tree algorithm. When an SRE is issued by a source route bridging end system, it will be forwarded among the source route bridges along the spanning tree; but, once it reaches the destination end system, the SRE will be returned by the same flooding concatenation mechanisms used with AREs on the way to the destination end system. Thus, an SRE fans out as it returns to discover all the possible routes and collects the RIFs exactly as an ARE does. Most Token-Ring drivers in end systems use only AREs in part because many source route bridges, particularly older boxes, do not run the spanning tree algorithm.

Another feature designed to reduce the overhead of route discovery in source route bridging is the incorporation in more sophisticated source route bridging implementations of a cache of MAC addresses that have previously been the subject of explorer discovery, along with the RIFs returned. This hybrid "learning" source route bridging has the advantage of significantly reducing the management overhead due to the proliferation of explorer traffic. As with caches in transparent bridges, it is important to age the cache entries so that end systems are not inadvertently given routes that are no longer correct or, as we just mentioned, were discovered at a particularly inopportune moment in terms of network performance.

14.5   Remote Bridging

The last type of L2 concatenation we want to look at is remote bridging. We start by noting that "bridging" in the context of serial protocols is different than "bridging" used in the context of LAN protocols, where it refers to one of several techniques (transparent, source route, and so on) for concatenating data links at layer 2. Used in the context of serial protocols, however, the term refers to a form of layer 2 tunneling, in which a LAN frame (L2 PDU) is encapsulated in a serial frame (L2 PDU) for transport across a data link.

Recall that we briefly considered tunneling in Chapter 12. As we saw there, a very wide variety of mechanisms have been devised that can be broadly characterized as tunneling, by which we mean encapsulation of a layer N PDU in a PDU that is *not* a layer $N - 1$ PDU. The former we referred to as the payload protocol, whereas the latter is the carrier protocol. Although there are numerous varieties of L2 tunneling and remote bridging, many of them proprietary implementations that were developed prior to PPP and which used HDLC or even SDLC, we confine ourselves to remote bridging using PPP as the carrier protocol.

Obviously, using PPP as the tunneling protocol for remote bridges would encourage standardization and interoperability of internetworking devices, a major PPP goal. Toward this end, the IETF published the Bridging Control Protocol for negotiating various encapsulation and bridging parameters. Note that several L2 tunneling protocols have been defined that use PPP as the carrier protocol, including Cisco's Layer 2 Forwarding (L2F) Protocol, Microsoft's Point-to-Point Tunneling Protocol (PPTP), and the Layer 2 Tunneling Protocol (L2TP).

Figure 14.5 illustrates the difference between local and remote bridges connecting two 802.3 Ethernet LANs. (The choice is arbitrary—other LANs such as Token Ring and FDDI can also be remotely bridged.)

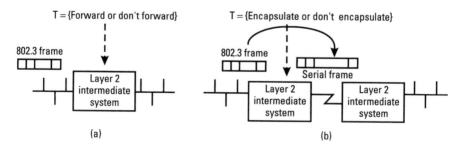

Figure 14.5   (a) Local versus (b) remote bridges.

Unlike the simple L2 intermediate system that is executing local bridging, remote bridging entails more than a question of forwarding: The decision is whether to encapsulate for tunneling across the WAN link, rather than merely just taking the frame to be bridged and reissuing it on the downstream LAN.

Beyond this, there are two types of remote bridges, depending on the visibility of the serial link to the bridging protocol. If the serial link is invisible (pure tunneling) then the two PPP link stations are called half-bridges, in that the two together combine to act as a single local bridge insofar as the bridging protocol is concerned. On the other hand, if the serial link is visible to the bridging protocol (which considers it to be merely another LAN data link) the two PPP link stations are called full bridges. Remote bridging in either case involves creating a virtual topology: With full bridges it logically appears to involve two bridges and three (LAN) data links, whereas with half-bridges it appears to involve one bridge and two (LAN) data links.

Remote bridging with PPP is managed with a special network control protocol called the Bridging Control Protocol (BCP). While it may seem unusual to call bridging a network protocol, it does indeed use the PPP link in the same way any true network protocol such as IP or IPX does. The protocol type code for bridging is 0x0031 and for BCP is 0x8031. BCP is used by PPP link stations to negotiate certain parameters that define remote bridging configurations. Table 14.1 lists the options that can be negotiated between two remote bridges using PPP to encapsulate their LAN traffic.

Options 1 and 2 in Table 14.1 involve the two PPP link stations exchanging a proposed 12-bit LAN segment number and a 4-bit bridge ID and negotiating agreement on one or the other but not both. The split or half-bridge approach requires that both sides of the PPP link agree on the bridge ID since the two remote bridges are considered halves of the same bridge. This is negotiated with Option 1, Bridge Identification.

Table 14.1

Bridging Control Protocol Options

Bridge identification	01
Line identification	02
MAC support	03
Tinygram compression	04
LAN identification	05
MAC address	06
Spanning tree protocol	07

If a full-bridge approach is used, then the PPP link is visible to the bridging mechanisms and is treated as a virtual LAN segment. It thus requires a LAN segment number that must be negotiated by the PPP link stations using BCP Option 2, Line Identification. The reason why both LAN segment number and bridge ID cannot be simultaneously agreed to is that full-bridges and half-bridges are mutually exclusive. With the former, the bridge IDs cannot be identical since the two remote bridges are considered separate entities, while with the latter the bridge IDs *must* be identical because the two remote bridges are considered parts of the same entity.

Option 03, MAC Support, allows PPP remote bridges to specify which types of the MAC addresses they will support. Option 04, Tinygram Compression, exploits the fact that many Ethernet frames, in order to meet the minimum PDU size of 64 octets, contain padding that can be removed prior to encapsulation in the PPP frame, thus reducing the overall traffic on the PPP link. Following deencapsulation at the receiving bridge, the padding is reinserted.

Option 05, LAN Identification, concerns the use of LAN IDs, not to be confused with the bridge IDs and LAN segment numbers discussed with regard to options 1 and 2, respectively. To understand LAN IDs, consider a pair of remote bridges (half or full) that is multiported, that is, each of which has two LAN interfaces, as in Figure 14.6, and where it is desired to remotely bridge LANs but at the same time not bridge the LANs locally, that is, keep the local traffic separate (perhaps for administrative or security reasons). One way to do this would be to use four remote bridges, two at each location, but this would entail two serial links as well as possible disruptions to other protocols that need to be routed rather than bridged.

The alternative is to define two *bridge groups*, in effect creating two virtual remote bridges executed/instantiated by the actual remote bridges. Keeping the traffic for the respective bridge groups separated requires that the actual remote bridges route frames originating in bridge group 1 only to LANs in bridge group 1, route frames originating in bridge group 2 only to LANs in bridge group 2, and so on. This requires some form of tagging, by which frames

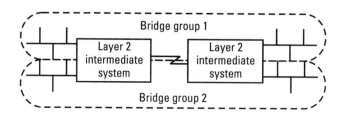

Figure 14.6 Multiple bridge groups.

can be identified as belonging to a given bridge group. This tag is the LAN ID, which is a 32-bit number and which is appended to the PPP frame header by the bridging protocol when it encapsulates the LAN frames. In effect, the LAN ID is a multiplexing field. The PPP link stations negotiating the parameters of the remote bridging protocol would include option 05 in their Link Configuration packets to indicate its use (multiple bridge groups) or not (monolithic remote bridging).

Option 06, MAC Address (not to be confused with option 03, MAC support), allows one of the remote bridges on a PPP link to either send or solicit a MAC address (Ethernet in canonical format). It is designed to support limited functionality Ethernet remote bridges that have just one Ethernet LAN port.

Finally, option 07, Spanning Tree Protocol, is used by the remote bridges to negotiate which if any bridging spanning tree protocol is to be used. This is primarily a concern with transparent bridging, although as we just mentioned spanning tree has been incorporated in source route bridging with SREs.

14.6 Summary

In this chapter we discussed the two dominant approaches to layer 2 (L2) concatenation, that is, bridging and L2 switching: the spanning tree/transparent method used in Ethernet networks and the source route method used in Token-Ring networks. We stressed the fact that the IEEE 802 MAC address space is flat and the pivotal role this played in the development of both transparent and source route bridging. This is because of the fact that with L2 relays network (i.e., L3) addresses never enter the picture. The end-to-end transporter is realized as a concatenation of MAC addresses, as it is with L3 concatenation, but with the difference that there is no way to associate a given address with a part of the network such as an IP network or SNA subarea.

We then moved on to examine how bridging began with promiscuous concatenation and how it is an example of model-free workload scheduling. When learning algorithms were added, however, the concatenation ceased to be model free; in fact, each transparent bridge/switch does maintain a topology model in the form of tables of addresses, one table for each transporter interface to which the bridge/switch is attached: taken together, these tables constitute a global topology model, albeit one distorted by the centrality of the bridge/switch in question. We then saw how simple learning bridges can fail in redundant topologies, and how an elementary routing protocol has been developed to allow bridges to exchange local topology information and reconstruct global topology to the extent of discovering where the other transparent bridges in the network are relative to each other. With the limited model, the transparent bridges superimpose a spanning tree on the network to prevent any looping of L2 PDUs.

In contrast to transparent bridging, we saw that source route concatenation relies on a very different technique in which a source end system initiates a search procedure that culminated in the discovery of all possible end-to-end concatenations (data links and intermediate systems) that could connect it to the destination end system it desired. The rationale for this approach was to allow less expensive bridges to be implemented, because as was explained the only function of a basic source route bridge is to read RIFs, add local ring and bridge numbers, and forward PDUs. However, we saw that to reduce the broadcast storms and other management overhead associated with source route bridging's discovery mechanisms, more sophisticated variants have been defined that employ spanning tree protocols and/or learning bridge-style caches.

Finally, we looked at remote bridging and saw how this is, in fact, a form of tunneling in which the carrier protocol is itself generally an L2 protocol such as HDLC or PPP. We explored the options offered by PPP's Bridging Control Protocol.

About the Author

Philippe Byrnes is a founder of Traphyx, a Silicon Valley startup focusing on traffic management and capacity planning tools. He has worked for IBM, 3Com, and Cisco Systems. He was educated at the University of Illinois, Urbana-Champaign and Stanford University, and is a member of the IEEE and ACM.

Index

fixed-length, 173
header, 147, 172, 307
illustrated, 147, 148
L2, 289
L3, 289
payload, 147
SNA, 352–57
trailer, 147
UDP, 339–40
Protocol machines
defined, 146
hierarchy, 147
Protocols
AHHP, 323–24
ANR, 371
APPN, 350, 369–70
ARPANET, 319–24
authentication, 233–36
BACP, 220, 251–52
BAP, 220, 251–52
BCP, 429, 441–42
BGP, 423–25
byte-oriented, 176
CHAP, 234, 235, 236
datagram, 392
data link, 146–49
defined, 146
distance vector routing, 409–16
EGP, 423
EIGRP, 416
end-to-end, 283–94, 302–14
GGP, 402, 414
HAP, 320
IGRP, 416
IMP-IMP, 321–23
IPCP, 238–40
IPXCP, 240–41
L2F, 440
L2TP, 440
LCP, 220, 227–33
link state routing, 416–23
LLC, 256, 261–62, 276
LQM, 226, 227, 236
MP, 245–51
NCP, 237–41
OSPF, 401, 417, 420–23
PAP, 234, 235, 236

PAR, 162
PPP, 217, 219–53
PPTP, 440
realization, 146
RIP, 401, 415–16
routing, 31–32, 388–89, 408–25
RTP, 371–72
SLIP, 221
SPAP, 234
transforming, 241–45
virtual circuit, 392–93
workload managers, 168
Proxy client, 22
Pulse Amplitude Modulation (PAM), 62
demodulators, 75–76
waveforms, 74
Pulse code modulation (PCM), 62

Quadrature Amplitude Modulation
(QAM), 62
defined, 55
scatter plot, 55
Queuing systems, 10

Random access radio channel
multiplexing, 258
Rapid Transport Protocol (RTP), 371–72
Reachability, 404
Reassembly lockup, 312
Received Line Signal Detector (RLSD), 123
Receivers
channel relationship, 46
decomposition of, with demodulator, 67
timing, 133
Reconstruction, 8
Redundancy
concurrent, 83
encoder induced, 86
Remote bridging, 440–43
with PPP, 441
See also Bridging
Request/Response Units (RUs), 356, 358
Requests For Service (RFSs), 4, 10, 19, 166
implicit, 12
mapping with composite servers, 22
ReSerVation Protocol (RSVP), 335
Response time, 6

Understanding Networking Technology: Concepts, Terms, and Trends, Second Edition, Mark Norris

Understanding Token Ring: Protocols and Standards, James T. Carlo, Robert D. Love, Michael S. Siegel, and Kenneth T. Wilson

Videoconferencing and Videotelephony: Technology and Standards, Second Edition, Richard Schaphorst

Visual Telephony, Edward A. Daly and Kathleen J. Hansell

Wide-Area Data Network Performance Engineering, Robert G. Cole and Ravi Ramaswamy

Winning Telco Customers Using Marketing Databases, Rob Mattison

World-Class Telecommunications Service Development, Ellen P. Ward

For further information on these and other Artech House titles, including previously considered out-of-print books now available through our In-Print-Forever® (IPF®) program, contact:

Artech House
685 Canton Street
Norwood, MA 02062
Phone: 781-769-9750
Fax: 781-769-6334
e-mail: artech@artechhouse.com

Artech House
46 Gillingham Street
London SW1V 1AH UK
Phone: +44 (0)20 7596-8750
Fax: +44 (0)20 7630-0166
e-mail: artech-uk@artechhouse.com

Find us on the World Wide Web at:
www.artechhouse.com